D1067640

Undergraduate Topics in Computer Science

Undergraduate Topics in Computer Science (UTiCS) delivers high-quality instructional content for undergraduates studying in all areas of computing and information science. From core foundational and theoretical material to final-year topics and applications, UTiCS books take a fresh, concise, and modern approach and are ideal for self-study or for a one- or two-semester course. The texts are all authored by established experts in their fields, reviewed by an international advisory board, and contain numerous examples and problems. Many include fully worked solutions.

For further volumes:
www.springer.com/series/7592

005.14 R44

José Bacelar Almeida · Maria João Frade ·
Jorge Sousa Pinto · Simão Melo de Sousa

Rigorous Software Development

An Introduction to Program Verification

 Springer

Nyack College - Bailey Library
One South Blvd.
Nyack, NY 10960

Dr. José Bacelar Almeida
Depto. Informática
Universidade do Minho
Campus de Gualtar
Braga 4710-057
Portugal
jba@di.uminho.pt

Dr. Maria João Frade
Depto. Informática
Universidade do Minho
Campus de Gualtar
Braga 4710-057
Portugal
mjf@di.uminho.pt

Dr. Jorge Sousa Pinto
Depto. Informática
Universidade do Minho
Campus de Gualtar
Braga 4710-057
Portugal
jsp@di.uminho.pt

Dr. Simão Melo de Sousa
Depto. Informática
Universidade Beira Interior
rua Marques d'Avila e Bolama
Covilhã 6201-001
Portugal
desousa@di.ubi.pt

Series editor
Ian Mackie

Advisory board
Samson Abramsky, University of Oxford, Oxford, UK
Chris Hankin, Imperial College London, London, UK
Dexter Kozen, Cornell University, Ithaca, USA
Andrew Pitts, University of Cambridge, Cambridge, UK
Hanne Riis Nielson, Technical University of Denmark, Lungby, Denmark
Steven Skiena, Stony Brook University, Stony Brooks, USA
Iain Stewart, University of Durham, Durham, UK

ISSN 1863-7310
ISBN 978-0-85729-017-5   e-ISBN 978-0-85729-018-2
DOI 10.1007/978-0-85729-018-2
Springer London Dordrecht Heidelberg New York

British Library Cataloguing in Publication Data
A catalogue record for this book is available from the British Library

© Springer-Verlag London Limited 2011
Apart from any fair dealing for the purposes of research or private study, or criticism or review, as permitted under the Copyright, Designs and Patents Act 1988, this publication may only be reproduced, stored or transmitted, in any form or by any means, with the prior permission in writing of the publishers, or in the case of reprographic reproduction in accordance with the terms of licenses issued by the Copyright Licensing Agency. Enquiries concerning reproduction outside those terms should be sent to the publishers.
The use of registered names, trademarks, etc., in this publication does not imply, even in the absence of a specific statement, that such names are exempt from the relevant laws and regulations and therefore free for general use.
The publisher makes no representation, express or implied, with regard to the accuracy of the information contained in this book and cannot accept any legal responsibility or liability for any errors or omissions that may be made.

Printed on acid-free paper

Springer is part of Springer Science+Business Media (www.springer.com)

Preface

It has become common sense that problems in software may have catastrophic consequences, ranging from economic to human safety aspects. This has led to the development of software norms that prescribe the use of particular development methods. These norms advise the use of *validation* techniques, and at the highest levels of certification, of *mathematical* validation.

Program verification is the area of computer science that studies mathematical methods for checking that a program conforms to its specification. It is part of the broader area of *formal methods*, which groups together very heterogeneous rigorous approaches to systems and software development. Program verification is concerned with properties of code, which can be studied in more than one way. From methods based on logic, in particular the combined use of a *program logic* and *first-order theories*, to other approaches like *software model checking*; *abstract interpretation*; and *symbolic execution*.

This book is a self-contained introduction to program verification using logic-based methods, assuming only basic knowledge of standard mathematical concepts that should be familiar to any computer science student. It includes a self-contained introduction to propositional logic and first-order reasoning with theories, followed by a study of program verification that combines theoretical and practical aspects—from a program logic to the use of a realistic tool for the verification of C programs, through the generation of verification conditions and the treatment of errors.

More specifically, we start with an overview of propositional (Chap. 3) and first-order (Chap. 4) logic, and then go on (Chap. 5) to establish a setting for the verification of sequential programs, building on the axiomatic semantics (based on Hoare logic) of a generic *While* language, whose behaviour is specified using *preconditions* and *postconditions*.

It is then shown how a set of first-order proof obligations, usually known as *verification conditions*, can be mechanically generated from a program and a specification that the program is required to satisfy (Chap. 6). Concrete programming languages can be obtained by instantiating the language of program expressions, which we illustrate with a language of integer expressions and then a language of integer-type arrays.

This setting is then adapted to cover the treatment of *safety properties* of programs, by explicitly incorporating in the semantics the possibility of runtime errors occurring. The set of generated verification conditions is extended to guard against that possibility (Chap. 7). This is illustrated in the concrete languages with arithmetic errors and out-of-bounds array accesses.

Finally the setting is extended to allow for the specification and verification of programs consisting of mutually-recursive procedures, based on annotations included in the code, usually known as *contracts* (Chap. 8). The idea here is that each procedure has its own public specification, called a contract in this context, and when reasoning individually about the correctness of an individual procedure one can assume that all procedures are correct with respect to their respective specifications.

The last part of the book illustrates the specification (Chap. 9) and verification (Chap. 10) of programs of a real-world programming language. We have chosen to use the ACSL specification language for C programs, and the Frama-C/Jessie tool for their verification. It is important to understand that these are being actively developed as this book is written, but we do believe that the core of the specification language and verification tools will remain unchanged.

Our emphasis is on the integrated presentation, and on making explicit the bits of the story that may be harder to grasp. Program verification already has a long history but remains a very active research field, and this book contains some material that as far as we can tell can only be found in research papers. The approach followed has logic at its core, and the theories that support the reasoning (which may include user-supplied parts) are always explicitly referred. The generation of verification conditions is performed by an algorithm that is synthesized from Hoare logic in a step-by-step fashion. The important topic of *specification adaptation* is also covered, since it is essential to the treatment of procedures. Finally, we develop a general framework for the verification of contract-based mutually-recursive procedures, which is an important step towards understanding the use of modern verification tools based on contracts.

The book will prepare readers to use verification methods in practice, which we find cannot be done correctly with only a superficial understanding of what is going on. Although verification methods tend to be progressively easier to use, software engineers can greatly benefit from an understanding of why the verification methods and tools are sound, as well as what their limitations are, and how they are implemented. Readers will also be capable of constructing their own verification platforms for specific languages—but it must be said at this point that one of the most challenging aspects, the treatment of heap-based dynamic data structures based on a memory model—is completely left out.

As mentioned before, program verification belongs to what are usually described as *formal methods* for the development of software. Chapter 1 motivates the importance of program verification, and explains how formal approaches to software development and validation are more and more important in the software industry. Chapter 2 offers an overview of formal methods that will hopefully give the reader a more general context for understanding the program verification techniques covered in the rest of the book.

Notes to the Instructor   The book assumes knowledge of the basic mathematical concepts (like sets, functions, and relations) that are usually taught in introductory courses on mathematics for computer science students. No knowledge of logical concepts is assumed.

The accompanying website for this book is http://www.di.uminho.pt/rsd-book, where teaching material, solutions to selected exercises, source code, and links to useful online resources can be found.

Acknowledgments   The authors are indebted to their students and colleagues at the departments of Informatics of the universities of Minho and Beira Interior, who have provided feedback and comments on drafts of several chapters. The idea of writing this book came up when the third author was on leave at the Department of Computer Science of the University of Porto (Faculty of Science), which kindly provided a welcoming environment for writing the first draft chapters.

Contents

Chapter 1
Introduction

This book is about the use of techniques and tools for the design and implementation of computer systems and software that are free from logical or functional flaws (in the sense of functional requirements). The word *rigorous* in the title of this book is justified by the fact that the arguments for such fault freeness have their roots in computer science, logic and mathematics rather than in empirical and statistical studies. In this sense this book will address concepts and techniques for *fault avoidance* rather than *fault tolerance* (which in itself represents a rich and very important area in computer system and software engineering).

1.1 A Formal Approach to Software Engineering

The importance of the role of information systems in essential sectors of contemporary societies is more and more prevalent. Transport, communications, health and energy are all representative examples. The rapid growth in terms of complexity of the computing infrastructures and software that underly and accompany the evolution of these systems has made it harder to have confidence on their reliability. Being able to establish that a system behaves as it is supposed to is now more delicate than ever.

Even though most software errors lead to no more than minor upsets, the situation is quite different when so-called *critical systems* are concerned. Classic examples of these include nuclear and medical equipment. The correct behaviour of the software components of these systems is essential. We nowadays live in a massively networked and digital era, and the emergence of new and sometimes revolutionary technologies only contributes to reinforce the need for the software to enjoy adequate correctness properties. Think for instance of the tremendous, but also invisible, introduction of information systems in the daily life of citizens (by means of ubiquitous and pervasive information systems or embedded and mobile systems).

The attention given by the media to the possibly bombastic effects of conception errors in information systems also has a role in creating a certain hype around the reliability of software. Governments, industrial consortia, and plain system users

J.B. Almeida et al., *Rigorous Software Development*,
Undergraduate Topics in Computer Science,
DOI 10.1007/978-0-85729-018-2_1, © Springer-Verlag London Limited 2011

have all decidedly become more conscious of this, and now increasingly demand that reliability be established. Let us look at a concrete and classic example of this phenomenon: in 1994 the popular science publication Scientific American published an article stating [12]:

> Despite 50 years of progress, the software industry remains years—perhaps decades—short of the mature engineering discipline needed to meet the needs of an information-age society.

Regrettably, the software industry has for a long time had a reputation for not being able to keep to any reliability or robustness commitments. In fact, the existence of bugs in commercial systems is so common that companies now systematically include a bug report system with each release. To mention some data, in 2002 the North-American Institute for Standards and Technologies estimated the cost of bugs in the American economy to ascend to 59 billion dollars.[1]

From the point of view of information system designers, product quality in terms of reliability is certainly a commercial concern. Goguen [5] lists a number of important contracts troubled or even terminated by reliability problems, such as the 8 billion dollars contract between IBM and the US Federal Aviation Agency regarding an air-traffic control system. In 1994 an error was discovered in the implementation of division operations by Pentium processors. Even though millions of processors had by then been sold, Intel was forced to exchange (free of charge) all the units produced [10]. Beyond the financial impact, the media emphasized the loss of confidence shown by Intel users (i.e. the computer manufacturing industry) that had a much broader and dramatic effect to the company.

It is extremely hard to certify in an absolute way the behaviour of a system. However, if absolute correctness may be illusory, reinforcing the degree of reliability and confidence that one can have in a product are still a central and well justified concern. For illustration purposes consider the following well-known facts in software engineering: the cost of maintaining software is about two thirds of its total cost, and a software specification fault is about 20 times more costly to repair if detected after production than before.

This state of affairs has been tackled by industry in a number of different ways, but the most popular weapon has certainly been program testing and simulation.

1.1.1 Test and Simulation-Based Reliability

A classic approach to ensuring the adequacy of a software system is *testing* or *simulation*. This consists in producing a set of data believed to be representative, giving it as input to the system (or a model thereof), and then comparing the outputs obtained with the set of expected outputs. Unfortunately, testing cannot be totally satisfying: for the vast majority of systems, the value space is immense (if not infinite), which renders exhaustive testing impracticable (if not impossible).

---

[1] http://www.nist.gov/public_affairs/releases/n02-10.htm.

A relative notion of reliability, let us call it *confidence*, is obtained from a judicious choice of entry values, which should affect the greatest possible number of components of (the control flow graph of) the system under test.[2] Of course, the complexity of the choice of input values is directly related to the complexity of the system, and for very complex systems reliability becomes very hard to guarantee. This is a particularly important point when one discusses how to ensure the robustness of a production-level system, since many problems are in fact triggered by an unexpected use of resources.

In spite of all this, as was already pointed out, the popularity of testing as a practice is huge in industry, due to several reasons. One reason is its acceptance by certification bodies, that typically consider its conclusions as valid, up to some considerations such as test coverage analysis.[3] In fact, *test engineering* is a reasonably well understood and mature discipline, whose expertise is now well integrated in industry.

1.1.2 An Alternative Approach: Formal Methods

Formal methods constitute an alternative, mathematical approach to reliability assurance, to which industry is progressively devoting more attention. In fact, not only is the need for formal methods increasingly more pressing, but the maturity of their underlying mechanisms is now more adequate for industrial use than it has been in the past, when it was extremely hard for non-specialists to be able to use formal methods tools. The perceived utility of formal methods is thus more and more indisputable.

At the heart of formal methods one finds a set of formalisms and mathematical tools for modeling and reasoning about systems (in particular, software systems and hardware systems). Formal methods offer the possibility of reasoning about the *complete* data space and set of configurations of a system, rather than samples of these sets. J. Rushby [9] defines formal methods as follows

> The term *formal methods* refers to the use of mathematical modeling, calculation and prediction in the specification, design, analysis and assurance of computer systems and software. The reason it is called *formal methods* rather than *mathematical modeling of software* is to highlight the character of the mathematics involved.

The deployment of formal methods is conditioned by the costs incurred, including the costs of the required expertise. While costs do remain high to this day, they are decreasing, and furthermore industry is more and more willing to support them: formal methods are slowly moving from a situation in which they were exclusively

---

[2]This quality measure is known as *test coverage*.

[3]A test suite has a good coverage, say 100% coverage, if all parts of the system are covered by its tests.

used in the academic milieu to being effectively used in the industrial context, particularly where critical systems are concerned. The reader will find a recent and detailed account of these issues in the survey of Woodcock et al. [13].

The present moment is particularly vibrant for the formal methods community, with the appearance of specific security requirements raised by the emergence of global computing (see below). An example of this is offered by the area of *language-based security*, which has recently seen the development of new techniques like *proof-carrying code* [8] and *verifying compilation* [7].

1.1.3 Requirements: Functional, Security, and Safety

Above we have used words like *reliability*, *robustness*, and *correctness*. Naturally, these qualifiers only make sense with respect to a given set of requirements. Requirements are usually divided into *functional* and *non-functional*, where the former broadly describe *what* a system does (in terms of the relation between inputs and outputs) while the latter class consists of properties related to *how* a system operates, including efficiency, quality of service, and maintenance issues.

One specific class of increasingly important functional requirements concerns the *security* of systems. This includes classic concerns such as authentication issues and limited access to functionality depending on users, or requirements relative to specific devices such as network firewalls or smart cards. The specific aspects of security requirements will not be dealt with in this book. We must however remark that security requirements, and in particular the advent of the *Common Criteria* international standard, have contributed to boosting the importance of formal methods (see below).

A second specific class of requirements concerns *safety* issues. In the context of modern programming languages these are usually understood to include issues that may cause a program to crash, such as the possibility of illegal memory accesses (*memory safety*). Compilers and language definitions have an important role to play in safety; for instance Java is clearly by design a safer language than C, since pointers are not free for the user to manipulate. As long as you trust the implementation of your Java compiler, you can trust your Java programs to execute in a memory-safe way.[4]

The expression "functional requirements" will be used for requirements that are not related to safety issues. Safety requirements will always be clearly identified as such. Interesting discussions on safety and security issues can be found in [6] and [11]. We quote from [6] on the difference between safety and security:

> To illustrate what we mean by safety and security consider gaining access to a building. The first concern would be our safety. For example, one of the hinges of the door might be broken, so that a person entering the building could get hurt opening the door. By *safety*,

---

[4]Incidentally, Java is also safer than C regarding *types*, since it is a strongly-typed programming language.

therefore we mean that nothing bad will happen. Our second concern would be the security of the building. This means that even if the hinges are in good condition, to make the building secure, a lock on the door is required to control access to the building. By *security*, therefore, we mean that there is access control to resources.

1.2 Formal Methods and Industrial Norms

We shall now consider the relevance of the use of formal methods in the software development process, and in particular its industrial adoption. Formal methods are not, in fact, the first or most natural choice. Its adoption by the industry depends on several factors. For the common practitioner, the fact that they are a good means to improve the software quality is seen as a quite philosophical or theoretical virtue, if not only a *beautiful* one. From our experience, the two most important factors for the adoption of formal methods are:

1. industry may be *required* to use formal methods; and
2. industry may *really benefit* from their use.

In this section we concentrate on the first factor; the second one will be left to Sect. 1.3.

Safety Critical System Standards   For obvious reasons, safety critical system design must follow strict rules and methodologies. The client that orders such a system requires that all reliability and security guarantees are provided by the designer. In this perspective, the different sectors of the safety critical system industry (electronics, railway, aerospace, medical systems, defence, etc.) have felt the need to regulate, in collaboration with standardisation bodies, the system development process.

In its general form, such standards propose a definition of safety (or reliability or even security, but let us focus on safety) by proposing a notion of *safety scale* specific to the targeted industry sector. To each level of this scale is associated a usage scenario for which the related safety is suitable.

In order to map a system (and its development process) to a safety measure (and thus to a specific level in the safety scale), the proposed standards define metrics and evaluation systems, as well as a validation and certification process. Thus, if a critical system is aimed to be used in a specific critical environment, then it must obey the rules of the corresponding level of the safety scale. Its compliance to this level must be validated and verified, and finally the system must be certified at this level by a *certification body* (i.e. the validation and verification process must be itself validated).

For instance the standard CENELEC EN 50128 (Railway Applications—Communications, Signaling and Processing Systems—Software for Railway Control and Protection Systems) provides a four-level safety scale, known as SIL (Safety Integrity Levels), where the level SIL4 establishes the highest level of safety integrity.[5] This level is targeted at systems that are intended to be used in the most critical scenarios in railway industry.

Similarly, in the areas of aerospace or defence industry, the standards DO-178B (Software Considerations in Airborne Systems and Equipment Certification) and ESA-GSW (Galileo Software Standard) define similar scales.[6] These standards introduce the use of various procedures and techniques to guarantee compliance to safety critical requirements. Examples include *Software Failure Modes, Effects and Criticality Analysis* (SFMECA) and *Software Error Effects Analysis* (SEEA), that are classically part of a *Reliability, Availability, Maintainability and Safety* (RAMS) analysis. These scales are known as *Design Assurance Levels* (DAL) and range from DAL E, the less critical, to DAL A, the most critical. In usage scenarios where a failure has catastrophic consequences, systems must be certified DAL A.

While to a large extent these standards consider as adequate, regarding the correctness of software, a suitable and heavily documented application of test and simulation, their latest versions have introduced formal methods, and recommend their use for certification at the highest levels. It is believed among the safety critical industrial community that the use of formal methods may become mandatory in the near future.

The Common Criteria   The standards considered above are dedicated to particular sectors of the critical systems industry. The one that we will now address in more detail has pioneered the use of formal methods in a normative context. This standard covers not only the safety critical area but the software industry in general.

In the age of global computing security issues have become of the utmost importance. Even though it is impossible to guarantee an absolute level of security, governments have shown to be more and more interested in taking measures to attain satisfying levels. Six countries (Canada, France, Germany, the Netherlands, the United Kingdom, and the USA) that possessed internal regulations on the security of information systems, have come together to reach a common approach for the evaluation of information systems security. This joint effort culminated in the ISO/IEC-15408 standard, known as *Common Criteria* [3].

Common criteria aim to establish a framework for a more organised reflection on the security of information systems. The main goal of the standard is to rationalise the requirements of such systems and the approaches to their attainment. More precisely, common criteria propose to establish a solid basis for improving *confidence* on systems adhering to the standard.

The undeniable advantage of this unified approach is the existence of a universal framework for system security, at the level of both product designers and potential users. Users may resort to a single source, the Common Criteria, for evaluating a product or a service according to their security requirements. From the client's point

---

[5]The standard IEC 61508 "Functional safety of electrical/electronic/programmable electronic safety-related systems" also uses a (slightly different) SIL classification.

[6]The definition of the GSW standard was in fact based on DO-178B.

of view, Common Criteria play an important role in allowing users to identify techniques and security requirements suited to their organisational needs. Developers on the other hand will use the Common Criteria to identify elementary security requirements to be applied in their products.[7] Since the set of requirements put forward by the Common Criteria is internationally accepted, system designers are relieved of having to evaluate the situation on a national basis for each country where they wish to disseminate their products.

A central notion is the rating assigned to a project or system to quantify the level of confidence on its security, depending on the scope, detail, and the characteristics of the tools and methods used in the evaluation. This is done through the so-called *Evaluation Assurance Levels* (EAL). System developers aim at a particular EAL. Evaluation laboratories work together with the developers to analyse security requirements, models, functionality, and the measures put to practice to reach that EAL. Accredited certification institutes analyse and validate the resulting evaluations.

Assurance levels range from EAL1 to EAL7. The reason for the emphasis we are placing on the Common Criteria here is that levels EAL5 to EAL7 explicitly require the use of formal methods—in fact EAL7 even requires the use of *formal proof tools*. Very few systems have been certified at the highest levels of Common Criteria, and in fact after its introduction, the standard has already been subject to reformulations that have recalibrated (among other aspects) the use of formal methods, in order to be make their application more flexible than in the original version of the standard. Nevertheless, Common Criteria still constitute a major opportunity and application context for formal methods.

Quis Custodiet Ipsos Custodes? To conclude on the relevance of standards in industrial practice, we must note that the design of a formal method, and the development of tool support for it, are not sufficient in themselves for the use of the method to be accepted in this normative context. For instance, in order to achieve a CENELEC SIL4 certification, the tools used in the design of the system, and also in the verification process, must be themselves SIL4-certified. In fact, a key point to the success of the industrial use of a given formal method lies precisely in this certification.

1.3 From Classic Software Engineering to Formal Software Engineering

Let us now contextualise formal methods as a discipline in the context of software engineering. This will help us approach the second topic raised at the beginning of the previous section: in order to use formal methods, industry must really be able to benefit from its use.

---

[7]Such a combination of requirements is called a *protection profile*.

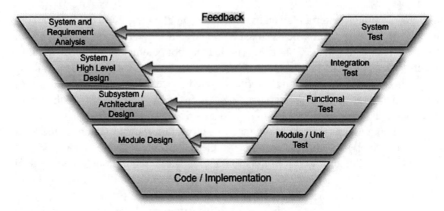

Fig. 1.1 The V software life cycle

This is, obviously, a complex and multifaceted issue, that we will go on exploring in a more detailed way in the next chapter. For now and as a first approach, we list some decisive key points that contribute to the adoption of a specific formal method:

- the underlying concepts and tools are mature and stable enough;
- the underlying tool support is suitable for an industrial use (say, it has good usability);
- the necessary expertise for its proper use can reasonably be acquired and mastered;
- its adoption can be integrated in the existing software engineering body of knowledge;
- the company understands and accepts the initial impact and cost of its integration and use.

Anyway, the adoption of formal methods cannot be seen in the same light as the adoption of a product or tool. Formal methods require a real strategic and methodological reflection, leading to deep modifications in the way of thinking, managing and producing software. We see such a change in the software development paradigm as a mandatory step for the success of the integration of formal methods in any industrial practice. Nevertheless, the shift cannot be too dramatic, since no software industry will ever start again from scratch. This justifies the importance of the last two items in the above list. We will argue below that there is indeed a fertile ground for the referred methodological shift.

In the discipline of software engineering several different notions of *software development cycle* are considered. Each one has its advantages and disadvantages. Let us consider for instance the *V cycle* depicted in Fig. 1.1, which is the development cycle most prevalent in the critical software industry.

This cycle proposes two distinct phases in the development of a software system. The first, corresponding to the left (descending) part of the V, is the *design and implementation* phase. The second part, corresponding to the right (ascending) part

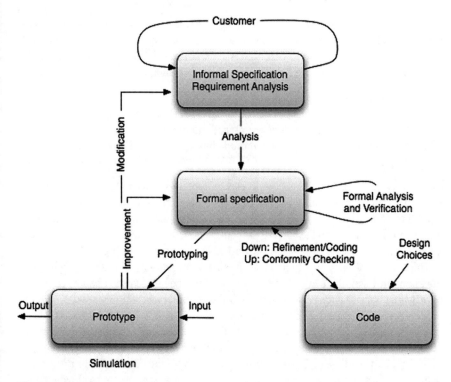

Fig. 1.2 The Balzer software life cycle

of the V, is the *validation and verification* phase, in which the validity of the results of the corresponding design stage is assessed.

In the descending phase, the model proposes an organisation of the software production process in stages, starting from a high-level view of the system, down to the code. Each stage proposes a more detailed look at the system than the previous; the requirements from the previous stage must then be redefined and adapted to the more finely-grained resulting structure. An important issue is that the new set of requirements should reflect the requirements of the previous stage. So the completion of each step should produce reports that underpin the work performed in the next step, as well as in the corresponding validation step in the ascending phase.

The ascending phase is dedicated to validation, taking as a basis the preparatory validation reports produced in the design phase. What is meant by validation here is typically testing and execution in a controlled environment. The reports produced at this stage must provide answers to the related preparatory reports.[8] If a problem is detected then the development process should go back to the related design stage.

---

[8]In a very simplistic way, the preparatory reports consist of lists of facts to be verified, and the reports of the validation phase are simply observations, based on the validation process, of whether or not those facts were successfully validated.

A natural question for us to ask is then, how should formal methods be integrated in such a development process? In the early 80's, Balzer and colleagues [2] proposed a notion of *software life cycle* that addresses precisely this issue. In their original view, the core of the software development process roughly comprises[9]

1. the *requirements* analysis;
2. the construction of a *formal model* or *formal specification*;
3. the development of a *functional prototype*; and finally
4. the development of the *implementation*, having as a basis the model and the prototype.

The novelty of Balzer's life cycle with respect to other software development cycles is that it explicitly advocates the use of formal methods at different stages, and in particular where the relationship between the requirements, the model, and the implementation is concerned (a topic to which we devote several sections of Chap. 2). An aggregate representation of the Balzer life cycle is shown in Fig. 1.2.

The Balzer software life cycle is, surprisingly, a relatively old idea. As was previously mentioned, the use of formal methods has been to a certain extent confined to very specific industry sectors, in which the necessary investments were seen as justified (typically, safety critical systems). Only recently, with the increased maturity of the formal methods concepts and tools, did it become feasible to apply these methods, and the Balzer life cycle, at a cost that industry is willing to pay.

An updated reading of Balzer's principles in the light of contemporary formal methods allows for their integration in the V cycle in one of the following two ways:

− *Downward application*: this form of integration results in the closest structure to the original. The formal specification phase of the Balzer life cycle encompasses the three design stages of the V life cycle. This is, for instance, the approach followed by B, VDM, and SCADE, that will be described in the next chapter.
− *Horizontal application*: each design stage can give rise to an application of the Balzer life cycle. For example during the design of the architecture of the system, the latter is translated into a formal specification which is then formally analysed and verified. Obviously, in this context, the application of the code production stage of the Balzer life cycle depends on the design stage under consideration.

The horizontal application to the implementation stage is a special case: from the code and requirements, a formal model is generated on which it is possible to conduct a formal analysis. In this configuration, the prototyping phase is not necessary.

For illustration purposes, we depict in Fig. 1.3 the development process defined by the CENELEC EN50128 standard, which has at its base a variant of the V life cycle. We can highlight, for example, the software module design phase that should provide the following three reports: the software module design specification, the

---

[9]Of course, other post-implementation phases such as testing, training, and maintenance may well be present in this process.

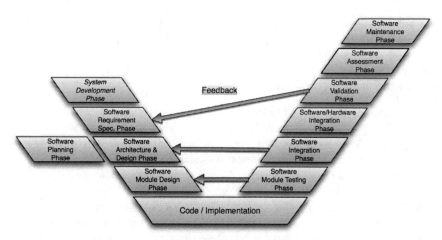

Fig. 1.3 The EN 50128 V software life cycle

software test module specification, and the software module verification report. The related test stage should provide a software module test report.

Without going into details, the software module verification report, for instance, should report on the soundness of the decomposition into modules. It should justify that the defined modules and requirements in fact contribute to the objectives stated in the previous step (i.e. the overall architecture of the target system and its requirements). In a horizontal application process of the Balzer principles, the software module verification report will present the conclusions of formally verifying the (formal) specification obtained from the module. This use of formal methods is in fact highly recommended in EN 50128, for SIL 3 and 4.

1.4 This Book

Books on formal methods tend to be either exhaustive, surveying all major approaches with little detail, or highly technical, covering all the advanced details of one specific method or tool. Our experience in teaching formal methods in a constrained context (usually two semester-long courses) tells us that neither of these is the best approach to introducing the discipline.

Taking the Balzer life cycle as a background basis, the book introduces formal methods as a computing discipline, with a broad-scope survey of the field, and then focuses on a particular approach that has proved to be effective (and unquestionably relevant for industry). The thread that runs through the book corresponds to fundamental, already historical ideas in software science, whose impact cannot be overestimated: the logical approach to the verification of programs, and in particular Hoare logic.

From the point of view of this book, these ideas stand at the heart of the principles underlying the approach to software development based on *contracts*, and it is also at the basis of model-oriented approaches like VDM and B. Hoare logic and Dijkstra's precondition calculus have been explored in textbooks [1, 4] as tools for program derivation; our approach focuses on the modern trend to use the formalisms in the context of program verification.

In a nutshell, the book includes:

1. An overview of the area of formal methods, covering different approaches and tools. This provides students with knowledge of the basic concepts and links to the most widely known and used formal methods.
2. A short introduction to computer science logic and logical methods. In particular the methods and tools for logical reasoning and its automation are introduced. This part forms the mathematical background for the rest of the book.
3. A detailed introduction to the verification of programs based on contracts. This includes the study of program logics for generic *While* programs, the generation of verification conditions, and the logical treatment of safety aspects. All of these are illustrated with many examples.
4. An illustration of how the ideas explained in the previous chapters are brought to practice, with the study of a specification language and a tool for the practical verification of safety and functional properties of C programs.

The prevailing techniques and tools explored in the book are widely-known. Excellent advanced books and research papers exist covering them, and the present book is no replacement for these. We include a substantial bibliography, ranging from fundamental textbooks on logic to research papers on verification techniques, and material covering the industrial use of formal methods.

References

1. Backhouse, R.: Program Construction—Calculating Implementations from Specifications. Wiley, New York (2003)
2. Balzer, R., Cheatham, T.E., Green, C.: Software technology in the 1990's: Using a new paradigm. IEEE Comput. **16**(11), 39–45 (1983)
3. Common Criteria. http://www.commoncriteria.org
4. Gries, D.: The Science of Programming. Springer, Secaucus (1987)
5. Goguen, J.: Hidden algebra for software engineering. In: Proceedings Combinatorics, Computation and Logic, vol. 21, Auckland, New Zealand, January 1999, pp. 35–59. Springer, Berlin (1999)
6. Hartel, P.H., Moreau, L.: Formalizing the safety of Java, the Java virtual machine, and Java card. ACM Comput. Surv. **33**(4), 517–558 (2001)
7. Hoare, C.A.R.: The verifying compiler, a grand challenge for computing research. In: Cousot, R. (ed.) VMCAI. Lecture Notes in Computer Science, vol. 3385, pp. 78–78. Springer, Berlin (2005)
8. Necula, G.C.: Proof-carrying code. In: Proceedings of POPL'97, pp. 106–119. ACM Press, New York (1997)
9. Rushby, J.: Formal methods and their role in the certification of critical systems. Technical Report SRI-CSL-95-1, Computer Science Laboratory, SRI International, Menlo Park, CA (March 1995)
10. The MathWorks: The Pentium papers. http://www.mathworks.com/company/pentium

11. Volpano, D.M., Smith, G.: Language issues in mobile program security. In: Vigna, G. (ed.) Mobile Agents and Security. Lecture Notes in Computer Science, vol. 1419, pp. 25–43. Springer, Berlin (1998)
12. Wayt Gibbs, W.: Trends in computing: Software's chronic crisis. Sci. Am. (September 1994)
13. Woodcock, J., Larsen, P.G., Bicarregui, J., Fitzgerald, J.: Formal methods: practice and experience. ACM Comput. Surv. **41**(4), 1–36 (2009)

Chapter 2
An Overview of Formal Methods Tools and Techniques

The goal of this chapter is to give an overview of the different approaches and tools pertaining to formal methods. We do not attempt to be exhaustive, but focus instead on the main approaches. After reading the chapter the reader will be familiar with the terminology of the area, as well as with the most important concepts and techniques. Moreover the chapter will allow the reader to contextualise and put into perspective the topics that are covered in detail in the book.

Why do we need an overview of formal methods? Why not just study *one* rigorous method for software development? This is a very pertinent and legitimate question. The behavioural essence of software is not captured by a unique unified mathematical theory. Such a general foundation is unlikely to exist.

Think for instance about the diversity of programming language paradigms and theories, and the resulting jungle of existing computer programming languages. Is there a definite paradigm (or, even, language) that makes obsolete all the other ones? Clearly not. Different languages will be chosen by different people to solve the same problem, and someone may well use different languages to solve different problems. Similarly, depending on the goals of the software designers and of the verification process, one may prefer a theory over another one, and even use more than one theory (and related formal methods techniques and tools), in the context of the development of a single system.

Even if the theory is fixed, several dialects and related tools may exist for it. Turning back to the programming language analogy, think for instance of the different existing C or Prolog dialects and compilers. A particular compiler may not even be significantly better than another, but its use will be justified for some users and some application scenarios. This is also a common situation with formal methods. Is it desirable? There is an open and vigorous debate about this issue; we simply point out that the potential user of formal methods should be aware of it and understand the different flavours available.

Our goal with the present overview is then to draw a map of this jungle of theories, techniques, and tools, to make some sense of it.

J.B. Almeida et al., *Rigorous Software Development*,
Undergraduate Topics in Computer Science,
DOI 10.1007/978-0-85729-018-2_2, © Springer-Verlag London Limited 2011

2.1 The Central Problem

Questions such as "What are formal methods?" or "What added value can be expected from the use of formal methods?" have been largely debated in the Software Engineering community [31, 36, 53, 88, 90]. In the following we sum up and discuss the main ideas.

The central problem of formal methods is to be able to guarantee the behaviour of a given computing system following some rigorous approach. At the heart of formal methods one finds the notion of *specification*. A specification is a model of a system that contains a description of its desired behaviour—*what* is to be implemented, by opposition to *how*. This specification may be totally abstract (in which case the model *is* the description of the behaviour), or it may be more operational, in which case the description is somehow contained, or implied, by the model. In this context, the central problem can be seen as being split in the following two aspects:

1. How to enforce, at the specification level, the desired behaviour? This is called the *model validation* problem.
2. How to obtain, from a specification, an implementation with the same behaviour? Or alternatively, given an implementation, how can it be guaranteed that it has the same behaviour as the specification? This is the *formal relation between specifications and implementations* problem.

The different families of formal methods cover a wide range of approaches to these two questions. This variety encompasses the study of specifications by *animation*, by *transformation*, or by *proving properties*. Analogously, implementations may be *derived* from specifications, or else they may be *guaranteed to be correct* with respect to them, either by *construction*, or by *verification*, i.e. by presenting a *formal proof*.

2.1.1 Some Existing Formal Methods Taxonomies

An interesting characterisation used to classify formal methods is the one that takes into account the ease of use and automation of the processes involved in their application. This gives rise to the taxonomy *lightweight* versus *heavyweight* usually found in the literature. Lightweight formal methods usually do not require deep expertise, by opposition to heavyweight formal methods, which are more complex, less automatic, but also more finely grained and powerful. They are typically confined to very specific application areas where their use and cost are justified.

Another existing taxonomy is related to the application of the Balzer software life cycle, as explained in the previous chapter. One speaks of *formally designed software* when formal methodologies and tools are applied in the horizontal fashion. The vertical application is referred to as the *correct-by-construction* approach or, when formal proof facilities are also provided in addition to the generation of code, as *formal development*.

2.1.2 This Overview

This chapter is based on a description of formal methods organised instead by layers of functionalities. We will first take a tour of the approaches to describing and analysing (formal) models; we will then cover the different existing proof mechanisms, and continue with a description of ways of formally relating models with programs, i.e. of approaching the second sub-problem identified above. Finally we will take a look at mechanisms for dealing with scalability issues. For each approach we will discuss in turn the key concepts and foundations involved and the corresponding tools.

The following central notions are common to a vast number of formal methods techniques and tools:

- The operational essence of the modelled systems is usually captured by some form of *transition system* (described either logically, relationally, or algebraically). The different mechanisms discussed below imply particular interpretations of *states*, *transitions*, and *state transformations*.
- The behavioural essence of the modelled systems is usually captured by some *program logic* (such as Hoare logic), a notion that stands at the heart of a large part of formal methods techniques and tools.

These notions are pervasive in the overview that follows.

2.2 Specifying and Analysing

The definition of a specification and the analysis of its behaviour may be carried out formally, i.e. in the context of some mathematical formalism. The advantages of this include the following:

- The formal nature of the specification language employed forces one to reason about and understand all the fine details of the specified system, and thus clarify potential hidden ambiguities. Several published case studies confirm such benefits (see Sect. 2.7). For instance the survey [61] reports on work by Don Syme, that allowed several non trivial errors to be found simply by writing down hand-built specifications in a formal way.
- The possibility to animate, or even execute, a specification—and thus to directly observe its behaviour—if an implementation of the underlying mathematical formalism exists. Such a functionality allows for the system to be *prototyped*. The straightforward advantage of having a prototype is that it is in general simpler to deploy than the system itself. A prototype makes possible the validation of a system without actually implementing it.
- A prototype obtained in this way is still a formal entity, amenable to being mathematically manipulated. One may thus reason about it (either by hand or with the assistance of a computer).

A specification essentially describes the manipulated data and how they evolve, i.e. the operations that transform them. The two main approaches to formal specification differ on the focus given to these two aspects:

— The behaviour of the modelled system can be expressed by focusing on its *operations*, available mechanisms (services), or actions that can be performed. In this view the crucial element is a clear definition of the modifications or changes performed by each operation on the *internal state* of the modelled system. Such specification languages are referred to as *state-based* or *model-based specification* languages.
— The behaviour of the target system can instead be expressed by focusing on the *manipulated data*, how they evolve, or the way in which they are related. This class of specifications includes *algebraic specifications*, sometimes also known as *axiomatic specifications*.

In what follows we will consider in turn the characteristics of the specification languages used for each of these approaches.

2.2.1 Model-Based Specification

The languages used for this class of specifications are characterised by the ability to describe the notion of internal state of the target system, and by their focus on the description of how the operations of the system modify this state. The underlying foundations are in discrete mathematics, set theory, category theory, and logic.

Abstract State Machines Proposed by Gurevich [58], abstract state machines (ASM), also called *evolving algebras*, form a specification language in which the notions of state and state transformation are central. A system is described in this formalism by the definition of states and by a finite set of (possibly non-deterministic) state transition rules, which describe the conditions (also called guards) under which a set of transformations (modifications of the machine's internal state) take place. These transitions are not necessarily deterministic: the formalism takes into account configurations in which several transitions are eligible for a certain state of the machine.

Given that the ASM formalism has the computational power of a Turing machine, it can be used as an executable specification language. The notion of execution of an ASM is the usual notion in the context of transition systems. For instance ASM_Gopher [94] is an implementation of this formalism that has served as the basis for a formalisation of the Java programming language and virtual machine.

Another specification methodology based on the notion of ASM is the B Method, accompanied by its B specification language [2]. The systems modelled in B are seen, as in many other formal methods, as transition systems. The basic unit is called an *abstract machine*, and specifications and programs are represented using a dedicated notation for these abstract machines. In a sense, the B modelling methodology

Fig. 2.1 An example abstract machine in B

```
MACHINE Car_status
SETS
    STATUS = {sold, available}
USES Cars
VARIABLES status
INVARIANT
    status ∈ CARS ↛ STATUS
INITIALISATION status := ∅
OPERATIONS
set_status(x, m) ≙
    PRE m ∈ STATUS ∧ x ∈ CARS
    THEN
        status(x) := m
    END
END;
```

is close to object-oriented modelling. Each machine defines the structure of its internal state (values), the properties that this state must always comply with (static properties called *machine invariants*), and the expected operations (the transitions). The expected properties are defined in a suitable first-order logic extended with a particular set theory. An important principle is that each specified operation *must preserve* the machine invariants (a property referred to as *internal consistency*).

Over the years the B method has given rise to more than a dozen implementations, including Atelier B [87] (a commercial product with a freely available version), BRILLANT [40, 41] (an open source platform), ProB [77] (also open source, includes an animator and a model checker), and Rodin [3] (an open source platform dedicated to the implementation of a recent and popular dialect called Event-B).

Let us give an example of a B specification. The abstract machine shown in Fig. 2.1 encapsulates the notion, for a car dealer, of a car being sold or available. Using the previously defined machine *Cars* that introduces the notion of *CARS*, the machine keeps track of the previously recorded cars and associated status by the use of a partial function from cars to status. At the level of code, one may expect this to be implemented by some form of database or container datatype.

This machine also illustrates how the internal state may evolve, by providing an operation that allows for the status of a car to be introduced or updated. More precisely, the specification ensures that if the operation *set_status* is executed in a state in which the condition $m \in STATUS \land x \in CARS$ holds, then the resulting state records the previous *status* updated, by mapping x to m.

Set and Category Theory   These two mathematical theories offer similar expressiveness at the level of specifications. States are described in terms of mathematical structures such as sets, relations, or functions. Transitions are expressed as *invariants* as well as preconditions and postconditions.

Z [93] and VDM [67] are classic examples of formal methods whose specification languages rely on set theory. These methods have been at the origin of many other systems. RAISE [56], for instance, is an evolution of VDM that also includes

Fig. 2.2   One task and one
resource

a concurrency specification layer *à la* CSP. The B specification method can also be seen as falling in this category, since it combines the use of ASMs with features inherited from both Z and VDM. We may also cite the emerging methodology Alloy and its related tool Alloy Analyser [65], which adapts and extends declarative languages like Z to bring in fully automatic (but partial) analysis. Specware [68] and Charity [38] on the other hand offer formalisms based on category theory.

Automata-Based Modelling   A different class of transition systems used for specification purposes are *automata* [5, 6, 74, 78, 86, 97, 100]. In this case it is the *concurrent behaviour* of the system being specified that stands at the heart of the model. The main idea is to define how the system reacts to a set of *stimuli* or events. A state of the resulting transition system represents a particular configuration of the modelled system. This formalism is particularly adequate for the specification of *reactive*, *concurrent*, or *communicating* systems, and also protocols. It is however less appropriate to model systems where the sets of states and transitions are difficult to express.

Let us consider for instance a system that allows two tasks to access a shared resource. For one task, the process of using this resource is depicted by the Büchi automaton shown in Fig. 2.2. The state N corresponds to the situation in which the task is idle. The access request is done via the event/transition P, after which the automaton reaches the state E, which corresponds to waiting for access to the resource. The transition T represents the authorisation of access, which allows the task to reach the state L, in which the task can perform any operation that requires access to the resource. The event F represents the release of the resource, leading back to state N.

Now if one wants to extend this simple model to two tasks (a and b) that compete for access to the same resource, this can be done with the automaton of Fig. 2.3, which is obtained by calculating the Cartesian product of the previous automaton with itself, and then removing states and transitions that are meaningless or simply unwanted.[1] We remark that we can easily convince ourselves that there is no deadlock in this schema, but more interestingly this can also be proved. One can

---

[1] The remaining transitions are usually called the *synchronisation set* over the Cartesian product.

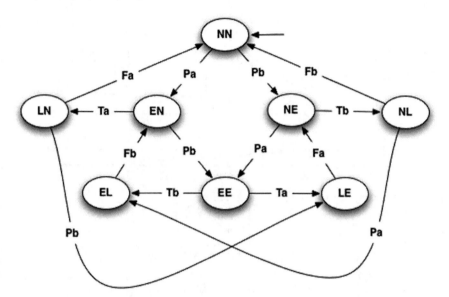

Fig. 2.3 Two tasks and one shared resource

also understand quite easily (and again, also prove) that this sharing policy is not fair. In fact it is possible that *a* asks for the resource followed by *b*, that repeatedly asks and is given access to it (this corresponds to the repeated execution of the loop *EN − EE − EL*). More on this simple example can be found in [19].

Modelling Languages for Real-Time Systems   Extending the simple automata framework gives rise to several interesting formalisms for the specification of real-time systems. When dealing with such systems, the modelling language must be able to cope with one or more physical concepts like time (or duration), temperature, inclination, altitude, etc. In fact, examples of real-time systems include for instance control systems that react in dynamic environments. Traditionally, time (usually modelled by means of clocks that follow a very specific progression law) is the dimension that has attracted most attention from researchers in the field.

For instance in the context of *synchronous concurrent models* [18, 59], the time flow is partitioned in discrete instants. An integer variable x can be treated, in fact, by considering the sequence of all the values that it takes at each discrete instant (e.g. $x = 1, 2, 3, 4, 5, 6, 7, \ldots$). The modelled systems progress (behave) according to successive atomic reactions upon the previous and actual values of the involved variables. Again, this kind of model of a system can be seen as an automaton where the states correspond to assignments of values to system variables, and where transitions correspond to system reactions.

Lustre [35] is a (textual) synchronous dataflow language, and SCADE [1] is a complete modelling environment that provides a graphical notation based on Lustre. Both provide a notation for expressing synchronous concurrency based on dataflow. Consider the SCADE example of Fig. 2.4. It introduces an operator (the unit of en-

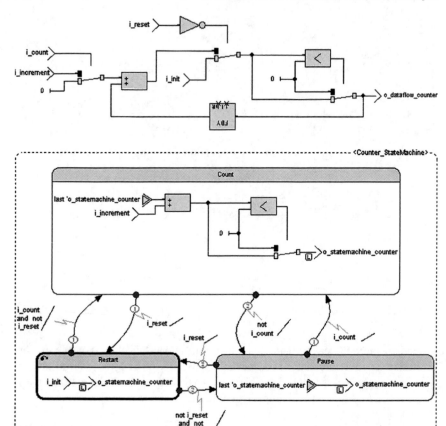

Fig. 2.4 Two counters in SCADE

capsulation in SCADE) that provides two alternative implementations of a counter: a dataflow-based counter and a state machine counter. Both counters run concurrently.

This operator has four input parameters:

- *i_init*: initial value of the counter;
- *i_increment*: value of the increment;
- *i_count*: counter switch;
- *i_reset*: reset the counter to the value of *i_init*;

and two output parameters:

- *O_statemachine_counter*: records the value of the state machine counter;
- *O_dataflow_counter*: records the value of the dataflow counter.

Other graphical formalisms have proved to be suitable for the modelling of real-time systems. One of the most popular is based on networks of *timed automata*.

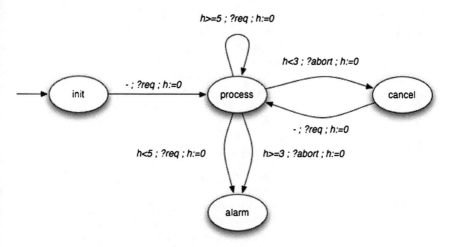

Fig. 2.5 A timed automaton for a requests processor

Basically, timed automata extend classic automata with clock variables (that evolve continuously but can only be compared with discrete values), communication channels, and guarded transitions. A theory of timed automata [7] provides the underlying foundations of this approach, which gave rise to model checking tools (see Sect. 2.3.3) like Uppaal [17] or Kronos [32].

The timed automaton represented in Fig. 2.5 models a control system that handles some processing request in a timely fashion. The control system reacts to two actions, a process request *req* and a cancel request *abort*. A process request is taken into account if a previous request has not taken place less than 5 time units before. Similarly an abort request is processed only if it is received at least three time units after a process request. If either of these two rules is not observed then the control system reaches an alarm state. The control involving time is done via the clock h. Only simple comparisons and a reset operation are possible. The transition $h \geq 5$; ?*req*; $h := 0$ stands for "this transition is possible when $h \geq 5$ and a process request is received; if the transition is chosen, then the clock h is reset to 0". This control system is supposed to be executed concurrently with a system (modelled by another timed automaton) that emits the required signals through the adequate communication channel (!*req* and !*abort*).

Hybrid automata [62] extend timed automata with the possibility to deal with other physical measures beyond time. Although the general theoretical context is very difficult (the problems that arise are easily undecidable), the existence of decidable fragments has enabled the creation of tools, such as Hytech [63]. Although essentially a deductive tool, we should also mention the recent Keymaera platform [85] for modelling and verification of hybrid systems, based on *differential dynamic logic* [84]. This is intended to model dynamic systems with interacting discrete and continuous behaviour, which is classically characterised by differential equations and discrete transitions. One advantage of the underlying theory is that it can handle such characterisations and provide proof mechanisms for them.

Fig. 2.6 Algebraic
specification of lists

Spec: *LIST$_0$ (ELT)*
Extends: *Nat$_0$*
Sorts: *list*
Operations:

nil	:	$\rightarrow$ *list*	// constructor
cons	: *elt* $\rightarrow$ *list*	$\rightarrow$ *list*	// constructor
length	: *list*	$\rightarrow$ *nat*	
hd	: *list*	$\rightarrow$ *elt*	
tl	: *list*	$\rightarrow$ *list*	
append	: *list* $\rightarrow$ *list*	$\rightarrow$ *list*	
rev	: *list*	$\rightarrow$ *list*	

Axioms: $xs, ys : list, \ x : elt$

$$length(nil) = 0$$
$$length(cons(s, xs)) = 1 + length(xs)$$
$$hd(cons(x, xs)) = x$$
$$tl(cons(x, xs)) = xs$$
$$append(nil, ys) = ys$$
$$append(cons(x, xs), ys) = cons(x, append(xs, ys))$$
$$rev(nil) = nil$$
$$rev(cons(x, xs)) = append(rev(xs), cons(x, nil))$$

2.2.2 Algebraic Specification

A second classic approach to specification is based on the use of multi-sorted algebras. A multi-sorted algebra consists of a collection of data grouped into sets (one
set for each datatype), a collection of functions on these sets corresponding to functions of the program being modelled, and a set of axioms specifying the basic properties of the functions in the algebra. Such a formalism allows one to abstract away
from the algorithms used to encode the desired properties, to concentrate on the
representation of data and on the input-output behaviour of functions. In addition to
multi-sorted algebras, the foundations of algebraic specification lie in mathematical
induction and equational logic.

An algebraic specification then consists in a series of sort declarations, function
signatures, and axioms that declare the basic behaviour of each function symbol.
CASL [22, 39], OBJ [54], Clear [33], Larch [55], and ACT-ONE [51] are all examples of tools based on algebraic specification languages. LOTOS [24] extends
the algebraic framework with CCS primitives, and thus allows for specifying and
reasoning about concurrent systems.

As an illustrative example, Fig. 2.6 introduces a container datatype for lists,
based on the two usual elementary operations on lists—the constructors *nil* (the
empty list) and *cons* (the operation that adds one element at the head of a list). This
datatype relies on some existing algebraic specification of natural numbers. The operations over the datatype are declared by stating their names and signatures; their
behaviour is expressed equationally in terms of their relation with the two constructors. Any implementation complying with this specification must ensure the stated
equational properties hold. Thus, such implementations will also comply with the
inferred properties.

An interesting point is the ability to infer additional and rich properties from the stated ones. We will address this issue in the sequel.

2.2.3 Declarative Modelling

An important class of specification languages includes logic-based languages, functional languages, rewriting languages, and languages for defining formal semantics. All of these rely on well-known mathematical foundations.

Logic programming languages, such as Prolog [95], propose an approach to modelling based on the notion of *predicate*. Data are represented with the help of some simple, but sufficiently expressive datatypes, such as lists, and operations are described by their behavioural properties, similarly to axioms in algebraic specifications. Prolog allows for specifications to be executed.

Functional languages on the other hand offer a specification framework in which the notion of *function* is the central element. The core of a functional language is the *λ-calculus* [10, 11], which has the same expressiveness as Turing machines and allows for the formulation of any operation, and even any datum, in terms of higher-order functions.

Languages like Scheme [50], SML [81], Haskell [96] or OCaml [75], and proof assistants such as ACL2 [69], Coq [79], PVS [91], HOL [57], Isabelle [83], and Agda [27], are all based on typed variants or extensions of the λ-calculus. These extensions are easier to use than the original calculus, and all propose a number of basic data types, powerful type construction operators (such as inductive types), and mechanisms for the definition of functions.

Execution in these languages relies on the notion of *reduction*, which resembles the notion of calculation in mathematics. These languages are all higher-order, and some (such as Coq) possess even richer extensions. This allows for great flexibility and expressiveness. The languages underlying proof assistant systems based on type theory and the Curry-Howard isomorphism [11, 92] may also be used as higher-order logical languages.

Rewriting systems [8, 21, 72] like ELAN [25] or SPIKE [26] offer languages that are very close to those used in algebraic specification, with the difference that axioms are replaced by equations that characterise the behaviour of function symbols in calculations.[2] As in the λ-calculus, execution relies on the notion of reduction, which is usually defined by sets of equations.

For illustration purposes, the Coq example shown in Fig. 2.7 introduces the abstract syntax of two simple languages. `expr` denotes simple arithmetic expressions with variables and assignment; `intr` corresponds to expressions of a very simple assembly language. The recursive function `compil` introduces a compilation schema, that is a translation of expressions to programs (lists of instructions). Once these notions have been introduced, one can define the operational semantics via the

---

[2]See http://rewriting.loria.fr/ for a list of many other rewriting systems.

```
Inductive expr : Set :=  │  Var  : ident -> expr
                         │  Num  : nat -> expr
                         │  Atrib: ident -> expr -> expr
                         │  Sum  : expr -> expr -> expr
                            ...

Inductive instr : Set :=  │  LOAD  : ident -> instr
                          │  STORE : ident -> instr
                          │  PUSH  : nat -> instr
                          │  DUP  : instr
                          │  ADD  : instr
                             ...

Definition program := (list instr).

Fixpoint compil (e : expr) {struct e} : program := match e with
        │  Var i => (LOAD i)::nil
        │  Num n => (PUSH n)::nil
        │  Atrib i e2 => app (compil e2) (DUP::(STORE i)::nil)
        │  Sum e1 e2 => (app (app (compil e1) (compil e2)) (ADD::nil))
           ...
end.
...

Theorem correctness : forall e:expr,
                  (extract_eval (aval e nil)) =
                  (extract_exec (exec (mkstate nil nil )(compil e))).
Proof.
...
```

Fig. 2.7 Two simple languages in Coq

notion of execution at source (`aval`, that depends on a variable environment) and
at machine level (`exec`, that depends on an execution environment). This allows us
to elegantly define the notion of compiler correctness in this very simple context:
for every expression `e`, executing `e` at source level or executing its compiled form
at machine level must always give the same result. This is what is stated by the
`correctness` theorem.

2.3 Specifying and Proving

Animation and execution are not sufficient to ensure a certain behaviour for a specification: it is essential to obtain a rigorous demonstration. Rather than simply constructing specifications and models one is interested in proving properties about
them. We are thus in the realm of *formal verification*.

Different notions of verification exist; Rushby [88] proposes a three-level classification of proof methods as follows:

1. The first level groups formal frameworks that do not offer any computer-based
   support for handling proofs. Demonstrations must then be carried out by hand;
   proofs are validated when reviewers are convinced of their contents.
2. In the second level we have frameworks that additionally offer a formal system
   allowing for a more rigorous formulation of demonstrations. For instance usage
   of natural language, which is possible in level 1, is not admissible in level 2. But
   demonstrations are still carried out by hand.

3. The third level consists of computer-based tools with support for proofs and for carrying out demonstrations. This level offers the highest degree of exactitude and guarantee. The undeniable advantage is that models can be both *expressed* and *reasoned about* in a formal (and possibly mechanical) way.

Whichever method is used, proving properties about specifications presupposes the use of some logical system. In what follows we first give in Sect. 2.3.1 an overview of logical concepts. Propositional logic and first-order logic will be the subject of Chaps. 3 and 4, so our goal here is to give the basic notions necessary for understanding the subsequent Sects. 2.3.2 to 2.3.4, in which we discuss different classes of computer-based proof tools that are often used in formal verification.

2.3.1 Logic in a Nutshell

Logic can be described as the study of the principles of reasoning. Reasoning about situations means constructing arguments about them. We are interested in doing this formally, so that the arguments are valid and can be defended rigorously. A formal logic is a language equipped with rules that allow one to establish when the truth of a given sentence can be concluded from the truth of other sentences.

Symbolic logic is the branch of mathematics devoted to formal logic, i.e. to the study of logical languages, their semantics, and their proof theory, and the way in which these are related.

A logic consists of:

– A *logical language* in which sentences are expressed. A logical language is a formal language having a precise, syntactic characterisation of well-formed sentences. A logical language consists of logical symbols, characterised by having a fixed interpretation, and non-logical ones, whose interpretations are not fixed. These symbols are combined together to compose well-formed formulas.
– A *semantics* that differentiates valid sentences from refutable ones. The semantics is defined in terms of the truth values of sentences. This is done using an interpretation function that assigns meaning to the basic components, given some domain of objects that our reasoning is concerned with.
– An *inference system* (or *proof system*) that supports the formalisation of arguments justifying the validity of sentences. The inference system is composed of a set of axioms (sentences of the logic that are accepted as true) and inference rules (that give ways of deriving the right conclusions from a given set of premises).

Of course, extreme care must be taken regarding the definition of an inference system, since it is expected that all the derived formulas are indeed justified semantically (this is usually called a *soundness criterion*), and that all consequences justified semantically are derivable by the system (a *completeness criterion*).

Propositional Logic, First-Order Logic, and Higher-Order Logic Let us briefly describe these three well-known logics. Each is a richer logic than the previous; all of them are widely used in the context of formal verification, so it is useful to have some understanding of their characteristics.

Propositional logic (also known as the *propositional calculus*) is the simplest of the three. The formulas of propositional logic are built from atomic propositions, which are sentences with no internal structure and which one can classify as being "true" or "false". Propositions are declarative sentences such as "Mary is the mother of John" or "$3 < 5$". Propositions are combined using Boolean operators that capture notions like "not", "and", "or", "implies", etc. In fact, the content of the propositions is not relevant. Propositional logic is not the study of the truth of individual formulas, but rather of the way in which the truth of one statement affects that of another.

First-order logic is a considerably richer logic than propositional logic. In addition to the symbols of propositional logic, a first-order language contains elements that allow us to reason about individuals of a given domain of discourse. These include functions, predicates, and quantification over individuals, dealing with the notions of "there exists" and "for all". There are two sorts of things involved in a first-order logic formula: *terms*, which are interpreted as individuals in the domain of discourse; and *formulas*, which are interpreted as truth values. First-order logic is also known as the *predicate calculus* in the sense that it is a calculus for reasoning about predicates such as "x is the mother of y" or "$x < x + 1$". While propositions are either true or false, predicates evaluate to true or false depending on the values given to their parameters (x and y in the previous examples). Quantification over individuals makes possible to express concepts such as "every person has a mother".

Higher-order logic is distinguished from first-order logic in several ways. It has both individual and relational variables, and both types of variables can be quantified, so quantifiers may apply to variables standing for predicates. There is a typing discipline to distinguish individuals from predicates; predicates can take as arguments both individual symbols and predicate symbols (these are higher-order predicates). Concepts such as "every property that holds for x also holds for y" can be naturally expressed in higher-order logic.

First-order logic is more expressive than propositional logic: it has more rules, that allow one to construct more complex formulas. In turn, higher-order logic is more expressive than first-order logic.

Classical versus Intuitionistic Logic There are two different branches of formal logic: the *classical* (based on the notion of *truth*) and the *intuitionistic* (based on the notion of *proof*). The classical branch of logic is based on the understanding that the truth of a statement is absolute: statements are either true or false. In a classical setting, "false" and "not true" mean the same thing. This is expressed by the *law of the excluded middle*, which states that $A \vee \neg A$ must hold independently of the meaning assigned to A. Classically, in order to prove a proposition A, it is valid to assume $\neg A$ and obtain a contradiction as result. This classic practice of proving a statement *by contradiction* is captured by an inference rule known as *reductio ad absurdum*.

The intuitionistic (or *constructive*) branch of logic rejects the law of the excluded middle. A statement A is "true" if we can prove it, or is "false" if we can show that if we have a proof of A we get a contradiction. If neither of these can be shown, then there exists no justification for the presumed truth of the disjunction $A \lor \neg A$. In an intuitionistic setting, judgements about a statement are based on the existence of a proof (or "construction") of that statement. Mathematicians are typically inclined to resort to classical reasoning, in spite of the fact that most standard mathematics fit within the framework of intuitionistic logic. In some cases the inability to use classical proof methods such as proofs by contradiction make reasoning much more difficult.

The guiding principle of intuitionistic logic is however very attractive to computer scientists, due to the algorithmic nature of constructive proofs. The importance of the relationship between logic and computer science cannot be overstated. Each field is of paramount importance to the other: logic plays a crucial role in computer science since it supplies the tools to formalise many important concepts in this field, in such a way that they can be reasoned about formally. On the other hand computers are useful tools for logic: formal logic makes it possible to calculate consequences at the symbolic level, and computers can be used to automate such symbolic calculations.

One of the most remarkable manifestations of this interplay between logic an computer science is the correspondence between systems of intuitionistic logic and typed lambda calculi, known as the *Curry-Howard isomorphism*. This correspondence establishes a connection between proof theory and type theory that is visible in many aspects: propositions correspond to types, proofs correspond to terms, the provability of a formula corresponds to the inhabitation of a type, proof normalization corresponds to term reduction, and so on. On a practical level, this correspondence give us ways to extract computer programs from constructive proofs, or even to view proofs as programs themselves. Moreover, the Curry-Howard isomorphism is at the core of some proof assistant tools (see also Sect. 2.4.2).

Propositional and first-order logic are essential tools for the verification of programs in the sense explained in Sect. 2.3.4, which is the main theme of this book. As such they will be covered in detail in Chaps. 3 and 4.

Temporal Logic The examples we gave to illustrate the expressive power of the different calculi where chosen because the truth value of the statements are static and cannot vary over time (i.e. they are always true or always false). Consider now the statement: "It is raining". Although the meaning of this statement is stable over time, its truth value can vary, sometimes it is true and sometimes it is false. One can capture this dependence by considering time as an object of discourse and making statements depend on a time variable, but the result is very clumsy.

The expression *temporal logics* is used to refer to logics that implicitly include a notion of time, providing a way to represent temporal information in the logical framework. In temporal logics the truth of a formula is not fixed in the semantics but depends on the point in time in which it its considered. Temporal logics have two kinds of operators: the usual logical connectives (such as "not", "and" or "implies") and temporal connectives (such as "eventually", "always" or "until"), allowing one to express statements like "It will be raining until the end of the race" or "It will eventually rain".

There exist many different sorts of temporal logics. Concerning the way time is viewed they are classified as *linear-time*, when time is represented by a sequence of time instants, or *branching-time*, when time is viewed as a tree of time instants, having the present instant as root, and with each branch corresponding to one possible future evolution. Temporal logics have found important applications in formal methods, in modeling behavioural aspects of the execution of computer systems.

2.3.2 Proof Tools

We have mentioned that computers can be used to calculate logical consequences. Let use discuss in some detail the ways in which this can be done.

Two opposing factors have an impact on the deductive behaviour of proof engines. On one hand, logical *expressiveness* permits studying complex properties and deductions; on the other hand, the *simplicity* of the logical formalism facilitates the automation of deductions. The two families of proof tools presented below represent choices that lead to different trade-offs between expressiveness and automation.

Automated Theorem Provers   These tools favour automation of deduction rather than expressiveness. Construction of proofs is automatic, once the proof engine has been adequately parameterised. Naturally these systems must rely on the decidability of at least a large fragment of the underlying theory. This is the case for *Horn clauses* as used in Prolog; for the first-order rewriting used for instance by ELAN; and for the fragment of first-order logic used by ACL2. Satisfiability Modulo Theory (SMT) solvers also fall in this category: tools like Yices [48], CVC3 [13], Z3 [46], or Alt-Ergo[3] [42] provide decision procedures for several theories of real numbers, integers, and of various data structures such as lists, arrays, bit vectors and so on. A historically important theorem prover with an almost unbeatable reputation in the program verification community is Simplify [47], an ancestor of modern SMT solvers.

Unlike model checkers (covered in Sect. 2.3.3), theorem provers may be able to employ techniques that allow for reasoning about infinite sets. Inductive reasoning techniques are an example of these.

Proof Assistants   Unlike theorem provers, proof assistants elect highly expressive (and thus undecidable) underlying logics, such as higher-order logic. There exist no decision procedures capable of proving arbitrary properties in such logics. If this

---

[3] Among others; see http://www.smtlib.org/ for a more complete list.

sounds restrictive, it must be emphasized that many properties need the power and elegance of higher-order logic to be adequately expressed and proved.

Proof assistants typically combine the following two modules:

- a *proof-checker*, responsible for verifying the well-formedness of the theories defined in the modeling process, and for checking the correctness of proofs;
- an interactive *proof development system*, to help (error-prone) users developing proofs. When the construction of a proof is finished, a proof script can be stored, describing that construction.

In most proof assistants proofs are *interactively constructed* by applying high-level proof-manipulation functions, usually known as *tactics*. Each tactic encodes a proof step. The proof state is usually represented as a stack of sequents (a pair of a sequence of hypotheses and a conclusion to be proved from them). A proof of a property ϕ is established by applying (in an appropriate order and with the right parameters) a set of tactics, in order to construct a proof tree linking axioms and theorems to the conclusion ϕ. In its simplest form, a tactic expresses a basic proof step, such as *modus ponens*. Tactics are however not restricted to such atomic steps—there is scope for complex reasoning and even for complete proofs, as is the case with tactics implementing decision procedures on decidable fragments of higher-order logic.

Two approaches are possible in the world of proof assistants. The first consists in giving users the possibility to define the logic in which they desire to express proofs. This logic is usually called the *object logic*. This axiomatic approach has been adopted for instance by the Isabelle system. The other approach is to offer a basic language that is sufficiently expressive to formulate most of mathematics. This integrated approach can be found in systems like Coq. In Coq proofs are first-class citizens of the language, at the same level as propositions and specifications. In these systems, proofs can be directly written by the user—but they are not in general necessarily easy to write. In practice, proof terms are more often generated as the result of a successful proof process, in addition to a proof script. Proof terms can be independently checked, which finds applications notably in the proof-carrying code techniques mentioned in Chap. 1. Other advantages of this approach will be discussed below.

To finish the section, let us remark that it is possible to combine automatic theorem proving with interactive proof facilities—this is the case in implementations of the B Method, whose (first order) proof mechanism allows for the interactive demonstration of lemmas that could not be proved mechanically.

2.3.3 Model Checking

Model checking [9, 19, 37] is a technique for the verification of finite-state (concurrent) systems (typically modelled by automata, see Sect. 2.2.1). It is one of the most widely used families of formal methods tools.

The idea of model checking is that the expected properties of the model are expressed by formulae of a *temporal logic*, and efficient symbolic algorithms are used to traverse the model in its entirety, so as to verify if all possible configurations validate those properties. The set of all states is called the model's *state space*. When a system possesses a finite state space, model-checking algorithms may in theory be used to realise the automatic demonstration of properties. If a property is not valid, a *counterexample* is exhibited.

A serious drawback of this approach is *state space explosion*: the transition graph typically grows exponentially on the size of the system, with the immediate consequence that no matter how efficient the checking algorithms are, the exploration of the state space eventually becomes impracticable.

Different techniques have been proposed that try to solve this problem. *Abstraction* is one such technique: a simplified version of the model is proposed, called an *abstract model*, whose state space may be explored within reasonable time. The abstract model must respect certain properties, in order to ensure that if a property is valid in it, then it is valid in the original model.

2.3.4 Program Logics and Program Annotation

To finish off with proof techniques, let us now consider tools based on *program annotations*. A program annotation is a formula placed together with the code of a program whose behaviour one wants to verify. The annotation of a program function, method, or piece of code in general, is supposed to indicate the conditions that should be met before the said code is executed, as well as describe the logical state of the program after its execution. The logical formalisms underlying this approach are program logics like *Hoare logic*. This family of formalisms is very diverse: even if the basis of the annotation languages is quite standard, the semantics of annotations is specific to the programming language at hand.

Note that this is a somewhat different setting with respect to what we have been considering in Sect. 2.3: we are no longer discussing properties of formal models of systems in general, but instead behavioural properties of *programs* (in particular source code) written as annotations. Roughly speaking, an annotation is a code-aware version of the requirements. The rationalisation of the code annotation methodology, coupled with its integration within the software engineering discipline, gave rise to a software development paradigm based on the notion of *contract* (a specific form of annotation), as pioneered in the Eiffel programming language [80], which implements the notion of *runtime* or *dynamic verification* of contracts.

This paradigm has nowadays become very popular, in fact almost every widespread programming language benefits from a contracts layer. Let us cite for instance the programming languages SPEC# [12] (which can be seen as a superset of C#) or SPARK [34], a carefully chosen subset of ADA targeted to the development of safety critical systems. Like Eiffel, both languages natively support the paradigm, but they additionally support the static verification of contracts. In the context of the Java programming language, different annotation systems exist that are based on the JML annotation language [66], such as Esc/Java [52], KeY [4] and Krakatoa [73].

The last programming language we consider here (and many are left out of this discussion) is C. Along with ADA, C is a popular choice in the safety critical industry, and the contract-based approach fits well the need for the static assurance of safety properties. A complete analysis and validation platform for C that provides a contracts layer is the Frama-C toolset [43], based on the ACSL annotation language [16], which in turn was inspired by JML. ACSL and the program verification functionality of Frama-C will be covered in Chaps. 9 and 10 of this book respectively. Another interesting contracts layer for C is provided by the VCC toolset [45]. The VCC approach allows for the verification of concurrent aspect of C programs.

Tools for statically checking the correspondence between the code and the annotations can be totally proof-based, but they can also associate the use of model-checking with a proof assistant. This is the case of the Loop [20] and Bandera [49] tools.

The undeniable advantage of the annotation-based approach is that it is the source-language implementation, and not some specification, that serves as the basis for the verification. In fact, a model is here constructed by taking as inputs a program and its annotations, together with an underlying model of the programming language. This approach is more and more seen as providing a satisfying alternative to the central problem of formal methods. This book covers the foundations of this approach, from Hoare logic to the generation of verification conditions for programs consisting of annotated routines.

2.4 Specifying and Deriving

We have to this point considered tools and techniques that address the first part of the central problem of formal methods. We now turn to the second part of this problem; given a specification that enjoys the desired properties, how to obtain an implementation whose behaviour matches the specification?

Again, different solutions exist to this problem. The solutions fit in two categories: either the specification is itself a program that can be directly executed (and the problem is immediately solved), or an implementation is produced from the specification, in which case the problem of the *correctness* of derivations must be dealt with.

One approach to dealing with this problem focuses on the derivation mechanisms, which can be restricted in appropriate ways, to ensure that the derived code satisfies the properties of the original specification. A second approach is to make the (not necessarily correct) derivation process generate a set of *proof obligations* such that, if all these obligations can be proved, this guarantees the correctness of the implementation with respect to the specification. Correction may then be ensured either manually, or preferably by machine, using level 3 formal verification tools (see Sect. 2.3). Formal verification is important even if the first approach is followed: in this case, it is the derivation process itself that has to be validated. If level 3 verification can be applied successfully to the derivation mechanism, then a universally valid procedure is obtained for all derivations.

Thus either the derivation mechanism or individual derivations must be subject to formal verification, which is in fact omnipresent at all stages of the central problem of formal methods. Both these approaches are sometimes referred to as *correct-by-construction* software development. We remark that some authors use this expression for the first approach only, when the derivation mechanism has been verified once and for all and specific derivations do not require proof. Other authors use it for development based on successive verified refinement steps, as will be described in Sect. 2.4.1.

We remark that in the program annotation approach discussed in Sect. 2.3.4, an internal model is deduced from the annotated code; the code is correct with respect to the annotations if the proof obligations that arise from this translation process and inspection of the model can be proved. So although the perspective is slightly different, the correctness properties still concern the relation between implementation and model.

2.4.1 Refinement

Refinement is the technique that synthesizes a program from a specification step by step, such that each step increases the degree of precision with respect to the initial specification. Each additional step represents an implementation choice, such as the choice of algorithm for implementing a given function, or the choice of a concrete datatype to implement an abstract type (say the implementation of a set as a linked list) or even the weakening of a precondition of an operation.

Individual refinement steps must be proved correct, i.e. the effect of the concrete specification must not contradict the effect of the abstract/refined specification, in order for the final program to enjoy the same properties as the original specification. Each step thus generates a number of *refinement proof obligations* that must be discharged. The good news is that the correctness of each individual step is in principle much easier to establish than the overall correctness.

This is the technique followed by approaches like Z, VDM, and B. This very special ability to link a high level view to the resulting code via a chain of design choices and proofs is particularly suitable (and has been used) in the context of a vertical application of the Balzer life cycle. Refinement is a very popular and successful example of application of formal methods in industry. It is well supported in terms of tools, and what is more it provides the simplest way to realize Balzer's vision of the software development process. As mentioned before, software developed through a chain of formally verified refinement steps is sometimes referred to as *correct-by-construction*.

In order to illustrate this concept, we consider in Fig. 2.8 the very classic example of a B machine that introduces the notion of a finite subset of the natural numbers with cardinal less than or equal to a given parameter (*maxelem*). Using the set-theoretic foundations of the method, the internal state includes the set in question,

Fig. 2.8 Finite sets of natural
numbers in B

MACHINE $Set1\,(maxelem)$
CONSTRAINTS $maxelem \in \mathbb{N}_1$
VARIABLES set
INVARIANT
  $set \subseteq \mathbb{N}$
  $\wedge\ card(set) \leq maxelem$
INITIALISATION $set := \emptyset$
OPERATIONS
$add(n) \;\widehat{=}\;$
  PRE $n \in \mathbb{N} - set\ \wedge\ card(set) < maxelem$
  THEN $set := set \cup \{n\}$
  END
END;

Fig. 2.9 Refinement of finite
sets in B

REFINEMENT $Set2$
REFINES $Set1$
VARIABLES $tab, index$
INVARIANT
  $index \in 0..maxelem$
  $\wedge\ tab \in 1..index \rightarrowtail \mathbb{N}$
  $\wedge\ ran(tab) = set$
INITIALISATION $index, tab := 0, \emptyset$
OPERATIONS
$add(n) \;\widehat{=}\;$
  PRE $n \in \mathbb{N} - ran(tab) \wedge index < maxelem$
  THEN $index := index + 1 || tab(index + 1) := n$
  END
END;

that is initialised to the empty set. The machine has an invariant (a property that must hold before and after the execution of any operation) stating that the variable *set* is indeed a subset of the natural numbers, and has a cardinality that is less than or equal to the parameter. The only provided operation adds a new element *n* to the set, provided that *n* is not already an element, and that the set can in fact be augmented (in terms of its cardinality).

This machine can be refined by the following design choice: we opt for a scalar representation of the recorded values instead of a set. The resulting machine *Set2* is shown in Fig. 2.9. Here, the injective function *tab* assumes the role of the variable *set*. The invariant of the refined machine now states, and this is an important point, the relation between *set* and *tab*. The variable *index* records the cardinal of the set and indeed is used as the index of the last added element, if one sees *tab* as an array. The proof obligations generated by any implementation of B will establish, when discharged, that the behaviour of machine *Set2* is indistinguishable from the behaviour of machine *Set1*.

2.4.2 Extraction

The *Calculus of Inductive Constructions* (CIC), that stands at the foundation of the Coq system, is an extension of the typed λ-calculus. By the Curry-Howard isomorphism [11, 92], this calculus is also a *constructive* minimal higher-order logic, with the consequence that every logical proposition can also be seen as a specification, and every proof expressible in the CIC can be seen as a program that obeys a specification. Indeed, this strong paradigm allows for the *extraction* of the computational contents of the proof (the program it contains) in a given programming language. As an example, let ϕ be the following theorem

$$\forall x, y \in \mathbb{N}. \exists q, r \in \mathbb{N}. (y = (q \times x + r) \ \wedge \ 0 \le r < x)$$

A proof of ϕ is, in this context, a function that takes two integers x and y and computes *the* pair (q, r) that testifies the validity of $(y = (q \times x + r) \ \wedge \ 0 \le r < x)$. This pair is of course $(y \div x, y \bmod x)$. In general the proof of a theorem of the form $\forall x. \exists y. (R \ x \ y)$, if it exists, is a function that maps x into a value y such that $(R \ x \ y)$ holds.

Coq is capable of performing extractions into the untyped functional language Scheme or into the typed functional languages Haskell and OCaml. The extraction of the program corresponding to the proof t of a property ϕ is done automatically and in a single step.

2.4.3 Execution

Specifying with the help of (declarative) programming languages is a *de facto* method for obtaining implementations. Logic-based languages like Prolog and functional languages like Scheme, ACL2, Haskell, SML, and OCaml can all be seen as offering scope for both specification and implementation in the same language.

The simple problem faced by these languages is their relevance in terms of industry-strength applications. It is arguable whether, say, Prolog can in fact be regarded by industry as a viable implementation language.

2.5 Specifying and Transforming

When discussing model checking we mentioned that it is sometimes desirable to transform a specification in order to make it amenable to manipulation by verification methods. This is true also outside the context of model checking: it is often useful, and even necessary, to construct variations of a specification in order to hide details that obscure the verification, or conversely to enrich the specification with an extra level of detail to take into account new behaviour. The main and general foundation for such transformations is the theory of *abstract interpretation* [44], which provides a framework for defining sound approximations.

In general, the possibility of constructing variations of the initial model allows for the *modularity* of formal verification. Behaviours may be decomposed in a number of different *views*, each of which concerns a well-defined part of the global model. In principle, studying each individual aspect of the model is easier than studying the global behaviour, and under certain conditions it may be equivalent in terms of the results obtained.

Surprisingly there is no popular and well established tool support for such transformations; in fact, even the definition of "tool-supported model transformation" is still an open issue. This is due in part to the fact that the abstraction problems, in their general form, are easily undecidable, and transformations are deeply tied to the properties to be proved and the modelling language used. Designing appropriate and sound approximations is not an easy task, and is usually undertaken in an ad-hoc fashion.

The JaKarTa toolset [14] for reasoning about JavaCard specifications is an early example of this approach. It provides a (rule-based) language and mechanisms for the specification of ad-hoc model transformations, based on their effects on the data structures of the model under analysis. For instance, when modelling the operational semantics of a virtual machine one may want to focus on the typing policy. In this case, the manipulated values by themselves are not relevant for the analysis. Given the specification of the effect of the transformation to be performed on the data manipulated by the virtual machine (forgetting the values, keeping the types), JaKarTa is able to automatically pass this transformation on to the operations of the machine, and to produce proof obligations (in Coq) that ensure the soundness of the transformation.

2.6 Conclusions

Clarke et al. [36] claim that no single tool seems to solve in a completely satisfying manner the central problem of formal methods. While proof assistants are solid tools for formal verification, they are hard to use and lack automation. Some tools propose a vertical approach, complete from specifications to programs, but they miss proof-support functionality. Hardly any proof tool offers support for transformation of specifications. Some efforts have been made to integrate functionally vertical specification methods with proof capabilities, such as B. Also, model checking modules have been proposed for both PVS and Coq [98], but the results cannot be considered to be entirely satisfying.

The obvious conclusion to draw from these observations is that in the current state of development, resorting to a *combination of methods and tools* is an appealing alternative. Code-oriented platforms like Key or Frama-C propose rich environments that integrate several tools. For instance, Frama-C allows the integration and interaction of several static analyses (slicing, value, interval, dead-code analysis, etc.) with deductive methods. The deductive facility itself allows for formal verification using several proof tools like Coq or SMT-solvers.

We finish the chapter with a discussion of the applicability of formal methods in industry.

2.6.1 Are Formal Methods Tools Ready for Industry?

After the discussion in the previous chapter and the overview of the present chapter we may now attempt to answer this question. We saw how both the horizontal and vertical application of the Balzer life cycle are addressed by formal methods tools— recall for instance the use of the correct by construction paradigm, or tool-supported approaches like SCADE or the B Method. In these last few years there has been a dramatic increase in the maturity of several tools, thus one can reasonably expect an even better context for formal methods in the coming years.

Nevertheless, even if it is now more reasonable, the use and application of formal methods still requires a solid knowledge of basic mathematics, and can still be considered to be challenging to the average software engineer (if not simply frightening or a waste of time). The reasons for this are multiple and complex, and include for instance

- the lack of adequate mathematical training;
- a software development context that is under the strong commitments of a reduced *time to market*; or
- the simple absence of proper planning, due to the development process being subject to constantly changing requirements.

The first argument is a fairly difficult foundational issue, but the variable geometry of the development process can at least in part be addressed by formal methods instruments; think for instance of the contract-based approach to software development. Nevertheless, there is undoubtedly a question of image at the heart of the problem. Any training or dissemination activity is a valuable contribution to the improvement of visibility, understanding, and acceptance of formal methods. This is especially important to demonstrate that, as we have already noted, formal methods are now sufficiently mature and usable.

The adequate use of most formal methods tools in an industrial context requires that the development team contains only one specialist in the field.[4] We refer the reader to [70] for a recent remark in this direction. A notable exception is the use of heavyweight formal methods (involving proof assistants, for instance) that clearly require specialised mathematical skills, but whose application is only justified in very specific contexts.

Nevertheless, while formal methods in general still have to improve their ability to cope with modularity and scalability, we have been seeing with increasing frequency the announcement of several *tours de force* in formal verification[5] which strengthen our belief that formal methods now possess all the arguments to change the state of affairs. As stated by Jim Woodcock in the context of the software verification grand challenge,

---

[4]In the same way that it takes only a single Linux guru in a team to disseminate and properly use this operating system.

[5]Consider for instance the published results on the formal verification of compilers [76], operating systems [71], avionic control systems [23] or cryptographic software [15], among many others.

1000000 of verified lines of code: you can't say any more it can't be done! Here, we've done it!

2.6.2 Is Industry Ready to Use Formal Methods?

An important aspect when considering the use of formal methods is that they are not a mere product. Using these methods is not like installing and applying an antivirus. As stated by J.-R. Abrial in several tutorials and documents about the B Method, adopting formal methods in a software company is more a strategical and method-ological issue than a technical one. We do not believe or advocate the widespread use of these methods in the software industry in general; their application should in-stead be considered when reliability, safety or security are a concern. Conscientious industrial applications of formal methods have already been conducted successfully in key areas, that have become flagship application areas.

Nevertheless, every software company has favoured and adopted some particular development process, and is unlikely to renounce it in favour of a completely new development process based on the use of formal methods. In order to adopt these methods, software companies have to reshape and adapt their in-house software design *savoir-faire*. This brings us again to the arguments stated in the previous chapter, and in particular to the Balzer life cycle.

Formal specification and verification are not easy or cheap, but the real cost has to be considered in the long term. One the other hand, their conclusions have to be taken with care: formal methods can only be used to specify or prove what was care-fully stated beforehand, and cannot be used to reason about what was not. Formally specifying and verifying a whole system is then unlikely to be feasible or even rea-sonable. The advisable practice is then to determine the important (or critical) parts of the system do be designed and validated, and to apply formal methods on these parts.

2.7 To Learn More

Formal methods are the subject of numerous books, surveys and technical overviews. Many of them have already been cited in this chapter. We highlight here some gen-eral popular references. The most widely cited references [28–30, 60] report on the use of formal methods in the general context of software engineering. More techni-cal surveys can be found in [36, 89] or in the more recent [64], dedicated to software verification. The latter special issue includes the already cited overview [99] that covers an important aspect barely touched in this chapter: the practice and industrial use of formal methods.

Several specialised books are also dedicated to formal methods, for instance [82] provides a nice introduction to the subject. [19] complements the previous reference by giving an overview of model checking tools.

References

1. Abdulla, P.A., Deneux, J.: Designing safe, reliable systems using scade. In: Proc. ISoLA 2004 (2004)
2. Abrial, J.-R.: The B-Book: Assigning Programs to Meanings. Cambridge University Press, Cambridge (1996)
3. Abrial, J.-R.: Modeling in Event-B System and Software Engineering. Cambridge University Press, Cambridge (2010)
4. Ahrendt, W., Baar, T., Beckert, B., Bubel, R., Giese, M., Hähnle, R., Menzel, W., Mostowski, W., Roth, A., Schlager, S., Schmitt, P.H.: The KeY tool. Softw. Syst. Model. **4**, 32–54 (2005)
5. Alur, R., Dill, D.: Automata-theoretic verification of real-time systems. In: Formal Methods for RealTime Computing. Trends in Software Series, pp. 55–82. Wiley, New York (1996)
6. Alur, R., Dill, D.: A theory of timed automata. Theor. Comput. Sci. **126**(2), 183–235 (1994)
7. Alur, R., Dill, D.L.: A theory of timed automata. Theor. Comput. Sci. **126**(2), 183–235 (1994)
8. Baader, F., Nipkow, T.: Term Rewriting and All That. Cambridge University Press, Cambridge (1998)
9. Baier, C., Katoen, J.-P.: Principles of Model Checking. MIT Press, Cambridge (2008)
10. Barendregt, H.P.: The Lambda Calculus, its Syntax and Semantics. Studies in Logic and the Foundations of Mathematics, vol. 103. North-Holland, Amsterdam (1984)
11. Barendregt, H.P.: Lambda calculi with types. In: Abramsky, S., Gabbay, D., Maibaum, T. (eds.) Handbook of Logic in Computer Science, vol. 2, pp. 117–310. Oxford University Press, New York (1992)
12. Barnett, M., Leino, K.R.M., Schulte, W.: The Spec# programming system: An overview. In: CASSIS: Construction and Analysis of Safe, Secure, and Interoperable Smart Devices, vol. 3362, pp. 49–69. Springer, Berlin (2004)
13. Barrett, C., Tinelli, C.: CVC3. In: Damm, W., Hermanns, H. (eds.) Proceedings of the 19th International Conference on Computer Aided Verification (CAV '07). Lecture Notes in Computer Science, vol. 4590, pp. 298–302. Springer, Berlin (2007)
14. Barthe, G., Courtieu, P., Dufay, G., de Sousa, S.M.: Tool-assisted specification and verification of typed low-level languages. J. Autom. Reason. **35**(4), 295–354 (2005)
15. Barthe, G., Grégoire, B., Béguelin, S.Z.: Formal certification of code-based cryptographic proofs. In: Shao, Z., Pierce, B.C. (eds.) POPL, pp. 90–101. ACM, New York (2009)
16. Baudin, P., Fillitre, J.-C., March, C., Monate, B., Moy, Y., Prevosto, V.: ACSL: ANSI/ISO C Specification Language. Preliminary Design (version 1.4). From the Frama-C website, http://frama-c.com (2010)
17. Behrmann, G., David, A., Larsen, K.G.: A tutorial on UPPAAL. In: Bernardo, M., Corradini, F. (eds.) Formal Methods for the Design of Real-Time Systems: 4th International School on Formal Methods for the Design of Computer, Communication, and Software Systems, SFM-RT 2004. LNCS, vol. 3185, pp. 200–236. Springer, Berlin (2004)
18. Benveniste, A., Caspi, P., Edwards, S.A., Halbwachs, N., Le Guernic, P., de Simone, R.: The synchronous languages 12 years later. Proc. IEEE **91**(1), 64–83 (2003)
19. Bérard, B., Bidoit, M., Finkel, A., Laroussinie, F., Petit, A., Petrucci, L., Schnoebelen, P.: Systems and Software Verification. Model-Checking Techniques and Tools. Springer, Berlin (2001)
20. van den Berg, J., Jacobs, B.: The LOOP compiler for Java and JML. In: Margaria, T., Yi, W. (eds.) Proceedings of TACAS'01. Lecture Notes in Computer Science, vol. 2031, pp. 299–312. Springer, Berlin (2001)
21. Bezem, M., Klop, J.W., de Vrijer, R. (eds.): Term Rewriting Systems. Cambridge Tracts in Theoretical Computer Science. Cambridge University Press, Cambridge (2002)
22. Bidoit, M., Mosses, P.D.: CASL User Manual. LNCS (IFIP Series), vol. 2900. Springer, Berlin (2004). With chapters by T. Mossakowski, D. Sannella, and A. Tarlecki
23. Blanchet, B., Cousot, P., Cousot, R., Feret, J., Mauborgne, L., Miné, A., Monniaux, D., Rival, X.: A static analyzer for large safety-critical software. CoRR, abs/cs/0701193 (2007)

24. Bolognesi, T., Brinksma, E.: Introduction to the ISO specification language Lotos. Comput. Netw. ISDN Syst. **14**(1), 25–59 (1987)

25. Borovanský, P., Kirchner, C., Kirchner, H., Moreau, P.-E., Ringeissen, C.: An overview of ELAN. In: Kirchner, C., Kirchner, H. (eds.) Proceedings of the International Workshop on Rewriting Logic and its Applications. Electronic Notes in Theoretical Computer Science, vol. 15. Pont-à-Mousson, France, September 1998. Elsevier, Amsterdam (1998)

26. Bouhoula, A., Kounalis, E., Rusinowitch, M.: SPIKE, an automatic theorem prover. In: Voronkov, A. (ed.) Proceedings of the International Conference on Logic Programming and Automated Reasoning (LPAR'92). Lecture Notes in Artificial Intelligence, vol. 624, pp. 460–462. Springer, Berlin (1992)

27. Bove, A., Dybjer, P., Norell, U.: A brief overview of Agda—a functional language with dependent types. In: TPHOLs '09: Proceedings of the 22nd International Conference on Theorem Proving in Higher Order Logics, pp. 73–78. Springer, Berlin (2009)

28. Bowen, J.P., Hinchey, M.G.: Seven more myths of formal methods. IEEE Softw. **12**(4), 34–41 (1995)

29. Bowen, J.P., Hinchey, M.G.: Ten commandments of formal methods. Computer **28**(4), 56–63 (1995)

30. Bowen, J.P., Hinchey, M.G.: Ten commandments of formal methods … ten years later. Computer **39**(1), 40–48 (2006)

31. Bowen, J.P., Stavridou, V.: Safety-critical systems, formal methods and standards. IEE/BCS Softw. Eng. J. **8**(4), 189–209 (1993)

32. Bozga, M., Daws, C., Maler, O., Olivero, A., Tripakis, S., Yovine, S.: Kronos: A model-checking tool for real-time systems (1998)

33. Burstall, R.M., Goguen, J.A.: An informal introduction to specification using CLEAR. In: Boyer, R.S., Moore, J.S. (eds.) The Correctness Problem in Computer Science, pp. 185–213. Academic Press, New York (1981)

34. Carré, B., Garnsworthy, J.: Spark—an annotated Ada subset for safety-critical programming. In: TRI-Ada '90: Proceedings of the Conference on TRI-ADA '90, pp. 392–402. ACM, New York (1990)

35. Caspi, P., Pilaud, D., Halbwachs, N., Plaice, J.A.: Lustre: A declarative language for real-time programming. In: POPL '87: Proceedings of the 14th ACM SIGACT-SIGPLAN Symposium on Principles of Programming Languages, pp. 178–188. ACM, New York (1987)

36. Clarke, E.M., Wing, M.J.: Formal methods: State of the art and future directions. ACM Comput. Surv. **28**(4), 626–643 (1996)

37. Clarke, E.M., Grumberg, O., Peled, D.A.: Model Checking. MIT Press, Cambridge (1999)

38. Cockett, R., Fukushima, T.: About Charity. Technical Report 92/480/18, University of Calgary (June 1992)

39. CoFI (The Common Framework Initiative): CASL Reference Manual. LNCS (IFIP Series), vol. 2960. Springer, Berlin (2004)

40. Colin, S., Petit, D., Mariano, G., Poirriez, V.: BRILLANT: An open source platform for B. In: Workshop on Tool Building in Formal Methods (held in conjunction with ABZ2010), February 2010

41. Colin, S., Petit, D., Poirriez, V., Rocheteau, J., Marcano, R., Mariano, G.: BRILLANT: An open source and XML-based platform for rigorous software development. In: SEFM '05: Proceedings of the Third IEEE International Conference on Software Engineering and Formal Methods, Washington, DC, USA, 2005, pp. 373–382. IEEE Computer Society, Los Alamitos (2005)

42. Conchon, S., Contejean, E., Kanig, J.: Ergo: A theorem prover for polymorphic first-order logic modulo theories (2006)

43. Correnson, L., Cuoq, P., Puccetti, A., Signoles, J.: Frama-C user manual. From the Frama-C website, http://frama-c.com (2010)

44. Cousot, P., Cousot, R.: Abstract interpretation: A unified lattice model for static analysis of programs. In: Proceedings of the 4th ACM Symposium on Principles of Programming Languages, pp. 238–252. ACM, New York (1977)

45. Dahlweid, M., Moskal, M., Santen, T., Tobies, S., Schulte, W.: VCC: Contract-based modular verification of concurrent C
46. De Moura, L., Bjørner, N.: Z3: An efficient smt solver. In: Conference on Tools and Algorithms for the Construction and Analysis of Systems (TACAS) (2008)
47. Detlefs, D., Nelson, G., Saxe, J.B.: Simplify: A theorem prover for program checking. J. ACM **52**(3), 365–473 (2005)
48. Dutertre, B., De Moura, L.: The Yices SMT solver. Technical report, SRI (2006)
49. Dwyer, M., Hatcliff, J., Joehanes, R., Laubach, S., Pasareanu, C., Visser, R.W., Zheng, H.: Tool-supported program abstraction for finite-state verification. In: Proceedings of ICSE'01 (2001)
50. Dybvig, R.K.: The Scheme Programming Language: ANSI Scheme, 2nd edn. Prentice-Hall International, Upper Saddle River (1996)
51. Ehrig, H., Fey, W., Hansen, H.: ACT ONE: An algebraic specification language with two levels of semantics. Technical Report 83–03, Technical University of Berlin, Fachbereich Informatik (1983)
52. Flanagan, C., Leino, K.R.M.: Houdini, an annotation assistant for ESC/Java. In: International Symposium on FME 2001: Formal Methods for Increasing Software Productivity. Lecture Notes in Computer Science, vol. 2021, pp. 500–517. Springer, Berlin (2001)
53. Formal methods ressources. http://www.afm.sbu.ac.uk/
54. Futatsugi, K., Goguen, J., Jouannaud, J.-P., Meseguer, J.: Principles of OBJ-2. In: Reid, B. (ed.) Proceedings 12th ACM Symp. on Principles of Programming Languages, pp. 52–66. Association for Computing Machinery, New York (1985)
55. Garland, S.J., Guttag, J.V., Horning, J.: An Overview of Larch. Lecture Notes in Computer Science, vol. 693, pp. 329–348. Springer, Berlin (1993)
56. George, C., Haxthausen, A.E., Hughes, S., Milne, R., Prehn, S., Pedersen, J.S.: The Raise Development Method. Prentice-Hall International, London (1995)
57. Gordon, M.J.C., Melham, T.F. (eds.): Introduction to HOL: A Theorem Proving Environment for Higher-Order Logic. Cambridge University Press, Cambridge (1993)
58. Gurevich, Y.: Evolving algebras 1993: Lipari guide. In: Börger, E. (ed.) Specification and Validation Methods, pp. 9–36. Oxford University Press, Oxford (1995)
59. Halbwachs, N.: Synchronous Programming of Reactive Systems. Kluwer Academic, Norwell (1993)
60. Hall, A.: Seven myths of formal methods. IEEE Softw. **7**(5), 11–19 (1990)
61. Hartel, P.H., Moreau, L.: Formalizing the safety of Java, the Java virtual machine, and Java card. ACM Comput. Surv. **33**(4), 517–558 (2001)
62. Henzinger, T.A.: The theory of hybrid automata. In: LICS '96: Proceedings of the 11th Annual IEEE Symposium on Logic in Computer Science, Washington, DC, USA, 1996, p. 278. IEEE Computer Society, Los Alamitos (1996)
63. Henzinger, T.A., Ho, P.-H., Wong-toi, H.: Hytech: A model checker for hybrid systems. Softw. Tools Technol. Transf. **1**, 460–463 (1997)
64. Hoare, C.A.R., Misra, J.: Preface to special issue on software verification. ACM Comput. Surv. **41**(4), 1–3 (2009)
65. Jackson, D.: Software Abstractions: Logic, Language, and Analysis. MIT Press, Cambridge (2006)
66. JML Specification Language. http://www.jmlspecs.org
67. Jones, C.B.: Software Development. A Rigorous Approach. Prentice-Hall International, Englewood Cliffs (1980)
68. Juellig, R., Srinivas, Y., Liu, J.: SPECWARE: An advanced environment for the formal development of complex software systems. In: Proceedings of AMAST'96. Lecture Notes in Computer Science, vol. 1101, pp. 551–554. Springer, Berlin (1996)
69. Kaufmann, M., Strother Moore, J.: ACL2: An industrial strength version of Nqthm. COMPASS—Proceedings of the Annual Conference on Computer Assurance, pp. 23–34 (1996). IEEE catalog number 96CH35960
70. Klein, G.: Correct os kernel? proof? done! USENIX ;login: **34**(6), 28–34 (2009)

71. Klein, G., Elphinstone, K., Heiser, G., Andronick, J., Cock, D., Derrin, P., Elkaduwe, D., Engelhardt, K., Kolanski, R., Norrish, M., Sewell, T., Tuch, H., Winwood, S.: sel4: Formal verification of an os kernel. In: Matthews, J.N., Anderson, T.E. (eds.) SOSP, pp. 207–220. ACM, New York (2009)

72. Klop, J.W.: Term-rewriting systems. In: Abramsky, S., Gabbay, D., Maibaum, T. (eds.) Handbook of Logic in Computer Science, vol. 2, pp. 1–116. Oxford Science Publications, New York (1992)

73. Krakatoa. http://www.lri.fr/marche/krakatoa/

74. Krauss, K.G.: Petri Nets Applied to the Formal Verification of Parallel and Communicating Processes. Lehigh University, Dissertation, Bethlehem, PA (1987)

75. Leroy, X., Doligez, D., Garrigue, J., Rémy, D., Vouillon, J.: The Objective Caml system, release 3.06 (2002). http://caml.inria.fr

76. Leroy, Xavier: Formal verification of a realistic compiler. Commun. ACM **52**(7), 107–115 (2009)

77. Leuschel, M., Butler, M.J.: ProB: an automated analysis toolset for the B method. Int. J. Softw. Tools Technol. Transf. (STTT) **10**(2), 185–203 (2008)

78. Mazzeo, A., Mazzocca, N., Russo, S., Savy, C., Vittorini, V.: Formal specification of concurrent systems: a structured approach. Comput. J. **41**(3), 145–162 (1998)

79. The Coq development team. The Coq proof assistant reference manual. LogiCal Project (2008). Version 8.2

80. Meyer, B.: Eiffel: The Language. Prentice Hall, Hemel Hempstead (1992)

81. Milner, R., Tofte, M., Harper, R., MacQueen, D.: The Definition of Standard ML (Revised). MIT Press, Cambridge (1997)

82. Monin, J.F.: Understanding Formal Methods. Springer, New York (2001)

83. Paulson, L.: Isabelle: A Generic Theorem Prover. Lecture Notes in Computer Science, vol. 828. Springer, Berlin (1994)

84. Platzer, A.: Logical Analysis of Hybrid Systems: Proving Theorems for Complex Dynamics. Springer, Heidelberg (2010)

85. Platzer, A., Quesel, J.-D.: KeYmaera: A hybrid theorem prover for hybrid systems. In: Armando, A., Baumgartner, P., Dowek, G. (eds.) IJCAR. LNCS, vol. 5195, pp. 171–178. Springer, Berlin (2008)

86. Reisig, W.: Petri nets and algebraic specifications. Theor. Comput. Sci. **80**, 1–34 (1991)

87. Requet, A.: An overview of Atelier B 4.0. In: Proceedings of the Conference The B Formal Method: From Research to Teaching'2008, Nantes (June 2008)

88. Rushby, J.: Formal methods and their role in the certification of critical systems. Technical Report SRI-CSL-95-1, Computer Science Laboratory, SRI International, Menlo Park, CA (March 1995)

89. Rushby, J.: Formal specification and verification for critical systems: Tools, achievements, and prospects. In: Suri, N., Walter, C.J., Hugue, M.M. (eds.) Advances in Ultra-Dependable Distributed Systems, pp. 282–296. IEEE Computer Society, Los Alamitos (1995)

90. Sannella, D.: A survey of formal software development methods. Technical Report ECS-LFCS-88-56, University of Edinburgh (July 1988)

91. Shankar, N., Owre, S., Rushby, J.M.: The PVS Proof Checker: A Reference Manual. Computer Science Laboratory, SRI International (February 1993)

92. Sørensen, M.H., Urzyczyn, P.: Lectures on the Curry-Howard Isomorphism. Studies in Logic and the Foundations of Mathematics, vol. 149. Elsevier, Amsterdam (2006)

93. Spivey, J.: An introduction to Z and formal specification. IEEE Softw. Eng. J. **4**(1), 40–50 (1989)

94. Stärk, R., Schmid, J., Börger, E.: Java and the Java Virtual Machine—Definition, Verification, Validation. Springer, Berlin (2001)

95. Sterling, L., Shapiro, E.: The Art of Prolog, 2nd edn. MIT Press, Cambridge (1994)

96. Thompson, S.: Haskell: The Craft of Functional Programming. Int. Comupt. Sci. Pearson Edn (1999)

 97. Vardi, M.Y., Wolper, P.: An automata-theoretic approach to automatic program verification. In: Symposium on Logic in Computer Science (LICS'86), pp. 332–345. IEEE Computer Society Press, Los Alamitos (1986)
 98. Verma, K.N., Goubault-Larrecq, J.: Reflecting BDDs in Coq. Technical Report RR3859, INRIA projet Coq (January 2000)
 99. Woodcock, J., Larsen, P.G., Bicarregui, J., Fitzgerald, J.: Formal methods: Practice and experience. ACM Comput. Surv. **41**(4), 1–36 (2009)
100. Wu, W., Saeki, M.: Specifying software architectures based on colored petri nets. IEICE Trans. Inf. Syst. **E83-D**(4), 701–712 (2000)

Chapter 3
Propositional Logic

Propositional logic is the basis for any study of logic. The sentences of propositional logic are built from a set of unstructured *atomic propositions* that are combined using a number of *logical connectives*. Logical connectives are Boolean operators whose names come from natural language, such as "*not*", "*and*", "*or*" and "*implies*", and they are given a formal meaning that mimics its usage in natural language. Such sentences, like "John plays tennis" or "$2 + 3 = 5$ and John has a car", can be said to have a truth value, i.e. to be true or false.

We are interested in reasoning about situations formally, so that valid arguments can be defended rigorously. The reader may ask why natural language is not suitable for carrying out logical reasoning. The main reason is that a sentence in natural language can be subject to different interpretations, depending on the context and implicit assumptions. This ambiguity of meaning makes natural language inadequate to carry out rigorous arguments. Another reason is that natural language is usually such a rich language that it cannot be formally described.

The sentences of propositional logic on the other hand are defined in a precise way by means of simple *syntactical rules*. The *semantics* of these sentences is given via an interpretation function that assigns a truth value (*true* or *false*) to every sentence. The syntax is also very relevant to the concept of *proof* (or *deduction*). When using logic as a proof system, we are not concerned with the meaning of the formulas manipulated, but with the syntax of these formulas. A proof system consists of a set of facts (*axioms*) and *inference rules* that allow us to deduce formulas from sets of formulas—the manipulation of formulas is purely mechanical, therefore appropriate to be performed by a computer. A *formal proof* is a sequence of applications of the rules of the proof system. An important property is that the set of provable formulas is exactly the same as the set of formulas shown to be true by semantic means, which legitimates the use of proof systems for establishing the validity of formulas.

This chapter is devoted to *classical* propositional logic, for which the relation between provability and validity will be explained in detail. We will also discuss the problem of deciding whether a formula is valid, and we will present some algorithms for this purpose. Due to space limitations some proofs will be omitted or left as exercises to the reader.

J.B. Almeida et al., *Rigorous Software Development*,
Undergraduate Topics in Computer Science,
DOI 10.1007/978-0-85729-018-2_3, © Springer-Verlag London Limited 2011

Propositional logic is the basis of more powerful logics. Most of the concepts presented here have counterparts in first-order logic, as will become evident in Chap. 4.

3.1 Syntax

The formal language of propositional logic is a precise language whose syntax can be completely described by simple formation rules, and whose semantics can be unambiguously defined. An important characteristic of this language is its conciseness and the way in which the logical structure of the sentences is emphasized.

Atomic propositions in themselves are not important; the study of propositional logic focuses on the relation between the truth of structured formulas and the truth of their constituent formulas. So, atomic formulas are denoted by letters without any mention of what these propositions actually say. The alphabet of a propositional language is organised into the following categories.

1. *Logical connectives*: $\perp$ (*absurdum*), $\neg$ (*negation*), $\wedge$ (*conjunction*), $\vee$ (*disjunction*) and $\rightarrow$ (*implication*).
2. *Auxiliary symbols*: "(" and ")".
3. *Propositions*: we assume a countable set **Prop** of propositional symbols. We let $P, Q, R, \ldots$ range over **Prop**.

Definition 3.1 The set of *formulas* of propositional logic is given by the abstract syntax:

$$\textbf{Form} \ni A, B, C ::= P \mid \perp \mid (\neg A) \mid (A \wedge B) \mid (A \vee B) \mid (A \rightarrow B)$$

An *atomic* formula is $\perp$ or a propositional symbol. *Structured* formulas are formed by combining other formulas with logical connectives. We let $A, B, C, \ldots$ range over **Form**, and read $\neg A$; $A \wedge B$; $A \vee B$ and $A \rightarrow B$ as *not A*; *A and B*; *A or B* and *A implies B*, respectively. Formulas in propositional logic are also referred to as *propositions*.

Some presentations of propositional logic include also connectives for *trueness* $\top$ and *equivalence* $\leftrightarrow$, but we prefer to see $\top$ as an abbreviation for $(\neg \perp)$ and $(A \leftrightarrow B)$ as an abbreviation for $((A \rightarrow B) \wedge (B \rightarrow A))$.

Remark 3.2 Care must be taken to distinguish between the formal language that is the object of our study (usually called the *object language*), and the (informal) language used to talk about the object of study (usually called the *meta-language*). We should always keep in mind this distinction to avoid confusion. For instance, the letters used to range over **Prop** and **Form** are variables of the meta-language, sometimes also called meta-variables.

Having considered parentheses as part of the syntax, we shall adopt some conventions to lighten the presentation of formulas:

- Outermost parentheses are usually dropped.
- Parentheses dictate the order of operations in any formula. In the absence of parentheses, we adopt the following convention about precedence. Ranging from the highest to the lowest precedence, we have respectively: $\neg$, $\wedge$, $\vee$ and $\rightarrow$.
- All binary connectives are right-associative.

Moreover, all meta-language symbols (i.e. symbols that do not belong to the syntax of formulas) bind loosely with respect to symbols of the syntax (e.g. $A \rightarrow B \in$ **Form** should be read $(A \rightarrow B) \in$ **Form**).

Example 3.3 Following our conventions for discarding brackets, $P \wedge Q \vee \neg R \rightarrow P \vee R \rightarrow Q$ stands for $(((P \wedge Q) \vee (\neg R)) \rightarrow ((P \vee R) \rightarrow Q))$, and $\neg P \wedge Q \wedge \neg R \vee \bot \vee \neg\neg Q$ stands for $(((\neg P) \wedge (Q \wedge (\neg R))) \vee (\bot \vee (\neg(\neg Q))))$.

When we characterise the syntax of a language by a grammar, we are in fact defining the set of sentences of the language by an inductive definition. Hence we have at our disposal the powerful tools of structural recursion to define functions on this set, and structural induction to perform proofs of properties on it. We illustrate the use of structural recursion with the definition of the set of subformulas of a given formula. Informally, given formulas A and B, we say that A is a *subformula* of B when A occurs syntactically within B.

Definition 3.4   The set of *strict subformulas* of A, $\mathsf{sSubF}(A)$, is defined inductively as follows

$$\mathsf{sSubF}(\bot) = \emptyset$$

$$\mathsf{sSubF}(P) = \emptyset$$

$$\mathsf{sSubF}(\neg A) = \{A\} \cup \mathsf{sSubF}(A)$$

$$\mathsf{sSubF}(A \wedge B) = \{A, B\} \cup \mathsf{sSubF}(A) \cup \mathsf{sSubF}(B)$$

$$\mathsf{sSubF}(A \vee B) = \{A, B\} \cup \mathsf{sSubF}(A) \cup \mathsf{sSubF}(B)$$

$$\mathsf{sSubF}(A \rightarrow B) = \{A, B\} \cup \mathsf{sSubF}(A) \cup \mathsf{sSubF}(B)$$

We say that B is a *subformula* of A if $B \in \mathsf{sSubF}(A)$ or $B = A$.

3.2 Semantics

Having described the syntax of propositional logic, we now describe the semantics which provides its meaning. The meaning is given by an *interpretation function* that maps formulas to *truth values* **T** and **F**, where **T** represents "true" and **F** represents "false" (hence, **T** $\neq$ **F**). This interpretation builds on the notion of *valuation*, which assigns to every propositional symbol a truth value. Recall that the only thing we assume about proposition symbols is that they represent "abstract" assertions that,

A	$\neg A$
F	**T**
T	**F**

A	B	$A \vee B$
F	**F**	**F**
F	**T**	**T**
T	**F**	**T**
T	**T**	**T**

A	B	$A \wedge B$
F	**F**	**F**
F	**T**	**F**
T	**F**	**F**
T	**T**	**T**

A	B	$A \rightarrow B$
F	**F**	**T**
F	**T**	**T**
T	**F**	**F**
T	**T**	**T**

Fig. 3.1 Truth tables for the connectives $\neg$, $\vee$, $\wedge$ and $\rightarrow$

in principle, can be either true or false. Thus, a valuation fixes a possible "scenario" (or "model"), in which some propositions are considered to be true, and others false.

Definition 3.5

1. A *valuation* is a function $\rho : \textbf{Prop} \rightarrow \{\textbf{F}, \textbf{T}\}$ that assigns truth values to propositional symbols.
2. Given a valuation ρ, the *interpretation function* $[\![\cdot]\!]_\rho : \textbf{Form} \rightarrow \{\textbf{F}, \textbf{T}\}$ is defined inductively as follows:

$$[\![\bot]\!]_\rho = \textbf{F}$$

$$[\![P]\!]_\rho = \textbf{T} \quad \text{iff} \quad \rho(P) = \textbf{T}$$

$$[\![\neg A]\!]_\rho = \textbf{T} \quad \text{iff} \quad [\![A]\!]_\rho = \textbf{F}$$

$$[\![A \wedge B]\!]_\rho = \textbf{T} \quad \text{iff} \quad [\![A]\!]_\rho = \textbf{T} \text{ and } [\![B]\!]_\rho = \textbf{T}$$

$$[\![A \vee B]\!]_\rho = \textbf{T} \quad \text{iff} \quad [\![A]\!]_\rho = \textbf{T} \text{ or } [\![B]\!]_\rho = \textbf{T}$$

$$[\![A \rightarrow B]\!]_\rho = \textbf{T} \quad \text{iff} \quad [\![A]\!]_\rho = \textbf{F} \text{ or } [\![B]\!]_\rho = \textbf{T}$$

The meaning of a structured formula is computed by combining the meanings of the subformulas according to a specific internal operation over the truth values, that captures the intended semantics of each connective. Concretely, we note that negation swaps the truth value (turns **F** into **T**, and vice-versa); conjunction returns **T** only if both arguments are **T**; disjunction returns **F** only if both elements are **F**; and implication returns **F** only if the antecedent is **T** and the consequent is **F** (i.e. it checks if truth is being preserved). These functions are normally specified in tabular form, giving rise to the *truth tables* presented in Fig. 3.1.

Example 3.6 Let ρ be a valuation such that $\rho(P) = \textbf{T}$, $\rho(Q) = \textbf{T}$ and $\rho(R) = \textbf{F}$. Then $[\![(P \rightarrow Q) \vee R]\!]_\rho = \textbf{T}$ (since $[\![P \rightarrow Q]\!]_\rho = \textbf{T}$), and $[\![\neg P \vee Q \rightarrow R \wedge P]\!]_\rho = \textbf{F}$ (since $[\![\neg P \vee Q]\!]_\rho = \textbf{T}$ and $[\![R \wedge P]\!]_\rho = \textbf{F}$).

Valuations assign truth values to all proposition symbols in **Prop**, but only the ones actually used in the construction of a formula do play a role in the computation of its meaning. This observation allows us to look at the interpretation of a formula A, constructed with proposition symbols $P_1, \ldots, P_n$, as a boolean function with n inputs (one for each proposition symbol used in A). Again, *truth tables* offer a convenient method for presenting the meaning of propositional formulas.

Example 3.7 Consider the formula $P \vee Q \rightarrow P \wedge Q$. The meaning of this formula can be computed by recursively combining the meaning of each of its subformulas by consulting the truth table of the corresponding connective (Fig. 3.1).

P	Q	$P \vee Q$	$P \wedge Q$	$P \vee Q \rightarrow P \wedge Q$
F	**F**	**F**	**F**	**T**
F	**T**	**T**	**F**	**F**
T	**F**	**T**	**F**	**F**
T	**T**	**T**	**T**	**T**

The resultant table is called the *formula's truth table*. We have chosen to include columns for each subformula of $P \vee Q \rightarrow P \wedge Q$, which facilitates the construction of the table. Nevertheless, the intermediate columns are only needed during the construction of the truth table, and are usually omitted in the final result.

Each row assigns a possible combination of truth values to the proposition symbols used in the formula. This gives rise to a total number of 2^n entries in the table (where n is the number of proposition symbols used in the formula). This turns the construction of truth tables impractical for (moderately) large values of n, a topic to which we will return in Sect. 3.5.

An alternative formulation of the semantics of propositional logic makes use of the notions of propositional model and validity relation.

Definition 3.8 A *propositional model* is a set of proposition symbols $\mathcal{M} \subseteq \textbf{Prop}$. The *validity relation* $\models \subseteq \mathcal{P}(\textbf{Prop}) \times \textbf{Form}$ is defined inductively by:

$$
\begin{aligned}
\mathcal{M} &\models P & \text{iff} \quad & P \in \mathcal{M} \\
\mathcal{M} &\models \neg A & \text{iff} \quad & \mathcal{M} \not\models A \\
\mathcal{M} &\models A \wedge B & \text{iff} \quad & \mathcal{M} \models A \text{ and } \mathcal{M} \models B \\
\mathcal{M} &\models A \vee B & \text{iff} \quad & \mathcal{M} \models A \text{ or } \mathcal{M} \models B \\
\mathcal{M} &\models A \rightarrow B & \text{iff} \quad & \mathcal{M} \not\models A \text{ or } \mathcal{M} \models B
\end{aligned}
$$

Atomic propositions are validated according to their membership in the model. The remaining clauses in the definition assign meaning to logical connectives.

The two formalisations are equivalent, since valuations are in bijection with propositional models: each valuation ρ defines a model $\mathcal{M} = \{P \in \textbf{Prop} \mid \rho(P) = \textbf{T}\}$. Conversely, a model $\mathcal{M}$ defines a valuation

$$
\rho(P) = \begin{cases} \textbf{T} & \text{if } P \in \mathcal{M}, \\ \textbf{F} & \text{otherwise.} \end{cases}
$$

A simple inductive proof establishes that both semantics assign equal meanings to formulas, i.e.

$$
\mathcal{M}_\rho \models A \quad \text{iff} \quad [\![A]\!]_\rho = \textbf{T}.
$$

The proof is left to the reader as Exercise 3.4. In the rest of the chapter we will follow this last formulation.

Definition 3.9

1. A formula A is said to be *valid in a model* $\mathcal{M}$ (or alternatively $\mathcal{M}$ is said to *satisfy* A), iff $\mathcal{M} \models A$. When $\mathcal{M} \not\models A$ the formula A is said to be *refuted* by the model $\mathcal{M}$.
2. A formula A is *satisfiable* iff there exists some model $\mathcal{M}$ such that $\mathcal{M} \models A$. It is *refutable* iff some model refutes A.
3. A formula A is *valid* (also called a *tautology*) iff every model satisfies A. A formula A is a *contradiction* iff every model refutes A.
4. Given two formulas A and B, A is said to be *logically equivalent* to B, denoted by $A \equiv B$, iff A and B are valid exactly in the same models.

More concisely, we can say that a formula is a contradiction if it is not satisfiable, and it is refutable if it is not a tautology. Note that every model refutes $\bot$ (since no rule allows to conclude that $\mathcal{M} \models \bot$) and therefore validates $\top$ (since $\top$ abbreviates $\neg\bot$). This justifies the reading of $\top$ and $\bot$ as *true* and *false* respectively. Finally, we remark that the symbols $\equiv$ and $\models$ belong to the meta-language (they are not part of the syntax of formulas).

Example 3.10 Consider the formula $P \vee Q \to P \wedge Q$, and models $\mathcal{M}_1 = \{P, Q\}$ and $\mathcal{M}_2 = \{Q\}$. We can check that:

- Since $\mathcal{M}_1 \models P$ and $\mathcal{M}_1 \models Q$, we have $\mathcal{M}_1 \models P \wedge Q$ and $\mathcal{M}_1 \models P \vee Q$ which allows us to conclude that $\mathcal{M}_1 \models P \vee Q \to P \wedge Q$.
- Since $\mathcal{M}_2 \models Q$ and $\mathcal{M}_2 \not\models P$, we have $\mathcal{M}_2 \models P \vee Q$ and $\mathcal{M}_2 \not\models P \wedge Q$. Hence $\mathcal{M}_2 \not\models P \vee Q \to P \wedge Q$.

From these observations, we can conclude that $P \vee Q \to P \wedge Q$ is neither a tautology nor a contradiction. It is both satisfiable and refutable.

Logical equivalence ($\equiv$) relates formulas with indistinguishable meanings (i.e. their meanings are the same in every model). The following proposition establishes that logical equivalence is indeed an equivalence relation (the proof is left to the reader as Exercise 3.9).

Proposition 3.11 $\equiv$ *is an equivalence relation on* **Form**, *i.e.*

1. *For every $A \in$ **Form**, $A \equiv A$.*   (*reflexivity*)
2. *If $A \equiv B$, then $B \equiv A$.*   (*symmetry*)
3. *If $A \equiv B$ and $B \equiv C$, then $A \equiv C$.*   (*transitivity*)

Figure 3.2 collects a set of notable equivalences of propositional formulas for future reference (A, B and C are arbitrary formulas). All of them are easily proved by constructing the corresponding truth tables for both sides of the equivalence.

Algebraic laws (boolean algebra):

$A \wedge B \equiv B \wedge A$	$A \vee B \equiv B \vee A$	(*commutativity*)
$A \wedge (B \wedge C) \equiv (A \wedge B) \wedge C$	$A \vee (B \vee C) \equiv (A \vee B) \vee C$	(*associativity*)
$A \wedge (A \vee B) \equiv A$	$A \vee (A \wedge B) \equiv A$	(*absorption*)
$A \wedge \neg A \equiv \bot$	$A \vee \neg A \equiv \top$	(*complement*)
$A \wedge (B \vee C) \equiv A \wedge B \vee A \wedge C$		($\wedge, \vee$-*distributivity*)
$A \vee (B \wedge C) \equiv (A \vee B) \wedge (A \vee C)$		($\vee, \wedge$-*distributivity*)

Interdefinability of operations:

$\neg(A \wedge B) \equiv \neg A \vee \neg B$	$\neg(A \vee B) \equiv \neg A \wedge \neg B$	(*De Morgan's laws*)
$A \to B \equiv \neg A \vee B$	$\neg A \equiv A \to \bot$	

Additional laws:

$\neg\neg A \equiv A$		(*double negation*)
$A \wedge \top \equiv A$	$A \vee \bot \equiv A$	(*neutral element*)
$A \wedge \bot \equiv \bot$	$A \vee \top \equiv \top$	(*absorption element*)

Fig. 3.2 Selected equivalences in propositional logic

To establish more elaborated equivalences, it is often convenient to reuse simpler equivalences as building blocks.

Proposition 3.12 *Let A, B and C be formulas such that B is a subformula of A and B $\equiv$ C. Then replacing in A some occurrence of B by C results in a formula that is logically equivalent to A.*

Proof By induction on the structure of A. □

The previous result supports the use of *equational reasoning* for establishing equivalences: we start with the left-hand side formula of a putative equivalence and try to reach the right-hand side by *rewriting* subformulas according to previously established equivalences. The soundness of this procedure is guaranteed by Propositions 3.11 and 3.12.

Example 3.13 We illustrate equational reasoning using the algebraic laws of Fig. 3.2 with a nontrivial equivalence: the double negation law ($\neg\neg A \equiv A$). We reason as follows:

$$\neg\neg A \equiv \neg\neg A \wedge \top, \qquad\qquad\qquad \text{by } (neutral\ element)$$
$$\equiv \neg\neg A \wedge (A \vee \neg A), \qquad\qquad \text{by } (complement)$$
$$\equiv \neg\neg A \wedge A \vee \neg\neg A \wedge \neg A, \quad \text{by } (distributivity)$$
$$\equiv \neg\neg A \wedge A \vee \bot, \qquad\qquad\quad \text{by } (complement)$$
$$\equiv \neg\neg A \wedge A \vee A \wedge \neg A, \qquad \text{by } (complement)$$
$$\equiv A \wedge (\neg\neg A \vee \neg A), \qquad\qquad \text{by } (commutativity)\ \text{and } (distributivity)$$
$$\equiv A \wedge \top, \qquad\qquad\qquad\quad \text{by } (complement)$$
$$\equiv A, \qquad\qquad\qquad\qquad\quad\ \text{by } (neutral\ element).$$

The equivalences in Fig. 3.2 also show that there is redundancy among the set of connectives adopted in the presentation of propositional language. In fact, the equivalences can be used to express propositional formulas using a restricted set of connectives. Examples of such sets are $\{\vee, \neg\}$, $\{\wedge, \neg\}$, $\{\rightarrow, \neg\}$ or $\{\rightarrow, \bot\}$ (see Exercise 3.14).

In addition to logical equivalence, the semantics of propositional logic induces other relevant relations between formulas. For example, given a model $\mathcal{M}$, if we have $\mathcal{M} \models P \wedge Q$ then we necessarily also have $\mathcal{M} \models Q \vee R$, because Q must be valid in $\mathcal{M}$. It is then reasonable to consider the second as a logical consequence of the first. This sort of argument is the aim of logical reasoning—in general, we are interested in describing logical consequences of a set of formulas. A semantic entailment relation, between sets of formulas and formulas, will be defined to capture this notion of logical consequence.

First let us show how the notions of validity, satisfiability, refutation and contradiction extend naturally to sets of formulas. We let $\Gamma, \Gamma', \ldots$ range over sets of formulas and adopt the convention that a comma-separated sequence of these sets denotes their union. Moreover, single formulas appearing in these sequences denote singleton sets, so for instance Γ, Γ', A, B denotes the set $\Gamma \cup \Gamma' \cup \{A\} \cup \{B\}$.

Definition 3.14 Let Γ be a set of formulas.

1. Γ is *valid in a model* $\mathcal{M}$ (or $\mathcal{M}$ *satisfies* Γ), iff $\mathcal{M} \models A$ for every formula $A \in \Gamma$. We denote this by $\mathcal{M} \models \Gamma$.
2. Γ is *satisfiable* iff there exists a model $\mathcal{M}$ such that $\mathcal{M} \models \Gamma$, and it is *refutable* iff there exists a model $\mathcal{M}$ such that $\mathcal{M} \not\models \Gamma$.
3. Γ is *valid*, denoted by $\models \Gamma$, iff $\mathcal{M} \models \Gamma$ for every model $\mathcal{M}$, and it is *unsatisfiable* iff it is not satisfiable, denoted by $\Gamma \models \bot$.

Definition 3.15 Let A be a formula and Γ a set of formulas. If every model that validates Γ also validates A, we say that Γ *entails* A (or A is a *logical consequence* of Γ).

We denote this by $\Gamma \models A$ and call $\models \subseteq \mathcal{P}(\textbf{Form}) \times \textbf{Form}$ the *semantic entailment* relation.

So if Γ entails A and $\mathcal{M} \models \Gamma$, then $\mathcal{M} \models A$. Note that we have adopted the same symbol for validity and semantic entailment. We will rely on the application context to disambiguate which relation is intended.

Clearly tautologies are entailed by the empty set. We will often omit the symbol $\emptyset$, and write $\models A$ for both assertions "A is valid" and "the empty set entails A". On the other hand, we observe that asserting $\Gamma \models \bot$ establishes that Γ is unsatisfiable (no model validates Γ). Note also that an alternative way of defining the logical equivalence of two formulas A and B would be to say that

$$A \equiv B \quad \text{iff} \quad A \models B \text{ and } B \models A.$$

Example 3.16 Consider the set of formulas $\Gamma = \{P \wedge \neg R, P \rightarrow Q\}$. Clearly Γ entails $R \vee Q$ because every model that satisfies Γ also satisfies P and Q, hence $R \vee Q$ is valid in such a model.

It is also easy to see that the formula $\neg Q \rightarrow P \wedge R$ is a logical consequence of Γ, since $\neg Q$ is always false in any model that satisfies Γ.

The following properties are often referred to as characterising an *abstract consequence relation* [1].

Proposition 3.17 *The semantic entailment relation satisfies the following properties*:

1. *For all $A \in \Gamma$, $\Gamma \models A$.*   (*inclusion*)
2. *If $\Gamma \models A$, then $\Gamma, B \models A$.*   (*monotonicity*)
3. *If $\Gamma \models A$ and $\Gamma, A \models B$, then $\Gamma \models B$.*   (*cut*)

Proof

1. Let $\mathcal{M}$ be a model such that $\mathcal{M} \models \Gamma$. Then $\mathcal{M}$ validates every formula in Γ and, since $A \in \Gamma$, $\mathcal{M} \models A$.
2. Let $\mathcal{M}$ be a model such that $\mathcal{M} \models \Gamma, B$. In particular, $\mathcal{M} \models \Gamma$. Hence, $\mathcal{M} \models A$ (since $\Gamma \models A$).
3. Let $\mathcal{M}$ be a model such that $\mathcal{M} \models \Gamma$. Since $\Gamma \models A$, we also have $\mathcal{M} \models A$, and hence $\mathcal{M} \models \Gamma, A$. Therefore $\mathcal{M} \models B$ follows from the hypothesis $\Gamma, A \models B$. $\square$

It is also instructive to see how semantic entailment relates with propositional connectives. The proof of the following proposition follows the same pattern of the previous result, and is left as Exercise 3.15.

Proposition 3.18

1. $\Gamma \models A$ *iff* $\Gamma, \neg A \models \bot$
2. $\Gamma \models \neg A$ *iff* $\Gamma, A \models \bot$
3. $\Gamma \models A \rightarrow B$ *iff* $\Gamma, A \models B$
4. $\Gamma \models A \wedge B$ *iff* $\Gamma \models A$ *and* $\Gamma \models B$
5. $\Gamma \models A \vee B$ *iff* $\Gamma \models A$ *or* $\Gamma \models B$

3.3 Proof System

So far we have taken the "semantic" approach to logic, with the aim of character-ising the semantic concept of model, from which validity and semantic entailment were derived. This, however, is not the only possible point of view.

Instead of adopting the view based on the notion of truth, we can think of logic as a codification of reasoning. This alternative approach to logic, called "deductive", focuses directly on the deduction relation that is induced on formulas, i.e. on what formulas are logical consequences of other formulas. We will now explore this per-spective by proposing a set of syntactical rules for deducing new formulas from sets of formulas (assumptions).

A *proof system* (or *inference system*) consists of a set of basic rules, known as *inference rules*, for constructing derivations. Such a derivation is a formal object that encodes an explanation of why a given formula—the *conclusion*—is deducible from a set of *assumptions*. The rules that govern the construction of derivations are called *inference rules* and consist of zero or more *premises* and a single *conclusion*. Derivations have a tree-like shape. We use the standard notation of separating the premises from the conclusion by a horizontal line, as in the following example:

$$\frac{prem_1 \qquad \cdots \qquad prem_n}{concl}$$

The intuitive reading of such a rule is "whenever the conditions prescribed by the premises $prem_1, \ldots, prem_n$ are met, we are allowed to conclude $concl$".

The proof system studied in this section is a formalisation of the reasoning used in mathematics, and was introduced by Gerhard Gentzen in the first half of the 20th century as a "natural" representation of logical derivations. It is for this reason called *natural deduction*. We choose to present the rules of natural deduction in sequent style. A *sequent* is a judgment of the form $\Gamma \vdash A$, where Γ is a set of formulas (the *context*) and A a formula (the *conclusion* of the sequent). A sequent $\Gamma \vdash A$ is meant to be read as "A can be deduced from the set of *assumptions* Γ", or simply "A is a consequence of Γ".

The inference rules of natural deduction were designed to render the intuitive meaning of the connectives as faithfully as possible. Each rule is in fact a schema (that is, a pattern containing meta-variables, each ranging over some phrase type). An *instance* of an inference rule is obtained by replacing all occurrences of each meta-variable by a phrase in its range. An inference rule containing no premises is called an *axiom schema* (or simply, an *axiom*).

Definition 3.19 The proof system $\mathcal{N}_{\mathsf{PL}}$ of *natural deduction* for propositional logic is defined by the rules presented in Fig. 3.3. A *derivation* (or *proof*) in $\mathcal{N}_{\mathsf{PL}}$ is in-ductively defined by the following clause:

– If

$$(\mathsf{R}) \ \frac{\Gamma_1 \vdash A_1 \qquad \cdots \qquad \Gamma_n \vdash A_n}{\Gamma \vdash A}$$

$$(\text{Ax}) \ \frac{}{\Gamma, A \vdash A} \qquad\qquad (\text{RAA}) \ \frac{\Gamma, \neg A \vdash \bot}{\Gamma \vdash A}$$

Introduction Rules:

$$(\text{I}_\to) \ \frac{\Gamma, A \vdash B}{\Gamma \vdash A \to B} \qquad\qquad (\text{I}_\neg) \ \frac{\Gamma, A \vdash \bot}{\Gamma \vdash \neg A}$$

$$(\text{I}_\wedge) \ \frac{\Gamma \vdash A \qquad \Gamma \vdash B}{\Gamma \vdash A \wedge B} \qquad (\text{I}_{\vee i}) \ \frac{\Gamma \vdash A_i}{\Gamma \vdash A_1 \vee A_2} \ i \in \{1, 2\}$$

Elimination Rules:

$$(\text{E}_\to) \ \frac{\Gamma \vdash A \qquad \Gamma \vdash A \to B}{\Gamma \vdash B} \qquad (\text{E}_\neg) \ \frac{\Gamma \vdash A \qquad \Gamma \vdash \neg A}{\Gamma \vdash B}$$

$$(\text{E}_{\wedge i}) \ \frac{\Gamma \vdash A_1 \wedge A_2}{\Gamma \vdash A_i} \ i \in \{1, 2\} \qquad (\text{E}_\vee) \ \frac{\Gamma \vdash A \vee B \qquad \Gamma, A \vdash C \qquad \Gamma, B \vdash C}{\Gamma \vdash C}$$

Fig. 3.3 System $\mathcal{N}_{\text{PL}}$ for classical propositional logic

is an instance of rule (R) of the $\mathcal{N}_{\text{PL}}$ proof system, and $\mathcal{D}_i$ is a derivation with conclusion $\Gamma_i \vdash A_i$ (for $1 \leq i \leq n$), then

$$(\text{R}) \ \frac{\overset{\displaystyle \mathcal{D}_1}{\Gamma_1 \vdash A_1} \quad \dots \quad \overset{\displaystyle \mathcal{D}_n}{\Gamma_n \vdash A_n}}{\Gamma \vdash A}$$

is a derivation with conclusion $\Gamma \vdash A$.

A sequent $\Gamma \vdash A$ is *derivable* in $\mathcal{N}_{\text{PL}}$ if it is the conclusion of some derivation.

The *deduction relation* $\vdash \in \mathcal{P}(\textbf{Form}) \times \textbf{Form}$ is the relation induced by derivability, i.e. one writes $\Gamma \vdash A$ (read "A can be deduced from Γ" or "Γ infers A") iff the sequent $\Gamma \vdash A$ is derivable in $\mathcal{N}_{\text{PL}}$. A formula that can be deduced from the empty context is called a *theorem*.

Although derivations are normally presented in inverted form (with the root shown at the bottom), it should be clear from the definition that they are tree-like structures with sequents at the nodes, and the conclusion at the root. Each node is related with its immediate descendents by an instance of a rule of $\mathcal{N}_{\text{PL}}$. The leaves of such a tree are necessarily instances of the axiom rule (Ax), since this is the only rule with no premises (it acts as the base case for the inductively defined set of derivations). The axiom rule merely asserts that deduction subsumes inclusion (c.f. inclusion property in Proposition 3.17).

Rule (RAA) is the *reductio ad absurdum* law. It captures a distinctive feature of classical logic, which is the ability to reason by contradiction: a formula A can be proved by establishing that assuming $\neg A$ leads to a contradiction.

The remaining rules are organised in two sets of *introduction* and *elimination* rules (with names prefixed by I and E respectively). The distinction is based on the observation that each of these rules either introduces a connective in the conclusion or removes it from a premise. In general, elimination rules explain the consequences of structured formulas, and introduction rules describe how such formulas can be derived. Together, they provide a sort of "behavioural description" of the logical connectives.

We briefly comment on some of these rules:

– Rule (E$_\rightarrow$), also known under the name of *modus ponens*, has a long tradition in the study of logic. It captures the conditional nature of implication: if one is able to deduce both an implication $A \rightarrow B$ and its antecedent A from a given set of assumptions Γ, then the consequent B can also be deduced from the same set of assumptions.
– Rules (I$_\rightarrow$) and (I$_\neg$) modify the assumption sets used in the conclusion and in the premise. This feature was in fact the main innovation of natural deduction with respect to previous proof systems, and allows to capture very clearly (and concisely) the meaning of these connectives:
  – (I$_\rightarrow$): proving an implication amounts to proving its conclusion assuming the antecedent as an additional assumption;
  – (I$_\neg$): proving the negation of A amounts to reaching a contradiction assuming A.
– Rule (E$_\vee$) also makes use of the ability to change the assumption set. It captures the fact that if a disjunction can be deduced from a given set of assumptions Γ, and some proposition can be deduced from the same set by additionally assuming either of the disjunct formulas, then that proposition can be deduced from Γ with no further assumptions.
– Contrary to the definition of validity, where $\bot$ was absent (no model validates $\bot$), in the $\mathcal{N}_{\mathsf{PL}}$ system $\bot$ appears explicitly. When $\bot$ can be deduced from Γ we say that Γ is *inconsistent*. Although all the elimination rules admit $\Gamma \vdash \bot$ in their conclusion, only one rule admits this without $\bot$ occurring in the premises, and is usually employed for deriving sequents of that form. This is (E$_\neg$), whose premises require that Γ infers both a formula A and its negation. In fact, rule (E$_\neg$) allows any formula to appear in its conclusion—this captures the *ex falso sequitur quodlibet* principle, stating that anything follows from a contradiction.

Example 3.20 Consider the derivation that establishes that $\neg P \rightarrow (Q \rightarrow P) \rightarrow \neg Q$ is a theorem:

$$
\cfrac{
  \cfrac{\text{(Ax)}\ \overline{\neg P, Q \to P, Q \vdash Q}}{}
  \qquad
  \cfrac{\text{(Ax)}\ \overline{\neg P, Q \to P, Q \vdash Q \to P}}{}
}{}
$$

$$
(E_\to)\ \cfrac{\neg P, Q \to P, Q \vdash Q \qquad (\text{Ax})\ \overline{\neg P, Q \to P, Q \vdash Q \to P}}{}
$$

(Ax) $\overline{\neg P, Q \to P, Q \vdash Q}$ (Ax) $\overline{\neg P, Q \to P, Q \vdash Q \to P}$

$(E_\to)$ $\dfrac{}{\neg P, Q \to P, Q \vdash P}$ (Ax) $\overline{\neg P, Q \to P, Q \vdash \neg P}$

$(E_\neg)$ $\dfrac{}{\neg P, Q \to P, Q \vdash \bot}$

$(I_\neg)$ $\dfrac{}{\neg P, Q \to P \vdash \neg Q}$

$(I_\to)$ $\dfrac{}{\neg P \vdash (Q \to P) \to \neg Q}$

$(I_\to)$ $\dfrac{}{\vdash \neg P \to (Q \to P) \to \neg Q}$

The well-formedness of the derivation can be easily verified by checking that each rule marked in the tree is indeed an instance of a rule of Fig. 3.3.

This example shows that even for such a reasonably simple formula, the size of the tree already poses a problem from the point of view of its representation. For that reason, we shall adopt an alternative format for presenting bigger proof trees: the root (conclusion) appears first, followed by its premises, numbered, in subsequent lines. Indentation is used to mark each level of the tree, and the rule name appears at the end of each line. In this alternative format, the previous derivation is represented as:

$$
\begin{array}{lll}
\vdash \neg P \to (Q \to P) \to \neg Q, & & (I_\to) \\
\quad \textbf{1.}\ \neg P \vdash (Q \to P) \to \neg Q, & & (I_\to) \\
\quad\quad \textbf{1.}\ \neg P, Q \to P \vdash \neg Q, & & (I_\neg) \\
\quad\quad\quad \textbf{1.}\ \neg P, Q \to P, Q \vdash \bot, & & (E_\neg) \\
\quad\quad\quad\quad \textbf{1.}\ \neg P, Q \to P, Q \vdash P, & & (E_\to) \\
\quad\quad\quad\quad\quad \textbf{1.}\ \neg P, Q \to P, Q \vdash Q, & & (\text{Ax}) \\
\quad\quad\quad\quad\quad \textbf{2.}\ \neg P, Q \to P, Q \vdash Q \to P, & & (\text{Ax}) \\
\quad\quad\quad\quad \textbf{2.}\ \neg P, Q \to P, Q \vdash \neg P, & & (\text{Ax}) \\
\end{array}
$$

This presentation style in fact corresponds to a popular strategy for constructing derivations. In *backward reasoning* one starts with the conclusion sequent and chooses to apply a rule that can justify that conclusion; one then repeats the procedure on the resulting premises. This process for building derivations is definitely not trivial, and will be addressed later in Sect. 3.6.

If one prefers to present derivations in a forward fashion, which corresponds to constructing derivations using the *forward reasoning* strategy, then it is customary to simply give sequences of judgments, each of which is either an axiom or follows from a preceding judgment in the sequence, by an instance of an inference rule. The last sequent of the sequence is the conclusion of the derivation. A derivation of $\vdash \neg P \to (Q \to P) \to \neg Q$ presented in the forward direction looks like this:

	Judgments	Justification
1.	$\neg P, Q \to P, Q \vdash Q$	(Ax)
2.	$\neg P, Q \to P, Q \vdash Q \to P$	(Ax)
3.	$\neg P, Q \to P, Q \vdash P$	$(E_\to)1, 2$
4.	$\neg P, Q \to P, Q \vdash \neg P$	(Ax)
5.	$\neg P, Q \to P, Q \vdash \bot$	$(E_\neg)3, 4$
6.	$\neg P, Q \to P \vdash \neg Q$	$(I_\neg)5$
7.	$\neg P \vdash (Q \to P) \to \neg Q$	$(I_\to) 6$
8.	$\vdash \neg P \to (Q \to P) \to \neg Q$	$(I_\to) 7$

No matter what presentation style is adopted for derivations, an important point is that it should be easy for anyone to validate a derivation by checking its well-formedness criterion, i.e. checking whether each step is a valid instance of a rule of the proof system.

Remark 3.21 Traditionally, presentations of the natural deduction system do not make use of sequents. Instead, leaves of the derivation trees can be either "open" or "closed", and each node is simply a formula (corresponding to the conclusion of the sequent)—the associated assumption set is implicitly assumed to be the set of formulas appearing in the open leaves of the tree. To cope with changes in the assumption set, some rules allow for formulas appearing in leaves to be closed, thus removing them from the assumption set.

As an example, the introduction rule for implication looks like this:

$$[A]_l$$

$$\vdots$$

$$(I_\to) \ \frac{B}{A \to B} \ l$$

Applying this rule, one transforms a derivation of B (with an arbitrary assumption set) into a derivation of $A \to B$, where any number of occurrences of the assumption A may be closed (signalled by the use of square brackets). The label l is just a mechanism for keeping track of which formulas are closed by which rules. As an example of a derivation in this sequent-free presentation, consider the following one from Example 3.20:

$$(I_\to) \ \frac{(I_\to) \ \frac{(I_\to) \ \frac{(I_\neg) \ \frac{(E_\neg) \ \frac{(E_\to) \ \frac{[Q]_3 \quad [Q \to P]_2}{P} \quad [\neg P]_1}{\bot}}{\neg Q} \ 3}{(Q \to P) \to \neg Q} \ 2}{\neg P \to (Q \to P) \to \neg Q} \ 1$$

Notice that the assumption set of the derivation is empty, since all leaves of the tree have been closed.

Despite being more verbose, the sequent style used in this book is more convenient for mathematical treatment. Also, the sequent notation is closer to the process of developing a proof in a computer with the help of a proof assistant.

As was the case with semantic entailment, the deduction relation also fits the requirements of an *abstract consequence relation*.

Proposition 3.22 *The deduction relation satisfies the following properties*:

1. *For all $A \in \Gamma$, $\Gamma \vdash A$.*   (*inclusion*)
2. *If $\Gamma \vdash A$, then $\Gamma, B \vdash A$.*   (*monotonicity*)
3. *If $\Gamma \vdash A$ and $\Gamma, A \vdash B$, then $\Gamma \vdash B$.*   (*cut*)

Proof

1. If $A \in \Gamma$, then (Ax) establishes that $\Gamma \vdash A$.
2. We construct a derivation of $\Gamma, B \vdash A$ by induction on the structure of $\Gamma \vdash A$: if the last rule is (Ax), then it can also be used to establish that $\Gamma, B \vdash A$ (since $A \in \Gamma$). For any other rule r, let $\Gamma_i \vdash C_i$ be the premises of r (with i ranging over the number of premises). By induction hypothesis we have that $\Gamma_i, B \vdash C_i$, and so applying rule r we obtain $\Gamma, B \vdash A$.
3. We have $\Gamma, A \vdash B$, so by rule $(I_\rightarrow)$ we have $\Gamma \vdash A \rightarrow B$. As we have $\Gamma \vdash A$, we conclude by applying rule $(E_\rightarrow)$. $\square$

Two other important concepts are those of *admissible rule* and *derivable rule* in a given proof system.

Definition 3.23 An inference rule is *admissible* in a formal system if every judgement that can be proved making use of that rule can also be proved without it (in other words the set of judgements of the system is closed under the rule).

So if a rule r is admissible in a formal system defined by a set of rules X, every derivation using the set of rules $X \cup \{r\}$ can be transformed in derivations using only the rules in X. In practical terms, this means that we can safely construct derivations using new rules that are not in the original system but are admissible in it.

Proposition 3.24 (Weakening) *The following rule, named* weakening, *is admissible in* $\mathcal{N}_{\mathsf{PL}}$

$$(W) \frac{\Gamma \vdash A}{\Gamma, B \vdash A}$$

Proof First assume that the last step of a given derivation is an instance of (W) and the derivation of the premise does not contain any other instances of this rule:

$$(W) \frac{\begin{array}{c} \mathcal{D} \\ \Gamma \vdash C \end{array}}{\Gamma', B \vdash C}$$

Then, by Proposition 3.22(2), we have that there must exist a weakening-free derivation of $\Gamma, B \vdash C$, since this sequent is guaranteed to be derivable. Any derivation containing a sole application of (W) as the last step can thus be transformed into one with the same conclusion, free from that rule. Now, if we have more occurrences of the weakening rule in a derivation we may eliminate each of these occurrences one by one, beginning with the uppermost instances of (W). $\square$

In most situations, eliminating from a derivation all instances of an admissible rule requires only local changes, in the sense that each occurrence is replaced by a derivation tree fragment in the original system.

Definition 3.25 An inference rule is said to be *derivable* in a proof system if the conclusion of the rule can be derived from its premises using the other rules of the system.

Of course, every derivable rule is also admissible, but the converse is not true. Weakening is an example of a rule that is admissible but not derivable.

Example 3.26 The following rule is derivable in $\mathcal{N}_{\mathsf{PL}}$:

$$\frac{\Gamma \vdash A \wedge B \qquad \Gamma, A, B \vdash C}{\Gamma \vdash C}$$

A derivation of it is

$$
\begin{array}{lll}
\Gamma \vdash C, & (\mathrm{E}_\rightarrow) \\
\quad \textbf{1.}\ \ \Gamma \vdash B, & (\mathrm{E}_{\wedge 2}) \\
\quad\quad \textbf{1.}\ \ \Gamma \vdash A \wedge B, & \text{premise} \\
\quad \textbf{2.}\ \ \Gamma \vdash B \rightarrow C, & (\mathrm{I}_\rightarrow) \\
\quad\quad \textbf{1.}\ \ \Gamma \vdash A, & (\mathrm{E}_{\wedge 1}) \\
\quad\quad\quad \textbf{1.}\ \ \Gamma \vdash A \wedge B, & \text{premise} \\
\quad\quad \textbf{2.}\ \ \Gamma \vdash A \rightarrow B \rightarrow C, & (\mathrm{I}_\rightarrow) \\
\quad\quad\quad \textbf{1.}\ \ \Gamma, A \vdash B \rightarrow C, & (\mathrm{I}_\rightarrow) \\
\quad\quad\quad\quad \textbf{1.}\ \ \Gamma, A, B \vdash C, & \text{premise}
\end{array}
$$

Lemma 3.27 *The following rule is admissible in* $\mathcal{N}_{\mathsf{PL}}$:

$$\frac{\Gamma, A \vdash B \qquad \Gamma, \neg A \vdash B}{\Gamma \vdash B}$$

Proof We show that this rule is derivable in the system $\mathcal{N}_{\mathsf{PL}} \cup (\mathrm{W})$:

$$
\begin{array}{lll}
\Gamma \vdash B, & (\mathrm{RAA}) \\
\quad \textbf{1.}\ \ \Gamma, \neg B \vdash \bot, & (\mathrm{E}_\neg) \\
\quad\quad \textbf{1.}\ \ \Gamma, \neg B \vdash \neg A, & (\mathrm{I}_\neg) \\
\quad\quad\quad \textbf{1.}\ \ \Gamma, \neg B, A \vdash \bot, & (\mathrm{E}_\neg) \\
\quad\quad\quad\quad \textbf{1.}\ \ \Gamma, \neg B, A \vdash B, & (\mathrm{W}) \\
\quad\quad\quad\quad\quad \textbf{1.}\ \ \Gamma, A \vdash B, & \text{premise} \\
\quad\quad\quad\quad \textbf{2.}\ \ \Gamma, \neg B, A \vdash \neg B, & (\mathrm{Ax}) \\
\quad\quad \textbf{2.}\ \ \Gamma, \neg B \vdash \neg\neg A, & (\mathrm{I}_\neg) \\
\quad\quad\quad \textbf{1.}\ \ \Gamma, \neg B, \neg A \vdash \bot, & (\mathrm{E}_\neg) \\
\quad\quad\quad\quad \textbf{1.}\ \ \Gamma, \neg B, \neg A \vdash B, & (\mathrm{W}) \\
\quad\quad\quad\quad\quad \textbf{1.}\ \ \Gamma, \neg A \vdash B, & \text{premise} \\
\quad\quad\quad\quad \textbf{2.}\ \ \Gamma, \neg B, \neg A \vdash \neg B, & (\mathrm{Ax})
\end{array}
$$

Then it is also admissible in that system, and Proposition 3.24 allows us to conclude its admissibility in $\mathcal{N}_{\mathsf{PL}}$. $\qquad\square$

The above result has an interesting corollary about consistency.

Proposition 3.28 *For every set* Γ *of formulas and formula* A, *if* $\Gamma \not\vdash \bot$ *then either* $\Gamma, A \not\vdash \bot$ *or* $\Gamma, \neg A \not\vdash \bot$.

Proof We prove the contra-positive: if $\Gamma, A \vdash \bot$ and $\Gamma, \neg A \vdash \bot$, then $\Gamma \vdash \bot$, by Lemma 3.27. $\qquad\square$

3.4 Soundness and Completeness

We dedicate this section to the study of the correspondence between the entailment and deduction relations. The first property of interest is the *soundness* of the system $\mathcal{N}_{PL}$, which means that deduction is compatible with the semantics, i.e. whenever we deduce a formula from an assumption set, then the formula is a logical consequence of that assumption set.

Theorem 3.29 (Soundness) *For every set Γ of formulas and formula A, if $\Gamma \vdash A$, then $\Gamma \models A$.*

Proof By induction on the derivation of $\Gamma \vdash A$. According to the last rule of this derivation we argue as follows:

− If the last step is

$$(\text{Ax}) \; \frac{}{\Gamma', A \vdash A}$$

Then $\Gamma', A \models A$ follows by Proposition 3.17(1).
− If the last step is

$$(\text{I}_\to) \; \frac{\Gamma, B \vdash C}{\Gamma \vdash B \to C}$$

By induction hypothesis we have $\Gamma, B \models C$ and by Proposition 3.18(3) $\Gamma \models B \to C$.
− If the last step is

$$(\text{E}_\to) \; \frac{\Gamma \vdash B \qquad \Gamma \vdash B \to A}{\Gamma \vdash A}$$

By induction hypotheses we have $\Gamma \models B$ and $\Gamma \models B \to A$. So by Proposition 3.18(3) we get $\Gamma, B \models A$ and by Proposition 3.17(3) $\Gamma \models A$.
− If the last step is

$$(\text{E}_\neg) \; \frac{\Gamma \vdash B \qquad \Gamma \vdash \neg B}{\Gamma \vdash A}$$

We have as induction hypotheses $\Gamma \models B$ and $\Gamma \models \neg B$, and thus by Proposition 3.18(1) $\Gamma, \neg B \models \bot$, and by Proposition 3.17(3) $\Gamma \vdash \bot$. By Proposition 3.17(2) we then have $\Gamma, \neg A \models \bot$, and we conclude $\Gamma \models A$ again by Proposition 3.18(1).
− If the last step is

$$(\text{I}_\neg) \; \frac{\Gamma, B \vdash \bot}{\Gamma \vdash \neg B}$$

By induction hypothesis $\Gamma, B \models \bot$, and by Proposition 3.18(2) $\Gamma \models \neg B$.
− If the last step is

$$(\text{RAA}) \; \frac{\Gamma, \neg A \vdash \bot}{\Gamma \vdash A}$$

By induction hypothesis $\Gamma, \neg A \models \bot$, and by Proposition 3.18(1) $\Gamma \models A$.

— If the last step is

$$(I_\wedge) \frac{\Gamma \vdash B \qquad \Gamma \vdash C}{\Gamma \vdash B \wedge C}$$

Induction hypotheses $\Gamma \models B$ and $\Gamma \models C$ and Proposition 3.18(4) yield $\Gamma \models B \wedge C$.

— If the last step is

$$(E_{\wedge i}) \frac{\Gamma \vdash A_1 \wedge A_2}{\Gamma \vdash A_i} \qquad i \in \{1, 2\}$$

By induction hypothesis $\Gamma \models A_1 \wedge A_2$ using Proposition 3.18(4), we have both $\Gamma \models A_1$ and $\Gamma \models A_2$.

— If the last step is

$$(I_{\vee i}) \frac{\Gamma \vdash A_i}{\Gamma \vdash A_1 \vee A_2} \qquad i \in \{1, 2\}$$

We have induction hypothesis $\Gamma \models A_i$. For both $i = 1$ and $i = 2$, using Proposition 3.18(5) we obtain $\Gamma \models A_1 \vee A_2$.

— If the last step is

$$(E_\vee) \frac{\Gamma \vdash B \vee C \qquad \Gamma, B \vdash A \qquad \Gamma, C \vdash A}{\Gamma \vdash A}$$

From induction hypothesis $\Gamma \models B \vee C$, by Proposition 3.18(5) we have either $\Gamma \models B$ or $\Gamma \models C$. Reasoning by cases with Proposition 3.17(3) and the remaining hypotheses $\Gamma, B \models A$ and $\Gamma, C \models A$, we obtain $\Gamma \models A$. $\qquad\qquad\square$

The converse property of soundness is called *completeness*. It states that the proof system is capable of deducing every possible semantic entailment, i.e. if a formula A is a logical consequence of a set of formulas Γ, then there exists a derivation in $\mathcal{N}_{PL}$ of the judgment $\Gamma \vdash A$.

Before proving the completeness of $\mathcal{N}_{PL}$, let us first prove the following auxiliary lemma relating unsatisfiability and inconsistency.

Lemma 3.30 *For every set Γ of formulas, if $\Gamma \models \bot$, then $\Gamma \vdash \bot$.*

Proof We prove the contra-positive formulation that any consistent set of formulas is satisfiable, i.e. if $\Gamma \nvdash \bot$, then $\Gamma \not\models \bot$. For this it suffices to exhibit a model that validates Γ. We will extend Γ in such a way that it remains consistent and fully characterises the desired model.

Let $\Gamma_0 = \Gamma$ and consider an enumeration $A_1, A_2, \ldots$ of the set of formulas **Form**. For each of these formulas define Γ_i to be $\Gamma_{i-1} \cup \{A_i\}$ if this is consistent, or $\Gamma_{i-1} \cup \{\neg A_i\}$ otherwise. Note that by Proposition 3.28 one of these sets is consistent. Let $\Gamma' = \bigcup_i \Gamma_i$ be the union of all these sets. Clearly, $\Gamma' \supseteq \Gamma$ and $\Gamma' \nvdash \bot$ (otherwise one of the sets Γ_i would be inconsistent, which contradicts their construction).

Note that for each formula A either A or $\neg A$ belongs to Γ'. We define the model $\mathcal{M}$ as the set of proposition symbols contained in Γ'. We claim that validity under

$\mathcal{M}$ is given by membership of Γ', that is:

$$\mathcal{M} \models A \quad \text{iff} \quad A \in \Gamma'. \tag{3.1}$$

Since $\Gamma \subseteq \Gamma'$, if our claim is true then clearly $\mathcal{M} \models \Gamma$, and we will be done.

To show claim (3.1) we proceed by induction on the structure of A. When A is $\bot$, we note that $\bot \notin \Gamma'$ since Γ' is consistent. If A is a proposition symbol P, then $\mathcal{M} \models P$ iff $P \in \mathcal{M}$ iff $P \in \Gamma'$. If A is $\neg B$, then $\mathcal{M} \models \neg B$ iff $\mathcal{M} \not\models B$, and by induction hypothesis, iff $B \notin \Gamma'$, iff $\neg B \in \Gamma'$. If A is $B \rightarrow C$, consider the following cases:

− Case $\mathcal{M} \models B \rightarrow C$. By definition, it must be the case that $\mathcal{M} \not\models B$ or $\mathcal{M} \models C$. By induction hypothesis, we have that either $\neg B \in \Gamma'$ or $C \in \Gamma'$. In any case we have $\Gamma' \vdash B \rightarrow C$:

$$\text{(E}_\neg) \frac{\text{(Ax)} \dfrac{}{\Gamma', B \vdash B} \quad \text{(Ax)} \dfrac{}{\Gamma', B \vdash \neg B}}{\text{(I}_\rightarrow) \dfrac{\Gamma', B \vdash C}{\Gamma' \vdash B \rightarrow C}} \qquad \text{(I}_\rightarrow) \frac{\text{(Ax)} \dfrac{}{\Gamma', B \vdash C}}{\Gamma' \vdash B \rightarrow C}$$

We can then conclude that $(B \rightarrow C) \in \Gamma'$ because $\Gamma', \neg(B \rightarrow C)$ is inconsistent as the following derivation shows:

$$\text{(E}_\neg) \frac{\text{(W)} \dfrac{\Gamma' \vdash B \rightarrow C}{\Gamma', \neg(B \rightarrow C) \vdash B \rightarrow C} \quad \text{(Ax)} \dfrac{}{\Gamma', \neg(B \rightarrow C) \vdash \neg(B \rightarrow C)}}{\Gamma', \neg(B \rightarrow C) \vdash \bot}$$

− Case $\mathcal{M} \not\models B \rightarrow C$. It must be the case that $\mathcal{M} \models B$ and $\mathcal{M} \not\models C$. By induction hypothesis $\{B, \neg C\} \subseteq \Gamma'$. The inconsistency of $\Gamma', B \rightarrow C$ is established by the following derivation

$$\text{(E}_\neg) \frac{\text{(E}_\rightarrow) \dfrac{\text{(Ax)} \dfrac{}{\Gamma', B \rightarrow C \vdash B} \quad \text{(Ax)} \dfrac{}{\Gamma', B \rightarrow C \vdash B \rightarrow C}}{\Gamma', B \rightarrow C \vdash C} \quad \text{(Ax)} \dfrac{}{\Gamma', B \rightarrow C \vdash \neg C}}{\Gamma', B \rightarrow C \vdash \bot}$$

Thus $B \rightarrow C \notin \Gamma'$. The remaining cases can be proved in a similar way (and are left to the reader as an exercise). □

Theorem 3.31 (Completeness) *For every set of formulas Γ and formula A, if $\Gamma \models A$ then $\Gamma \vdash A$.*

Proof If $\Gamma \models A$ then, by Proposition 3.18(2), $\Gamma, \neg A \models \bot$. By Lemma 3.30 we get $\Gamma, \neg A \vdash \bot$. Hence $\Gamma \vdash A$ follows by applying rule (RAA). □

The proof of the following theorem illustrates the application of soundness and completeness results. It shows how moving forward and backward between the semantics and proof-theoretic worlds can provide additional insight into the nature of validity in propositional logic.

Theorem 3.32 (Compactness) *A (possibly infinite) set of formulas Γ is satisfiable if and only if every finite subset of Γ is satisfiable.*

Proof We prove the following equivalent statement: $\Gamma \models \bot$ if and only if there exists a finite subset $\Gamma' \subseteq \Gamma$ such that $\Gamma' \models \bot$.

The right-to-left implication is immediate since if $\mathcal{M}$ is a model such that $\mathcal{M} \models \Gamma$, then necessarily $\mathcal{M} \models \Gamma'$. To prove the left-to-right implication, we assume $\Gamma \models \bot$. By completeness we have $\Gamma \vdash \bot$, thus there exists a derivation of this sequent in $\mathcal{N}_{\mathsf{PL}}$. Since a derivation is a finite tree, it can use only finitely many assumptions from Γ. Let Γ' be this finite set of assumptions; then $\Gamma' \vdash \bot$ holds, and $\Gamma' \models \bot$ follows by soundness. $\square$

3.5 Validity Checking: Semantic Methods

We have seen two different ways in which the validity status of formulas can be settled: directly using the definition of validity (the semantic method), or by exploiting soundness and completeness theorems to transfer the problem to the proof-theoretic side (the deductive method). However, we have not given any concrete algorithmic solutions to establishing the validity of propositional formulas. In this section we will consider the semantic method in more detail; the deductive method will be the subject of Sect. 3.6.

Given a formula A, there are two obvious decision problems regarding its validity status:

— *Validity problem* (VAL): Given a formula A, is A valid?
— *Satisfiability problem* (SAT): Given a formula A, is A satisfiable?

We will see that both of these problems are *decidable*, i.e. for each one it is possible to devise an *algorithm* that terminates on all inputs and answers "yes" if the input formula is valid (respectively satisfiable), and "no" otherwise. In fact these are *dual problems*, and are indeed two faces of a same coin since A is satisfiable if and only if $\neg A$ is not valid. Unfortunately these are also inherently complex problems, and it is believed that there exist no polynomial time algorithms for solving them.

Recall that a formula A is a tautology if $\mathcal{M} \models A$ for every model $\mathcal{M}$, and it is satisfiable if there exists a model $\mathcal{M}$ such that $\mathcal{M} \models A$. What distinguishes VAL from SAT is then the quantifier involved in each definition, since otherwise they are both based on the same underlying assertion (the validity of formulas in a given model). The assertion $\mathcal{M} \models A$ (for a given model $\mathcal{M}$ and formula A) is unproblematic, since Definition 3.8 can be easily turned into a recursive function (on the structure of A) that checks this assertion. Moreover, the infinitude of the universe of quantification $\mathcal{P}(\mathbf{Prop})$ can be overcome by noting that only propositional symbols used in a formula play a role in its validity—formally,

$$\mathcal{M} \models A \quad \text{iff} \quad \mathcal{M} \cap \mathsf{pSymbs}(A) \models A$$

where $\mathsf{pSymbs}(A)$ denotes the set of propositional symbols used in A (c.f. Exercise 3.5). These considerations lead to the truth tables already mentioned in Sect. 3.2. In fact, a procedure for VAL (resp. SAT) can trivially be designed that simply constructs the truth table for the given formula and checks if every (resp. some) row is assigned to the truth value **T**. The problem with this naive approach to VAL and SAT is the complexity of the resulting solution. The size of the truth table for a formula with n proposition symbols is 2^n, which turns the procedure impractical even for moderately sized formulas (tens of propositional symbols).

The question is then whether it is possible to devise better decision procedures for handling these problems. The answer to this question is clearly affirmative—the structure of logical validity allows for much better algorithms (even if they still lie in the same complexity class).

Remark 3.33 It is believed that both SAT and VAL cannot be solved by polynomial-time algorithms, i.e. these problems are not members of the complexity class **P**. However, this is actually a conjecture, for which no formal proof exists (at the time of writing of this book). In fact, SAT is an **NP**-complete problem, which means that any problem in **NP** can be encoded as a SAT problem.[1] VAL on the other hand is a **co-NP**-complete problem. So, if a polynomial-time algorithm to solve SAT or VAL were ever found, this would settle the $\mathbf{P} = \mathbf{NP}$ question (one of the most important unsolved problems in theoretical computer science).

To overcome the limitation of the truth tables method, consider the following strategy:

1. one first preprocesses the input formula to a restricted syntactic class, preserving the property under evaluation (validity for VAL, and satisfiability for SAT);
2. an efficient method is then applied to check the validity of formulas in this restricted class.

Of course, both steps should be kept "reasonably effective" since they are intended to be run in sequence—as we shall see, sometimes an apparently simple method for the second step relies on a compromisingly inefficient first step.

3.5.1 Normal Forms in Propositional Logic

To address the first step in the strategy outlined, we introduce *normal forms*: syntactical classes of formulas (i.e. formulas with a restricted "shape") that can be considered to be representative of the whole set of formulas. The idea is that we associate to a normal form a *normalization procedure* that, for any formula, computes a formula of this restricted class that is *equisatisfiable* with the original, i.e. it is satisfiable if and only if the original formula is.

In propositional logic there are three important normal forms that we now define. In the following a proposition symbol and its negation will be called *literals*.

---

[1]More rigorously, any problem in **NP** is reducible in polynomial-time to SAT.

Definition 3.34 (Normal Forms) Given a propositional formula A, we say that it is in:

- *Negation normal form* (NNF), if the implication connective is not used in A, and negation is only applied to atomic formulas (propositional symbols or the constant $\bot$);
- *Conjunctive normal form* (CNF), if it is a conjunction of disjunctions of literals, i.e. $A = \bigwedge_i \bigvee_j l_{ij}$, for literals l_{ij};
- *Disjunctive normal form* (DNF), if it is a disjunction of conjunctions of literals, i.e. $A = \bigvee_i \bigwedge_j l_{ij}$, for literals l_{ij}.

A formula in NNF is a formula built up from literals, the constants $\bot$ and $\top$ (i.e. $\neg\bot$), disjunctions and conjunctions. In the definition of CNF (resp. DNF), $\bot$ (resp. $\top$) is considered to be the empty disjunction (resp. the empty conjunction).[2]

Example 3.35 Let A be $(\neg P \vee Q) \wedge \neg R \vee (P \wedge Q)$, B be $(P \vee \neg Q) \wedge (Q \vee R \vee P) \wedge \neg P$, and C be $(R \wedge P) \vee \neg Q \vee (P \wedge \neg R \wedge Q)$. Then, A, B and C are NNFs, B is a CNF and C a DNF.

For checking the validity or satisfiability of formulas we will be primarily interested in CNFs and/or DNFs. Thus our perspective on NNFs is as an intermediate step needed to compute CNFs/DNFs. Since DNF and CNF are dual concepts, we will restrict our attention to the latter class in the rest of this section. The interested reader may transfer the results and methods to the former class by dualisation.

Let us first consider the normalisation procedure for NNF. We will use what is known as a *rewriting system*. Consider the following set of equivalences taken from Fig. 3.2.

$$A \to B \equiv \neg A \vee B$$

$$\neg\neg A \equiv A$$

$$\neg(A \wedge B) \equiv \neg A \vee \neg B$$

$$\neg(A \vee B) \equiv \neg A \wedge \neg B$$

Given an arbitrary formula, we repeatedly replace any subformula that is an instance of the left-hand-side (LHS) of one of these equivalences by the corresponding right-hand-side (RHS). This process eventually terminates since the RHSs are "closer" to normal forms than the LHSs (intuitively, the negation symbol is pushed towards the atoms—this argument is formalised in Exercise 3.20).

Moreover, by Propositions 3.11 and 3.12, the resultant formula is equivalent to the original, and is clearly in NNF (otherwise at least one of the above rules is

---

[2]It is not uncommon to find in the literature slightly different definitions for CNF and DNF, for instance allowing for instance constants $\bot$ and $\top$ to appear inside conjunctions/disjunctions, disallowing repetitions of literals, etc.

applicable). It is also worth mentioning that the result is not significantly bigger than the original formula—the number of binary connectives and atomic propositions is preserved, and the number of negations is at most increased by the number of binary connectives. This algorithm is indeed efficient since it can be encoded as a function with running time linear in the size of the input formula.

In order to reach a CNF, one should further apply the distributivity property of $\vee$ over $\wedge$ to pull conjunctions outside disjunctions, and remove occurrences of $\bot$ and $\top$ as conjuncts and disjuncts. One additionally considers the following equivalences taken from Fig. 3.2

$$A \vee (B \wedge C) \equiv (A \vee B) \wedge (A \vee C) \qquad (A \wedge B) \vee C \equiv (A \vee C) \wedge (B \vee C)$$

$$A \wedge \bot \equiv \bot \qquad \bot \wedge A \equiv \bot \qquad A \wedge \top \equiv A \qquad \top \wedge A \equiv A$$

$$A \vee \bot \equiv A \qquad \bot \vee A \equiv A \qquad A \vee \top \equiv \top \qquad \top \vee A \equiv \top$$

The repeated and exhaustive application of these equivalences to a formula in NNF leads to an equivalent CNF. The termination argument is slightly more evolved, but still holds (again, see Exercise 3.20 for details).

Example 3.36 Let us compute the CNF of $((P \to Q) \to P) \to P$. The first step is to compute its NNF by transforming implications into disjunctions and pushing negations to proposition symbols:

$$
\begin{aligned}
((P \to Q) \to P) \to P &\equiv \neg((P \to Q) \to P) \vee P \\
&\equiv \neg(\neg(P \to Q) \vee P) \vee P \\
&\equiv \neg(\neg(\neg P \vee Q) \vee P) \vee P \\
&\equiv \neg((P \wedge \neg Q) \vee P) \vee P \\
&\equiv (\neg(P \wedge \neg Q) \wedge \neg P) \vee P \\
&\equiv ((\neg P \vee Q) \wedge \neg P) \vee P
\end{aligned}
$$

To reach a CNF, distributivity is then applied to pull the conjunction outside:

$$((\neg P \vee Q) \wedge \neg P) \vee P \equiv (\neg P \vee Q \vee P) \wedge (\neg P \vee P).$$

The problem with the CNF translation is that it has an exponential worst-case running time, since the distributive equivalences duplicate formulas and the resulting formula can thus be exponentially bigger than the original formula. The following formula illustrates this bad behaviour.

$$(P_1 \wedge Q_1) \vee (P_2 \wedge Q_2) \vee \cdots \vee (P_n \wedge Q_n)$$
$$\equiv (P_1 \vee (P_2 \wedge Q_2) \vee \cdots \vee (P_n \wedge Q_n)) \wedge (Q_1 \vee (P_2 \wedge Q_2) \vee \cdots \vee (P_n \wedge Q_n))$$
$$\equiv \cdots$$
$$\equiv (P_1 \vee \cdots \vee P_n) \wedge$$
$$(P_1 \vee \cdots \vee P_{n-1} \vee Q_n) \wedge$$
$$(P_1 \vee \cdots \vee P_{n-2} \vee Q_{n-1} \vee P_n) \wedge$$
$$(P_1 \vee \cdots \vee P_{n-2} \vee Q_{n-1} \vee Q_n) \wedge$$
$$\cdots \wedge$$
$$(Q_1 \vee \cdots \vee Q_n)$$

Note that the original formula has $2 \cdot n$ literals, while the corresponding CNF has 2^n disjunctive clauses, each with n literals. This example reveals that, in practice, it is not reasonable to reduce a formula in its equivalent CNF as part of a VAL procedure.

There are alternative conversions to CNF that avoid this exponential growth. The trick is that instead of producing an equivalent formula, one can produce formulas that are *equisatisfiable* with the original formula, i.e. the resultant formula is satisfiable whenever the original formula is. This is of course a weaker requirement than equivalence; these conversions are adequate for solving the SAT problem, but not VAL.

These alternative conversions compute what is called the *definitional CNF* of a formula, and rely on the introduction of new proposition symbols that act as names for subformulas of the original one. In the previous example, one might consider proposition symbols $R_1, \ldots, R_n$ that act as names for each conjunction in the original formula. Additional disjunctive clauses are added to ensure that these new propositional symbols are tied to their corresponding subformulas. For the example at hand, one would use for instance R_i to stand for $P_i \wedge Q_i$, and add the formula $(\neg R_i \vee P_i) \wedge (\neg R_i \vee Q_i)$ to state that R_i implies both P_i and Q_i. This would result in the formula

$$(R_1 \vee \cdots \vee R_n) \wedge (\neg R_1 \vee P_1) \wedge (\neg R_1 \vee Q_1) \wedge \cdots \wedge (\neg R_n \vee P_n) \wedge (\neg R_n \vee Q_n)$$

Now let $\mathcal{M}$ be any model satisfying this CNF. Then, necessarily $\mathcal{M} \models R_i$ (for some i), which implies that $\mathcal{M} \models P_i$ and $\mathcal{M} \models Q_i$. It is then the case that $\mathcal{M}$ witnesses the satisfiability of the original formula, thus the satisfiability of the converted formula implies the satisfiability of the original formula. The converse implication is obviously not true, since propositions R_i are not taken in consideration in the original formula.

3.5.2 Validity of CNF Formulas

Having considered the translation of arbitrary formulas into CNF, let us now concentrate on the validity and satisfiability problems for this restricted class of formulas.

The strict shape of CNFs make them particularly suited for checking validity problems. Recall that CNFs are formulas with the following shape (each l_{ij} denotes a literal):

$$(l_{11} \vee l_{12} \vee \cdots \vee l_{1k}) \wedge \cdots \wedge (l_{n1} \vee l_{n2} \vee \cdots \vee l_{nj})$$

The inner disjuncts are usually called simply *clauses*. Associativity, commutativity and idempotence of both disjunction and conjunction (following equivalences in Fig. 3.2) allow us to treat each CNF as a set of sets of literals S

$$S = \{\{l_{11}, l_{12}, \ldots, l_{1k}\}, \ldots, \{l_{n1}, l_{n2}, \ldots, l_{nj}\}\}$$

In what follows we will often adopt this view. An empty inner set will be identified with $\bot$, and an empty outer set with $\top$. We note the following:

- Since a CNF is a conjunction of clauses, it is a tautology if and only if all the clauses are tautologies. If a clause $c \in S$ is a tautology, it can be removed from S without affecting its validity status, i.e. $S \equiv S \setminus \{c\}$.
- A clause c on the other hand is a disjunction of literals, and it is a tautology precisely when there exists a proposition symbol P such that $\{P, \neg P\} \subseteq c$ (otherwise it would be possible to build an assignment that refutes c, see Exercise 3.7). A clause c such that $\{P, \neg P\} \subseteq c$ for some P is said to be *closed*.

These considerations provide us with a simple criterion for checking validity:

A CNF is a tautology if and only if all its clauses are closed.

s an example, this criterion allows us to conclude that the formula $((P \rightarrow Q) \rightarrow P) \rightarrow P$ of Example 3.36 is a tautology, since all the clauses of its CNF are closed.

Unfortunately, the potential exponential growth introduced by the transformation of formulas into their equivalent CNFs compromises the applicability of this simple criterion for VAL. As explained before, to overcome this limitation one should instead consider transformations that preserve only satisfiability, and focus instead on the SAT problem.

3.5.3 Satisfiability of CNF Formulas

We will now consider one of the most important methods to check satisfiability of CNFs: the *Davis-Putnam-Logemann-Loveland procedure (DPLL)*. DPPL is an algorithm for verifying if a particular CNF is a contradiction. It incrementally constructs a model compatible with a CNF—if no such model exists, the formula is signaled as a contradiction. Otherwise it is satisfiable.

The basic observation underlying DPLL is that if we fix the interpretation of a particular proposition symbol, we are able to simplify the corresponding CNF accordingly. Consider a proposition symbol P, a CNF S, a clause $c \in S$, and a model $\mathcal{M}$:

- If $P \in \mathcal{M}$,

– if $P \in c$ then $\mathcal{M} \models c$ since c is a disjunction containing P. Thus $\mathcal{M} \models S$ iff $\mathcal{M} \models S \setminus \{c\}$. In short, clauses containing P can be ignored.

– $\mathcal{M} \models c$ iff $\mathcal{M} \models c \setminus \{\neg P\}$, because $\mathcal{M} \not\models \neg P$. In short, $\neg P$ can be removed from every clause in S.

— Analogously if $P \notin \mathcal{M}$ (i.e. $\mathcal{M} \models \neg P$):

– if $\neg P \in c$ then $\mathcal{M} \models S$ iff $\mathcal{M} \models S \setminus \{c\}$;

– $\mathcal{M} \models c$ iff $\mathcal{M} \models c \setminus \{P\}$.

These observations can be summarised as follows. Let

$$\mathsf{split}^P(S) = \{c \setminus \{\neg P\} \mid c \in S, P \notin c\}$$

$$\mathsf{split}^{\neg P}(S) = \{c \setminus \{P\} \mid c \in S, \neg P \notin c\}$$

Note that neither P nor $\neg P$ occur in any clause of $\mathsf{split}^P(S)$ or $\mathsf{split}^{\neg P}(S)$. Informally we can see $\mathsf{split}^P(S)$ (resp. $\mathsf{split}^{\neg P}(S)$) as a simplification of S assuming P is satisfied (resp. not satisfied).

Then for a CNF S and proposition symbol P,

$$S \equiv (P \to \mathsf{split}^P(S)) \wedge (\neg P \to \mathsf{split}^{\neg P}(S))$$

Recursively applying this simplification for every symbol occurring in the CNF is the heart of the DPLL algorithm.

Definition 3.37 (DPLL Algorithm) Let S be a CNF. The DPLL algorithm is defined recursively by

$$\mathrm{DPLL}(S) = \begin{cases} \mathbf{F} & \text{if } S = \top \\ \mathbf{T} & \text{if } \bot \in S \\ \mathrm{DPLL}(\mathsf{split}^P(S)) \text{ and } \mathrm{DPLL}(\mathsf{split}^{\neg P}(S)) & \text{otherwise} \end{cases}$$

where the proposition symbol P chosen in the recursive step is any proposition symbol appearing in S. The CNF S is a contradiction if $\mathrm{DPLL}(S) = \mathbf{T}$, and satisfiable otherwise.

Example 3.38 Figure 3.4 shows the recursion tree for the execution of DPLL on the CNF $(\neg P \vee \neg Q \vee \neg R) \wedge (\neg Q \vee \neg R) \wedge Q \wedge R$. Since all the leaves are tagged with $\mathbf{T}$, the formula is a contradiction.

Observe that the behaviour of the algorithm is highly dependent on the order in which the proposition symbols are chosen. In fact, actual implementations pay particular attention to how the next symbol is selected, in order to maximize the efficiency of the algorithm. Moreover, additional optimisations and heuristics are often explored to avoid unnecessary branches during execution (see Exercise 3.21).

Fig. 3.4 Recursion tree for
an execution of DPLL

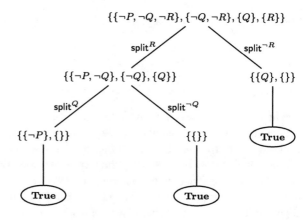

3.6 Validity Checking: Deductive Methods

Instead of directly using the semantic definition to establish the validity of a formula, one can rely on the correspondence given by the soundness and completeness theorems to conduct the argumentation on the deductive side. Checking the validity of a formula A then amounts to the construction of a derivation tree whose conclusion is the sequent $\vdash A$.

In general, when we face the problem of constructing a derivation for a sequent $\Gamma \vdash A$ (hereafter called a *proof goal*), two general strategies are possible, already mentioned in Sect. 3.3:

— in *forward reasoning*, assumptions and axioms are logically combined by inference rules to reason towards the goal;
— in *backward reasoning*, inference rules are directly applied to the goal, possibly generating new subgoals (when the rule has premises).

Forward reasoning arguably stands closer to the usual mathematical discourse, but needs to be properly oriented towards the proof goal—this is often a difficult task that poses serious challenges when mechanisation is desired.

The deductive method is also typically used in a computer-assisted interactive fashion, guided by the user. Proof assistants are tools that help the user construct derivations, applying inference rules in a stepwise fashion, and often providing facilities for partially automating the process of rule application. Backward reasoning is also typically employed when using proof assistants. The interactive construction of derivations is particularly important for other more expressive logics than propositional logic, when decidability is definitely lost.

Let us start with a small example in order to understand the difficulties raised by the process of proof construction. For concreteness, assume one is willing to show that $P \rightarrow (P \rightarrow Q) \rightarrow Q$ is a tautology, by exhibiting a derivation with conclusion $\vdash P \rightarrow (P \rightarrow Q) \rightarrow Q$. The first step is to identify the last rule used in the derivation. Notice that since the formula is an implication, the most natural choice is the rule $(I_\rightarrow)$.

$$\vdash P \rightarrow (P \rightarrow Q) \rightarrow Q, \qquad (I_\rightarrow)$$
$$\textbf{1.} \quad P \vdash (P \rightarrow Q) \rightarrow Q, \quad (I_\rightarrow)$$
$$\textbf{1.} \quad P, P \rightarrow Q \vdash Q, \qquad ???$$

The first rule still leaves an implication at the conclusion of the sequent, so the same argument leads to a second application of rule $(I_\rightarrow)$. Then we reach a point where we have an unstructured formula at the right-hand side of the sequent, but we are still not able to close the tree using the (Ax) rule. From a purely syntactic point of view, any rule from the elimination group or (RAA) admits as conclusion the current proof goal. In this particular example it is not difficult to anticipate that we should choose rule $(E_\rightarrow)$, since its premises are readily available in the context for being canceled by rule (Ax). Thus:

$$\textbf{1.} \quad P, P \rightarrow Q \vdash Q, \qquad (E_\rightarrow)$$
$$\textbf{1.} \quad P, P \rightarrow Q \vdash P, \qquad (Ax)$$
$$\textbf{2.} \quad P, P \rightarrow Q \vdash P \rightarrow Q, \quad (Ax)$$

This example shows that even though we have adopted the backward reasoning strategy, we still have to resort to some form of forward reasoning at specific points (at least mentally) to choose appropriate rule instances allowing us to proceed with the construction. Moreover, we note that the natural deduction rules exhibit different behaviour regarding the backward derivation construction process. In particular, introduction rules are much better behaved than elimination rules (or (RAA)), since they exhibit the following desirable features:

Syntactically directed: the applicability of a rule is constrained by the syntactic shape of the elements appearing in its conclusion. This is a desirable feature as it always allows to narrow the candidate rules for proof construction. In the above example for instance, the application of rule $(I_\wedge)$ is not possible, since it would require the conclusion of the current goal sequent to be a conjunctive formula.

Subformula property: a rule is said to exhibit the *subformula property* when all the formulas that appear in the premises are subformulas of formulas appearing in the conclusion of the rule. In particular, all the meta-variables that appear in the premises also appear in the conclusion. This is desirable since when the rule is applied backwards all the premises are fully determined by the conclusion, and there is no need to guess any formulas.

Simpler premises: the sequents appearing in the premises are simpler than those in the conclusion. This provides simple evidence of progress during the derivation-construction process.

None of the elimination rules exhibit these properties. In fact, they share some kind of dual properties that make them more suitable for forward reasoning. The (RAA) rule is, to this respect, similar to elimination rules.

Example 3.39 We illustrate the above discussion with a well-known non-trivial example: a derivation of the *excluded middle* tautology $A \vee \neg A$. Note that it is of no use to apply the well-behaved rules $(I_{\vee i})$ here, since neither of the resulting sequents $\vdash A$ nor $\vdash \neg A$ are provable. In fact there is a single viable rule for starting the derivation, which is (RAA). This corresponds to a very specific proof strategy known as *proof by contradiction*.[3]

$$\vdash A \vee \neg A, \qquad\qquad \text{(RAA)}$$

$\quad$ **1.** $\neg(A \vee \neg A) \vdash \bot,$ $\qquad$ (E$_\neg$)
$\qquad$ **1.** $\neg(A \vee \neg A) \vdash ?_1,$ $\qquad$???
$\qquad$ **2.** $\neg(A \vee \neg A) \vdash \neg ?_1,$ $\quad$???

The second rule has to introduce $\bot$ in the conclusion; the natural choice for this is rule (E$_\neg$). However, we need to choose a formula for filling the hole $?_1$, corresponding to a formula that appears in the premises of the rule and not in the conclusion. For that, note that the formula $\neg(A \vee \neg A)$ is equivalent to $\neg A \wedge \neg\neg A$ (using the De Morgan's laws of Fig. 3.2), which suggests $\neg A$ as a candidate for filling the hole:

$\quad$ **1.** $\neg(A \vee \neg A) \vdash \bot,$ $\qquad$ (E$_\neg$)
$\qquad$ **1.** $\neg(A \vee \neg A) \vdash \neg A,$ $\qquad$???
$\qquad$ **2.** $\neg(A \vee \neg A) \vdash \neg\neg A,$ $\quad$???

To prove the first premise we apply (I$_\neg$), followed again by (E$_\neg$)

$\quad$ **1.** $\neg(A \vee \neg A) \vdash \neg A,$ $\qquad$ (I$_\neg$)
$\qquad$ **1.** $\neg(A \vee \neg A), A \vdash \bot,$ $\qquad$ (E$_\neg$)
$\qquad\quad$ **1.** $\neg(A \vee \neg A), A \vdash ?_2,$ $\qquad$???
$\qquad\quad$ **2.** $\neg(A \vee \neg A), A \vdash \neg ?_2,$ $\quad$???

Observing that from A we easily get $A \vee \neg A$ makes this last formula the right candidate for filling the hole $?_2$:

$\quad$ **1.** $\neg(A \vee \neg A), A \vdash \bot,$ $\qquad$ (E$_\neg$)
$\qquad$ **1.** $\neg(A \vee \neg A), A \vdash A \vee \neg A,$ $\qquad$ (I$_{\vee 1}$)
$\qquad\quad$ **1.** $\neg(A \vee \neg A), A \vdash A,$ $\qquad$ (Ax)
$\qquad$ **2.** $\neg(A \vee \neg A), A \vdash \neg(A \vee \neg A),$ $\quad$ (Ax)

---

[3]Proofs by contradiction are not possible in the intuitionistic setting mentioned in Sect. 2.3.1. The rule (RAA) is absent from proof systems for intuitionistic logic.

The other branch is similar, leading to the following complete derivation.

$$
\begin{array}{ll}
\vdash A \vee \neg A, & \text{(RAA)} \\
\quad 1. \ \neg(A \vee \neg A) \vdash \bot, & \text{(E}_\neg) \\
\quad\quad 1. \ \neg(A \vee \neg A) \vdash \neg A, & \text{(I}_\neg) \\
\quad\quad\quad 1. \ \neg(A \vee \neg A), A \vdash \bot, & \text{(E}_\neg) \\
\quad\quad\quad\quad 1. \ \neg(A \vee \neg A), A \vdash A \vee \neg A, & \text{(I}_{\vee 1}) \\
\quad\quad\quad\quad\quad 1. \ \neg(A \vee \neg A), A \vdash A, & \text{(Ax)} \\
\quad\quad\quad\quad 2. \ \neg(A \vee \neg A), A \vdash \neg(A \vee \neg A), & \text{(Ax)} \\
\quad\quad 2. \ \neg(A \vee \neg A) \vdash \neg\neg A, & \text{(I}_\neg) \\
\quad\quad\quad 1. \ \neg(A \vee \neg A), \neg A \vdash \bot, & \text{(E}_\neg) \\
\quad\quad\quad\quad 1. \ \neg(A \vee \neg A), \neg A \vdash A \vee \neg A, & \text{(I}_{\vee 2}) \\
\quad\quad\quad\quad\quad 1. \ \neg(A \vee \neg A), \neg A \vdash \neg A, & \text{(Ax)} \\
\quad\quad\quad\quad 2. \ \neg(A \vee \neg A), \neg A \vdash \neg(A \vee \neg A), & \text{(Ax)}
\end{array}
$$

In essence we can say that constructing derivations in the natural deduction system is pretty much like playing a game: there are rules that must be followed, and some moves are almost mechanical while others demand judicious choices. Of course, experience plays a fundamental role in the process, and users rapidly develop strategies to overcome the most challenging patterns originated by the elimination rules.

It is also possible to use appropriate admissible rules in the construction of derivations. Good rules for this purpose would share the good properties of introduction rules, but operate on the assumption set, and thus behave like elimination rules. Examples include the two following admissible rules, both admissible in system $\mathcal{N}_{\mathsf{PL}}$:

$$
\frac{\Gamma, \neg A, \neg B \vdash C}{\Gamma, \neg(A \vee B) \vdash C} \qquad\qquad \frac{}{\Gamma, A, \neg A \vdash \bot}
$$

These rules make the previous example trivial (apart from the choice of rule (RAA)). We should also mention that the difficulties posed by natural deduction are overcome by other alternative deductive systems. We will return to this point in Sect. 3.7.

3.7 To Learn More

In addition to natural deduction, other proof systems that formalise logical derivations include *Hilbert systems* and Gentzen's *sequent calculus*.

The *sequent calculus* was, like the natural deduction system, introduced by Gentzen. It shares many features with natural deduction, like a varying set of assumptions, multiple inference rules, a single axiom scheme. What distinguishes it from natural deduction is the absence of elimination rules, which are replaced instead by introduction rules that act on the context. This feature makes the system more amenable to the study of its meta-theoretical properties, and easier to use when constructing derivations. In particular, the sequent calculus has the property that all rules enjoy the subformula property (see Sect. 3.6), and is thus much more suited for backward proof construction than natural deduction.

Hilbert systems, introduced by David Hilbert in the beginning of the 20th century, are characterised by leaving the assumption set unchanged in any derivation. They are based on a very small set of inference rules (a single one, for propositional logic) and substantial number of axiom schemes. The preferred strategy for building derivations in these systems is by forward reasoning (one starts with known facts and derives new facts), but it is normally awkward to use when compared with natural deduction or sequent calculus.

A detailed account of Hilbert systems and of the sequent calculus can be found for instance in [2].

In Sect. 3.5 we have introduced the semantic approach to both the validity and the satisfiability problems in propositional logic. Even though both problems are decidable, they are inherently complex. Our goal was simply to give the reader a basic understanding of some practical aspects of the validity of propositional formulas. We did not even study a concrete method for definitional CNF conversion (i.e. equisatisfiability-based conversion of formulas to CNF). For a deeper discussion of the issues involved in such a conversion and in its implementation see [3]. The reader should keep in mind that the choice of algorithms and appropriate implementation decisions are critical for achieving a decision procedure that can be used in practice.

As a mater of fact several implementations exist and have been successfully used in industry, in the context of hardware and software formal verification. A particular class of tools that are close to the computational approach exposed in this chapter are the so called *SAT solvers* [5]—an important family of these solvers are in fact based on the DPLL algorithm. The *satisfiability library* SATlib[4] [4] is an online resource that proposes, as a standard, a unified notation and a collection of benchmarks for performance evaluation and comparison of tools. Such a uniform test-bed has been serving as a framework for regular tool competitions organised in the context of the regular SAT conferences.[5] The resource collects problem instances, algorithms, and empirical characterisations of the performance of algorithms. All of this demonstrates the vitality of the research in this area.

Finally, given our emphasis on the semantic method and the importance of SAT solvers, the reader may have wondered why we bothered to mention the deductive approach, and whether it is useful at all, since the particular proof system studied in this chapter is not even particularly good for being mechanically applied. The deductive method is however often the only method applicable in full generality, since it can be used interactively, guided by human choices. This is a very important point to understand. Note that propositional logic is a very narrow fragment of logical reasoning, which has the remarkable feature that there exists a decision procedure for establishing the validity (or refutation) of every formula. We will in the next chapter discuss a richer logic (and there are many others!) for which such a procedure does not exist. In the absence of decidability, semantic proof methods will try to proceed by refutation, but may well fail in producing a conclusive answer.

---

[4]http://www.satlib.org/.

[5]http://www.satcompetition.org.

For such logics, the importance of the user-guided deductive method cannot be overstated, since it allows to establish in a definitive way the validity of a formula. We remark that the drawbacks identified in the formulation of the natural deduction system end up being diluted in problems that are inherent to the proof construction process. While it is true that having an appropriate formulation of the formal system can facilitate the constrution of proofs in certain scenarios, it may never replace human genius in every situation. The bottom line is that this process necessarily appeals to the *creativity* of the user, as hinted in Sect. 3.3.

For other classic references on propositional logic, see Sect. 4.8.

3.8 Exercises

3.1. Define a function pSymbs : **Form** → $\mathcal{P}(\mathbf{Prop})$ that computes the set of proposition symbols that occur in a formula.

3.2. Define the function subF : **Form** → $\mathcal{P}(\mathbf{Form})$ that computes the subformulas of a given formula. Consider the non-strict notion of subformula, i.e. a formula is considered a subformula of itself.

3.3. Prove that for every formula A, the set of proposition symbols pSymbs(A) coincides with subF(A) ∩ **Prop**.

3.4. Prove the equivalence between the semantics of propositional logic given by Definitions 3.5 and 3.8, i.e. given a valuation ρ and a model $\mathcal{M} = \{P \in \mathbf{Prop} \mid \rho(P) = \mathbf{T}\}$, we have:

$$\text{for every formula } A, \quad \mathcal{M} \models A \quad \text{iff} \quad [\![A]\!]_\rho = \mathbf{T}$$

3.5. Show that only the proposition symbols used in a formula A are relevant to its meaning, i.e. for every propositional model $\mathcal{M}$,

$$\mathcal{M} \models A \quad \text{iff} \quad (\mathcal{M} \cap \mathsf{pSymbs}(A)) \models A$$

where pSymbs is the function that computes the set of proposition symbols used in formula A (see Exercise 3.1).

3.6. Show that the following formulas are tautologies:
   (a) $((A \to B) \to A) \to A$ (*Peirce's Axiom*)
   (b) $(A \to B \to C) \to (A \to B) \to A \to C$ (*S axiom*)
   (c) $A \to B \to A$ (*K axiom*)

3.7. Let $B = A_1 \vee \cdots \vee A_n$ where A_i is either a proposition symbol or its negation. Show that B is a tautology if and only if there exist j, k such that $A_j = \neg A_k$.

3.8. Recall that $A \leftrightarrow B$ abbreviates $A \to B \wedge B \to A$. Construct a truth table for $A \leftrightarrow B$ and prove that

$$A \leftrightarrow B \quad \text{iff} \quad A \equiv B$$

3.9. Prove Proposition 3.11, i.e. that $\equiv$ is an equivalence relation on **Form**.

3.10. Let $A, B, C \in \mathbf{Form}$. Use semantic arguments (e.g. truth tables) to show that:

(a) $A \to B \to C \not\equiv (A \to B) \to C$ (*implication is non-associative*)
(b) $A \land B \land C \equiv (A \land B) \land C$ (*conjunction is associative*).

3.11. Use equational reasoning to derive the neutral and absorption element laws from the algebraic laws of Fig. 3.2.

3.12. When using equational reasoning for establishing equivalences, a valuable result is the following *indirect equivalence lemma*:

> Let A and B be propositional formulas. If $A \land \neg B \equiv \bot$ and $A \lor \neg B \equiv \top$, then $A \equiv B$.

Derive the above lemma from the algebraic laws of Fig. 3.2.

3.13. Derive the *De Morgan's laws* from the algebraic laws of Fig. 3.2.
(Hint: use the lemma proved in the previous exercise)

3.14. A set of connectives is called a *minimal functionally complete* set if every formula of propositional logic is equivalent to one that uses only connectives from that set. Show that the following sets are minimal functionally complete sets of connectives.

(a) $\{\lor, \neg\}$ (b) $\{\land, \neg\}$

(c) $\{\to, \neg\}$ (d) $\{\to, \bot\}$

3.15. Prove Proposition 3.18, i.e. that
(a) $\Gamma \models A$ iff $\Gamma, \neg A \models \bot$
(b) $\Gamma \models \neg A$ iff $\Gamma, A \models \bot$
(c) $\Gamma \models A \to B$ iff $\Gamma, A \models B$
(d) $\Gamma \models A \land B$ iff $\Gamma \models A$ and $\Gamma \models B$
(e) $\Gamma \models A \lor B$ iff $\Gamma \models A$ or $\Gamma \models B$

3.16. Some presentations of the system $\mathcal{N}_{\mathsf{PL}}$ replace rule (RAA) with

$$(\mathrm{E}_{\neg\neg}) \; \frac{\Gamma \vdash \neg\neg A}{\Gamma \vdash A}$$

(a) Show that rule $(\mathrm{E}_{\neg\neg})$ is admissible in $\mathcal{N}_{\mathsf{PL}}$.
(b) Let $\mathcal{N}'_{\mathsf{PL}}$ be the system which replaces (RAA) with $(\mathrm{E}_{\neg\neg})$. Show that (RAA) is derivable in $\mathcal{N}'_{\mathsf{PL}}$.

3.17. Show that the following rules are admissible in $\mathcal{N}_{\mathsf{PL}}$.
(Hint: actually, they are derivable in $\mathcal{N}_{\mathsf{PL}} \cup \{W\}$)
(a) If $\Gamma, \neg A \vdash B$, then $\Gamma, \neg B \vdash A$.
(b) If $\Gamma, A, B \vdash C$, then $\Gamma, A \land B \vdash C$.
(c) If $\Gamma, \neg A, \neg B \vdash \bot$, then $\Gamma \vdash A \lor B$.
(d) If $\Gamma, \neg A, \neg B \vdash C$, then $\Gamma, \neg(A \lor B) \vdash C$.

3.18. Prove the following sequents in $\mathcal{N}_{\mathsf{PL}}$.
(Hint: use admissible rules from the previous exercises)
(a) $(A \to C) \land (B \to D) \vdash (A \land B) \to (C \land D)$
(b) $(A \land B) \to C \vdash (\neg C \land A) \to \neg B$
(c) $\neg A \lor (B \to C) \vdash (A \land B) \to C$
(d) $(A \lor B) \land C \vdash (A \land C) \lor (B \land C)$

3.19. Compute equivalent CNFs and DNFs for the following formulas.
   (a) $A \vee (A \rightarrow B) \rightarrow A \vee \neg B$
   (b) $(A \rightarrow B \vee C) \wedge \neg (A \wedge \neg B \rightarrow C)$
   (c) $(\neg A \rightarrow \neg B) \rightarrow (\neg A \rightarrow B) \rightarrow A$
   What can you say about their validity status? (i.e. which are valid, contradiction, or satisfiable formulas).

3.20. In order to demonstrate the termination of normalisation procedures based on rewriting systems, one first defines a *measure function* $\#(\cdot) : \mathbf{Form} \rightarrow \mathbb{N}$ such that the following requirements are met:

   – the measure is *monotone* with respect to subformulas, i.e. if B is a subformula of A, then for any C such that $\#(C) < \#(B)$, it must be the case that $\#(A[C/B]) < \#(A)$ (where $A[C/B]$ stands for the substitution of occurrences of B by C in A);
   – applying each of the rewriting equations to a formula strictly decreases its measure.

   The well-foundedness of the relation $<$ on the natural numbers ensures that the normalisation procedure is terminating.
   (a) Define a measure and prove termination for the NNF normalisation procedure.
   (b) Show that the following measure is adequate for demonstrating termination of the normalisation procedure from NNF into CNF.

$$\#(\bot) = \#(\top) = \#(P) = \#(\neg P) = 3$$

$$\#(A \wedge B) = \#(A) + \#(B)$$

$$\#(A \vee B) = \#(A)^{\#(B)}$$

3.21. A *unit-clause* is a clause of a CNF that consists of a single literal. Let S be a CNF that contains a unit-clause $\{l\}$. The following CNF is called the *unit-propagation* of l in S

$$\mathsf{unitPropagate}(l, S) = \{c \setminus \{-l\} \mid c \in S, l \notin c\}$$

   where $-l$ denotes the opposite literal of l (i.e. if l is $\neg P$ then P else $\neg P$). Show that, when $\{l\} \in S$,

$$\mathsf{DPLL}(S) = \mathsf{DPLL}(\mathsf{unitPropagation}(l, S))$$

   thus avoiding an unnecessary branching on l.

References

1. Aczel, P.: Schematic consequence. In: Gabbay, D.M. (ed.) What Is a Logical System? Studies in Logic and Computation, pp. 261–272. Springer, Berlin (1994)

2. Goubault-Larrecq, J., Mackie, I.: Proof Theory and Automated Deduction. Applied Logic Series, vol. 6. Kluwer Academic, Dordrecht (1997)
3. Harrison, J.: Handbook of Practical Logic and Automated Reasoning. Cambridge University Press, Cambridge (2009)
4. Hoos, H.H., Stutzle, T.: Satlib: an online resource for research on sat. In: Walsh, T., Gent, I.P., v. Maaren, H. (eds.) SAT 2000, pp. 283–292. IOS Press, Amsterdam (2000)
5. Prasad, M.R., Biere, A., Gupta, A.: A survey of recent advances in sat-based formal verification. Int. J. Softw. Tools Technol. Transf. (STTT) 7(2), 156–173 (2005)

Chapter 4
First-Order Logic

First-order logic augments the expressive power of propositional logic as it links the logical assertions to properties of objects of some non-empty universe—the *domain of discourse*. This is achieved by allowing the propositional symbols to take arguments that range over elements of the domain of discourse. These are now called predicate symbols and are interpreted as relations on a domain. The terms of the domain of discourse are made up of variables, constants, and functions applied to other terms. First-order logic also expands the lexicon of propositional logic with the symbols $\forall$ and $\exists$ for "for all" and "there exists" along with various symbols denoting variables, constants and functions.

First-order logic then admits individual variables that range over the domain of discourse, predicates such as $sports(x)$ to represent the claim that the variable x is a sport, $flower(x)$ indicating that x is a flower, $older(x, y)$ stating that x is older than y, or $likes(x, y)$ expressing the claim that x likes y. The first-order language allows us to express that "John likes all sports" by $\forall x \,.\, sports(x) \rightarrow likes(John, x)$; "John's mother likes flowers" by $\forall x \,.\, flower(x) \rightarrow likes(mother(John), x)$; "John's mother does not like some sports" by $\exists y \,.\, sport(y) \wedge \neg likes(mother(John), y)$; "Peter only likes sports" by $\forall x \,.\, likes(Peter, x) \rightarrow sports(x)$; or "every mother is older than her children" by $\forall x \,.\, older(mother(x), x)$. Note that a function is used to refer to "John's mother", i.e., $mother(John)$ represents an object of the domain of discourse.

This chapter is devoted to *classical* first-order logic. Our presentation will be similar to the one conducted for propositional logic. We first define the syntax of first-order logic followed by its semantics. Next we define a proof system for it and present the fundamental theoretical results of soundness and completeness. We also discuss the decision problems related to this logic. The remaining sections of the chapter cover variations and extensions of first-order logic, as well as first-order theories.

J.B. Almeida et al., *Rigorous Software Development*,
Undergraduate Topics in Computer Science,
DOI 10.1007/978-0-85729-018-2_4, © Springer-Verlag London Limited 2011

4.1 Syntax

The alphabet of a first-order language is organised into the following categories.

- *Logical connectives*: $\bot$, $\neg$, $\wedge$, $\vee$, $\rightarrow$, $\forall$ and $\exists$. The symbols $\forall$ and $\exists$ are called the *universal* and *existential* quantifier respectively.
- *Auxiliary symbols*: ".", "(" and ")".
- *Variables*: we assume a countable infinite set $\mathcal{X}$ of variables that represent arbitrary elements of an underlying domain. We let $x, y, z, \ldots$ range over $\mathcal{X}$.
- *Constants*: we assume a countable set $\mathcal{C}$ of constants that represent specific elements of an underlying domain. We let $a, b, c, \ldots$ range over $\mathcal{C}$.
- *Functions*: we assume a countable set $\mathcal{F}$ of function symbols. We let $f, g, h, \ldots$ range over $\mathcal{F}$. Each function symbol f has a fixed arity $\mathrm{ar}(f)$, which is a positive integer.
- *Predicates*: we assume a countable set $\mathcal{P}$ of predicate symbols. We let $P, Q, R, \ldots$ range over $\mathcal{P}$. Each predicate symbol P has a fixed arity $\mathrm{ar}(P)$, which is a non-negative integer. Predicate symbols with arity 0 play the role of propositions.

We assume that all these sets are disjoint. $\mathcal{C}$, $\mathcal{F}$ and $\mathcal{P}$ are the non-logical symbols of the language. These three sets constitute the *vocabulary* $\mathcal{V} = \mathcal{C} \cup \mathcal{F} \cup \mathcal{P}$.

Remark 4.1 There are different conventions for dealing with equality in first-order logic. Here we follow the approach of considering equality symbol (=) as a non-logical symbol. We will thus be working with what are usually known as "*first-order languages without equality*". An alternative approach considers the equality symbol as a logical symbol with a fixed interpretation. This alternative approach results in "*first-order languages with equality*", and will be studied in Sect. 4.6.1.

There are two sorts of things involved in a first-order logic formula: *terms*, which denote the objects that we are reasoning about; and *formulas*, which denote truth values.

Definition 4.2 The set of *terms* of a first-order language over a vocabulary $\mathcal{V}$ is given by the following abstract syntax

$$\mathbf{Term}_{\mathcal{V}} \ni t, u \ ::= \ x \mid c \mid f(t_1, \ldots, t_{\mathrm{ar}(f)})$$

The set of variables occurring in t is denoted by $\mathsf{Vars}(t)$. A *ground term* t is a term without variables, i.e. $\mathsf{Vars}(t) = \emptyset$.

Variables and constants are the first building blocks of terms. More complex terms are built with function symbols using previously built terms. Note that functions may be nested.

Definition 4.3 The set of *formulas* of a first-order language over a vocabulary $\mathcal{V}$ is given by the abstract syntax

$$\mathbf{Form}_{\mathcal{V}} \ni \phi, \psi, \theta \quad ::= \quad P(t_1, \ldots, t_{\mathrm{ar}(P)}) \mid \bot \mid (\neg\phi) \mid (\phi \wedge \psi) \mid (\phi \vee \psi)$$
$$\mid (\phi \to \psi) \mid (\forall x. \phi) \mid (\exists x. \phi)$$

An *atomic formula* has the form $\bot$ or $P(t_1, \ldots, t_{\mathrm{ar}(P)})$. More complex formulas are constructed with the logical connectives. As for terms, we define *ground formulas* as formulas without variables, i.e. quantifier-free formulas ϕ such that all terms occurring in ϕ are ground terms.

As was done in propositional logic, we adopt some syntactical conventions to lighten the presentation of formulas:

- Parentheses dictate the order of operations in any formula. In the absence of parentheses we have, ranging from the highest precedence to the lowest: $\neg$, $\wedge$, $\vee$ and $\to$. Finally we have that $\to$ binds more tightly than $\forall$ and $\exists$.
- Outermost parentheses are usually discarded.
- Binary connectives are right-associative.
- Nested quantifications such as $\forall x. \forall y. \phi$ are abbreviated to $\forall x, y. \phi$.
- Occasionally, sequences of variables $x_1, \ldots, x_n$ are represented in vector notation $\overline{x}$. $\forall \overline{x}. \phi$ denotes the nested quantification $\forall x_1, \ldots, x_n. \phi$.

Example 4.4 The formula

$$\exists y. P(x, y) \to \neg R(y) \vee Q(y, x) \to Q(x, x)$$

stands for

$$(\exists y. (P(x, y) \to (((\neg R(y)) \vee Q(y, x)) \to Q(x, x))))$$

and the formula

$$\forall x. \exists z. P(x, z) \wedge R(x) \to \forall y. Q(x, y)$$

stands for

$$(\forall x. (\exists z. ((P(x, z) \wedge R(x)) \to (\forall y. Q(x, y)))))$$

A formula ψ that occurs syntactically within a formula ϕ is called a *subformula* of ϕ. In a quantified formula $\forall x. \phi$ or $\exists x. \phi$, x is the *quantified variable* and ϕ is the *scope* of the quantification. Occurrences of the quantified variable within the respective scope of quantification are said to be *bound*. Variable occurrences that are not bound are said to be *free*. For example, in the formula $\exists x. P(x, y)$ the occurrence of x is bound, but the occurrence of y is free. Free and bound variables of a term t are variables that occur respectively free and bound in t.

Definition 4.5 The set of *free variables* of a formula θ, denoted $\mathsf{FV}(\theta)$, is defined inductively as follows:

- if θ is atomic, then $\mathsf{FV}(\theta)$ is the set of all variables occurring in θ
- if θ is $\neg\phi$, then $\mathsf{FV}(\theta) = \mathsf{FV}(\phi)$
- if θ is $\phi \wedge \psi$ or $\phi \vee \psi$ or $\phi \to \psi$, then $\mathsf{FV}(\theta) = \mathsf{FV}(\phi) \cup \mathsf{FV}(\psi)$
- if θ is $\forall x.\, \phi$ or $\exists x.\, \phi$, then $\mathsf{FV}(\theta) = \mathsf{FV}(\phi) \setminus \{x\}$

Definition 4.6 The set of *bound variables* of a formula θ, denoted $\mathsf{BV}(\theta)$, is defined inductively as follows:

- if θ is atomic, then $\mathsf{BV}(\theta) = \emptyset$
- if θ is $\neg\phi$, then $\mathsf{BV}(\theta) = \mathsf{BV}(\phi)$
- if θ is $\phi \wedge \psi$ or $\phi \vee \psi$ or $\phi \to \psi$, then $\mathsf{BV}(\theta) = \mathsf{BV}(\phi) \cup \mathsf{BV}(\psi)$
- if θ is $\forall x.\, \phi$ or $\exists x.\, \phi$, then $\mathsf{BV}(\theta) = \mathsf{BV}(\phi) \cup \{x\}$

Note that a variable can have both free and bound occurrences within the same formula. Hence, the sets of free and bound variables of a formula are not necessarily disjoint.

Example 4.7 Let θ be the formula $\exists x.\, Q(x, y) \wedge (\forall y.\, P(y, x))$. The variable x occurs only as a bound variable, but the variable y occurs free in $Q(x, y)$ and bound in $\forall y.\, P(y, x)$. So, $\mathsf{FV}(\theta) = \{y\}$ and $\mathsf{BV}(\theta) = \{x, y\}$.

Definition 4.8 A *sentence* (or *closed* formula) is a formula without free variables. If $\mathsf{FV}(\phi) = \{x_1, \ldots, x_n\}$, the *universal closure* of ϕ is the formula $\forall x_1, \ldots, x_n.\, \phi$ and the *existential closure* of ϕ is the formula $\exists x_1, \ldots, x_n.\, \phi$.

The meaning of a bound occurrence of a variable will ultimately be determined by the corresponding quantifier that binds it. On the other hand, free occurrences are to be thought of as unspecified elements. In the next section we will address the semantics of first-order logic, which will formally support this reading. For now, we will introduce a concept that, while being of a syntactical nature, actually captures this intuition: the *substitution* of free variables by terms. More than one definition are possible; we adopt here the more general notion of simultaneous substitution of a set of variables.

We start by introducing the following auxiliary notation for function update.

Definition 4.9 Let X, Y be sets and f a function with domain X and codomain Y. For $x \in X$ and $y \in Y$, the *patching* of f in x to y (written $f[x \mapsto y]$) is the function defined as follows, also with domain X and codomain Y:

$$(f[x \mapsto y])(z) = \begin{cases} y & \text{if } z = x \\ f(z) & \text{otherwise} \end{cases}$$

Definition 4.10 (Substitution) A *substitution* is a mapping $\sigma : \mathcal{X} \to \mathbf{Term}_{\mathcal{V}}$ from variables to terms such that $\sigma(x) = x$ for all but a finite number of variables. The *substitution domain* of σ is the finite set $\mathsf{dom}(\sigma) = \{x \in \mathcal{X} \mid \sigma(x) \neq x\}$. The *range* of σ is the direct image of $\mathsf{dom}(\sigma)$ under σ, i.e. $\mathsf{rng}(\sigma) = \{\sigma(x) \mid x \in \mathsf{dom}(\sigma)\}$. The *empty substitution* is the substitution with empty substitution domain, which injects variables into terms.

The notation $[t_1/x_1, \ldots, t_n/x_n]$, for pairwise distinct $x_1, \ldots, x_n$, or in vector notation $[\overline{t}/\overline{x}]$, denotes the substitution whose domain is contained in $\{x_1, \ldots, x_n\}$ that maps each x_i to t_i ($i \in \{1, \ldots, n\}$) and any other variable to itself.

Definition 4.11 The *application of a substitution σ to a term t* is denoted by $t\sigma$ and is defined recursively by:

$$x\sigma = \sigma(x)$$

$$c\sigma = c$$

$$f(t_1, \ldots, t_{ar(f)})\sigma = f(t_1\sigma, \ldots, t_{ar(f)}\sigma)$$

When t' is $t\sigma$, t' is said to be an *instance* of t. If additionally t' is a ground term, then it is a *ground instance* and σ is called a *ground substitution*.

Notice that, in general, the result of

$$t[t_1/x_1, \ldots, t_n/x_n]$$

is the simultaneous substitution of $t_1, \ldots, t_n$ for $x_1, \ldots, x_n$ in t. This differs from the application of the corresponding singleton substitutions in sequence,

$$((t[t_1/x_1])\ldots)[t_n/x_n]$$

since in the latter case each substitution may introduce new occurrences of variables that are affected by subsequent substitutions. To give a concrete example, consider the substitution application

- $f(x, y)[y/x, z/y] = f(y, z)$, and
- $(f(x, y)[y/x])[z/y] = f(y, y)[z/y] = f(z, z)$.

Having defined the application of a substitution to a term, we now consider the much subtler notion of application of a substitution to a first-order formula.

Definition 4.12 The *application of a substitution σ to a formula ϕ*, written $\phi\sigma$, is defined recursively by:

$$\bot\sigma = \bot$$

$$P(t_1, \ldots, t_{ar(P)})\sigma = P(t_1\sigma, \ldots, t_{ar(P)}\sigma)$$

$$(\neg\phi)\sigma = \neg(\phi\sigma)$$

$$(\phi \odot \psi)\sigma = (\phi\sigma) \odot (\psi\sigma)$$

$$(Qx.\phi)\sigma = Qx.(\phi(\sigma[x \mapsto x]))$$

where $\odot \in \{\wedge, \vee \rightarrow\}$ and $Q \in \{\forall, \exists\}$ is a quantifier.

Notice that the effect of the patching $\sigma[x \mapsto x]$ is to remove the variable x from the substitution domain of σ. This way, only free occurrences of variables are affected by the application of substitutions. To exemplify this, let ϕ be $\forall x. P(x, y)$. The occurrence of x is bound, and the occurrence of y is free in ϕ. We have

- $(\forall x. P(x, y))[t/x] = \forall x. P(x, y)$
- $(\forall x. P(x, y))[t/y] = \forall x. P(x, t)$

There is however a problem with this definition. Consider the application of the substitution $[g(x)/y]$ to ϕ:

- $(\forall x. P(x, y))[g(x)/y] = \forall x. P(x, g(x))$

Both occurrences of the variable x are now bound (we say that the occurrence of x in $g(x)$ has been *captured* by the quantifier). This is unacceptable since the intuitive meaning of the formula $\forall x. P(x, y)$ should not depend on the choice of the name of the quantified variable; its meaning should be the same of, say, $\forall z. P(z, y)$, in which no variable capture would occur when the same substitution is applied (in the next section we will see that the semantic interpretations of $\forall x. P(x, y)$ and $\forall z. P(z, y)$ are indeed the same).

Instead of fixing our definition to work around variable capture, we characterise the situations in which the use of Definition 4.12 is safe. Observe that captured variables always occur in the terms being substituted, and are bound by quantifiers of the formula where the substitution is applied.

Definition 4.13 The notion of a term *free for a variable in a formula* is defined inductively as follows. Let θ be a formula, t a term, and x a variable.

- If θ is atomic, then t is free for x in θ.
- If θ is $\neg\phi$ and t is free for x in ϕ, then t is free for x in θ.
- If θ is either $\phi \wedge \psi$ or $\phi \vee \psi$ or $\phi \rightarrow \psi$, and t is both free for x in ϕ and free for x in ψ, then t is free for x in θ.
- If θ is either $\forall y. \phi$ or $\exists y. \phi$, and either
    - $x \notin \mathsf{FV}(\theta)$, or
    - $y \notin \mathsf{Vars}(t)$ and t is free for x in ϕ,
  then t is free for x in θ.

A substitution σ is said to be *free for θ* if $\sigma(x)$ is free for x in θ, for all $x \in \mathrm{dom}(\sigma)$.

Note that t is free for x in θ if no free occurrence of x in θ is in the scope of $\forall y$ or $\exists y$ for some variable y occurring in t.

Clearly, when σ is free for ϕ, no variable capture possibly occurs in $\phi\sigma$. In the rest of this chapter, whenever a substitution is applied as in $\phi\sigma$, we will add the proviso that σ is free for ϕ. Fortunately, we will see that such a requirement can always be met, no matter what formula and substitution we start with.

4.2 Semantics

Before formally defining the semantics of first-order logic, let us first analyse a small example. Consider the following sentence:

$$\forall x. \exists y.\ P(y, f(x))$$

This first-order sentence says that for all x there exists y such that $P(y, f(x))$ holds. Whether this sentence is true or not depends on what the variables represent and what the function f and the predicate P are. For example, suppose the variables are natural numbers, f is defined by $f(x) = x + 1$ and P is defined by the relation "less than". Then the above sentence is true since there always exists a natural number less than its successor. But if f is defined by $f(x) = x^2$ then the sentence is false since there is no natural number less than 0. However, if we assume that variables represent real numbers, the sentence will be true.

The previous discussion shows that, in order to assign meaning to formulas, we first need to fix how the symbols of the vocabulary are interpreted. This will be role of first-order *structures*.

Definition 4.14 Given a vocabulary $\mathcal{V}$, a $\mathcal{V}$-*structure* $\mathcal{M}$ is a pair $\mathcal{M} = (D, I)$ where D is a nonempty set called the *interpretation domain*, and I is an *interpretation function* that assigns constants, functions and predicates over D to the symbols of $\mathcal{V}$ as follows:

- for each constant symbol $c \in \mathcal{C}$, the interpretation of c is a constant $I(c) \in D$;
- for each $f \in \mathcal{F}$, the interpretation of f is a function $I(f) : D^{\mathrm{ar}(f)} \to D$;
- for each $P \in \mathcal{P}$, the interpretation of P is a function $I(P) : D^{\mathrm{ar}(P)} \to \{\mathbf{F}, \mathbf{T}\}$. In particular, 0-ary predicate symbols are interpreted as truth values.

$\mathcal{V}$-structures are also called *models* for $\mathcal{V}$.

Occasionally, we will be interested in extending a vocabulary $\mathcal{V}$ with new symbols. The corresponding $\mathcal{V}$-structures shall be related accordingly.

Definition 4.15 Let $\mathcal{V}$ and $\mathcal{V}'$ be vocabularies and $\mathcal{M}$ be a $\mathcal{V}$-structure.

- $\mathcal{V}'$ is said to be an *expansion* of $\mathcal{V}$ if $\mathcal{V} \subseteq \mathcal{V}'$.
- A $\mathcal{V}'$-structure $\mathcal{M}'$ is an *expansion* of $\mathcal{M}$ if $\mathcal{M}'$ has the same interpretation domain as $\mathcal{M}$ and interprets the symbols of $\mathcal{V}$ in the same way as $\mathcal{M}$.

When $\mathcal{M}'$ is an expansion of $\mathcal{M}$, $\mathcal{M}$ is said to be a *reduct* of $\mathcal{M}'$.

Example 4.16 Let $\mathcal{V} = \{c, f, P\}$, where c is a constant, f is a unary function, and P is a binary predicate. $\mathcal{M}_1$, $\mathcal{M}_2$ and $\mathcal{M}_3$, described below, are examples of $\mathcal{V}$-structures.

- $\mathcal{M}_1$ is defined such that its domain is the set of natural numbers $\mathbb{N}$, the constant c is interpreted as 0, f is interpreted as the function such that $f(x)$ is $x + 1$ (the successor function) and the predicate P is interpreted as the relation such that $P(x, y)$ is $x < y$.

– $\mathcal{M}_2$ is defined such that its domain is the set of real numbers $\mathbb{R}$, c is interpreted as 1, $f(x)$ as x^2, and $P(x, y)$ as $x < y$.
– $\mathcal{M}_3$ is defined such that its domain is $\mathbb{N}$, c is interpreted as 0, $f(x)$ as x^2 and $P(x, y)$ as $x \geq y$.

As an example of an expansion, consider $\mathcal{V}' = \mathcal{V} \cup \{g, Q\}$ where g is a binary function and Q a unary predicate, and the $\mathcal{V}'$-structure $\mathcal{M}'$ that expands $\mathcal{M}_1$ interpreting $g(x, y)$ as $x \times y$, and $Q(x)$ as $x > 10$. $\mathcal{M}_1$ is then a reduct of $\mathcal{M}'$.

$\mathcal{V}$-structures address the interpretation of symbols in $\mathcal{V}$, but this is not enough when we are dealing with formulas containing free variables. Consider as an example the formula $\exists y.\ P(y, f(x))$. Even in a concrete $\mathcal{V}$-structure that fixes the interpretation domain and the interpretation of f and P, it is not possible to assign any sensible meaning to the formula, since it is not known how the free occurrence of x is interpreted. This is accomplished by a new interpretation layer as follows.

Definition 4.17 An *assignment* for a domain D is a function $\alpha : \mathcal{X} \to D$ from the set of variables to the domain D. The set of all assignments for a domain D is denoted by Σ_D.

The semantics of first-order logic may now be presented. The idea is that together a $\mathcal{V}$-structure and an assignment α interpret terms as elements of the interpretation domain and formulas as truth values.

Definition 4.18 Let $\mathcal{M} = (D, I)$ be a $\mathcal{V}$-structure. A functional, denoted by $[\![\cdot]\!]_\mathcal{M}$, is associated to $\mathcal{M}$ mapping every term $t \in \mathbf{Term}_\mathcal{V}$ to a function $[\![t]\!]_\mathcal{M} : \Sigma_D \to D$ and every formula $\phi \in \mathbf{Form}_\mathcal{V}$ to a function $[\![\phi]\!]_\mathcal{M} : \Sigma_D \to \{\mathbf{F}, \mathbf{T}\}$, as given below.

– $[\![t]\!]_\mathcal{M} : \Sigma_D \to D$ is defined inductively as follows:

$$\begin{aligned}
[\![x]\!]_\mathcal{M}(\alpha) &= \alpha(x) \\
[\![c]\!]_\mathcal{M}(\alpha) &= I(c) \\
[\![f(t_1, \ldots, t_{\mathrm{ar}(f)})]\!]_\mathcal{M}(\alpha) &= I(f)([\![t_1]\!]_\mathcal{M}(\alpha), \ldots, [\![t_{\mathrm{ar}(f)}]\!]_\mathcal{M}(\alpha))
\end{aligned}$$

We read $[\![t]\!]_\mathcal{M}(\alpha)$ as the *value of t with respect to $\mathcal{M}$ and α*.

– $[\![\phi]\!]_\mathcal{M} : \Sigma_D \to \{\mathbf{F}, \mathbf{T}\}$ is defined inductively as follows:

$$\begin{aligned}
[\![\bot]\!]_\mathcal{M}(\alpha) &= \mathbf{F} \\
[\![P(t_1, \ldots, t_{\mathrm{ar}(P)})]\!]_\mathcal{M}(\alpha) &= I(P)([\![t_1]\!]_\mathcal{M}(\alpha), \ldots, [\![t_{\mathrm{ar}(P)}]\!]_\mathcal{M}(\alpha)) \\
[\![\neg\phi]\!]_\mathcal{M}(\alpha) = \mathbf{T} &\quad \text{iff } [\![\phi]\!]_\mathcal{M}(\alpha) = \mathbf{F} \\
[\![\phi \wedge \psi]\!]_\mathcal{M}(\alpha) = \mathbf{T} &\quad \text{iff } [\![\phi]\!]_\mathcal{M}(\alpha) = \mathbf{T} \text{ and } [\![\psi]\!]_\mathcal{M}(\alpha) = \mathbf{T} \\
[\![\phi \vee \psi]\!]_\mathcal{M}(\alpha) = \mathbf{T} &\quad \text{iff } [\![\phi]\!]_\mathcal{M}(\alpha) = \mathbf{T} \text{ or } [\![\psi]\!]_\mathcal{M}(\alpha) = \mathbf{T} \\
[\![\phi \to \psi]\!]_\mathcal{M}(\alpha) = \mathbf{T} &\quad \text{iff } [\![\phi]\!]_\mathcal{M}(\alpha) = \mathbf{F} \text{ or } [\![\psi]\!]_\mathcal{M}(\alpha) = \mathbf{T} \\
[\![\forall x.\ \phi]\!]_\mathcal{M}(\alpha) = \mathbf{T} &\quad \text{iff } [\![\phi]\!]_\mathcal{M}(\alpha[x \mapsto a]) = \mathbf{T} \text{ for all } a \in D \\
[\![\exists x.\ \phi]\!]_\mathcal{M}(\alpha) = \mathbf{T} &\quad \text{iff } [\![\phi]\!]_\mathcal{M}(\alpha[x \mapsto a]) = \mathbf{T} \text{ for some } a \in D
\end{aligned}$$

We read $[\![\phi]\!]_\mathcal{M}(\alpha)$ as the *value of ϕ with respect to $\mathcal{M}$ and α*.

An important observation to be made at this point is that first-order validity subsumes propositional validity: when we look for the clauses that assign meaning to propositional connectives ($\neg, \wedge, \vee, \rightarrow$), we find a perfect match to those of Definition 3.5, with the notion of valuation in propositional logic playing the role of both the $\mathcal{V}$-structure and assignment in first-order logic. This observation allows for a trivial encoding of propositional logic in first-order logic: one considers a vocabulary with only 0-adic predicate symbols that play the role of propositional symbols in formulas. However, a much more interesting embedding results from allowing arbitrary first-order formulas (over an arbitrary vocabulary) to play the role of proposition symbols. We will later use such an encoding to transfer results from propositional logic to first-order logic.

Clearly the interest of first-order logic resides in what it adds to the expressive power of propositional logic. To this respect we note that when the interpretation domain D of $\mathcal{M}$ is infinite, evaluating $\forall x.\, \phi$ or $\exists x.\, \phi$ may require testing infinitely many values (all the truth values $[\![\phi]\!]_{\mathcal{M}}(\alpha[x \mapsto a])$, for $a \in D$). In fact, universal and existential quantifications can be read as generalisations of the conjunction and disjunction connectives as follows

$$[\![\forall x.\, \phi]\!]_{\mathcal{M}}(\alpha) = \bigwedge_{a \in D} [\![\phi]\!]_{\mathcal{M}}(\alpha[x \mapsto a])$$

$$[\![\exists x.\, \phi]\!]_{\mathcal{M}}(\alpha) = \bigvee_{a \in D} [\![\phi]\!]_{\mathcal{M}}(\alpha[x \mapsto a])$$

Example 4.19 Recall Example 4.16 where three structures $\mathcal{M}_1, \mathcal{M}_2, \mathcal{M}_3$ were defined for the vocabulary $\{c, f, P\}$.

1. Let $\phi_1 = P(f(x), f(f(c)))$. Then $[\![\phi_1]\!]_{\mathcal{M}_1}(\alpha) = \mathbf{T}$ if $\alpha(x) = 0$, since ϕ_1 is interpreted as $0 + 1 < 0 + 1 + 1$ in $\mathbb{N}$. However $[\![\phi_1]\!]_{\mathcal{M}_1}(\alpha') = \mathbf{F}$ if $\alpha'(x) = 4$, as ϕ_1 is interpreted $4 + 1 < 0 + 1 + 1$ which does not hold in $\mathbb{N}$. Using $\mathcal{M}_2$ instead and assignments α and α' such that $\alpha(x) = 0.5$ and $\alpha'(x) = \sqrt{3}$, we have $[\![\phi_1]\!]_{\mathcal{M}_2}(\alpha) = \mathbf{T}$ and $[\![\phi_1]\!]_{\mathcal{M}_2}(\alpha') = \mathbf{F}$ since ϕ_1 is interpreted in $\mathbb{R}$ as $0.5^2 < (1^2)^2$ and $\sqrt{3}^2 < (1^2)^2$, respectively.

2. Let $\phi_2 = \forall x. P(x, f(x))$ and $\phi_3 = \exists x. \forall y. P(f(y), x)$. For any assignment α we have:

   - $[\![\phi_2]\!]_{\mathcal{M}_1}(\alpha) = \mathbf{T}$, $[\![\phi_2]\!]_{\mathcal{M}_2}(\alpha) = \mathbf{F}$ and $[\![\phi_2]\!]_{\mathcal{M}_3}(\alpha) = \mathbf{F}$.
   - $[\![\phi_3]\!]_{\mathcal{M}_1}(\alpha) = \mathbf{F}$, $[\![\phi_3]\!]_{\mathcal{M}_2}(\alpha) = \mathbf{F}$ and $[\![\phi_3]\!]_{\mathcal{M}_3}(\alpha) = \mathbf{T}$.

3. Let $\phi_4 = \forall x, y\,.\, P(x, y) \rightarrow P(f(x), f(y))$. For any assignment α we have:

   - $[\![\phi_4]\!]_{\mathcal{M}_1}(\alpha) = \mathbf{T}$, since ϕ_4 is interpreted in $\mathcal{M}_1$ as "for any $x, y \in \mathbb{N}$, if $x < y$ then $x + 1 < y + 1$".
   - $[\![\phi_4]\!]_{\mathcal{M}_2}(\alpha) = \mathbf{F}$, because ϕ_4 is interpreted in $\mathcal{M}_2$ as "for any $x, y \in \mathbb{R}$, if $x < y$ then $x^2 < y^2$". It is easy to find elements of $\mathbb{R}$ to assign to x and y such that the interpretation of the formula in the scope of the quantification is $\mathbf{F}$. For instance, let x be -5 and y be 2.

4. Let $\phi_5 = \forall x.\ \exists y.\ P(y,x) \wedge P(c, f(y))$. For any assignment α we have $[\![\phi_5]\!]_{\mathcal{M}_2}(\alpha) = \mathbf{T}$.

We now proceed with the definition of validity and related notions.

Definition 4.20 Let $\mathcal{V}$ be a vocabulary and $\mathcal{M}$ a $\mathcal{V}$-structure.

1. Given a formula ϕ and an assignment α, we say that $\mathcal{M}$ *satisfies* ϕ *with* α, denoted by $\mathcal{M}, \alpha \models \phi$, iff $[\![\phi]\!]_{\mathcal{M}}(\alpha) = \mathbf{T}$.
2. Given a formula ϕ and a $\mathcal{V}$-structure $\mathcal{M}$, we say that $\mathcal{M}$ *satisfies* ϕ (or that ϕ *is valid in* $\mathcal{M}$), denoted by $\mathcal{M} \models \phi$, iff for every assignment α, $\mathcal{M}, \alpha \models \phi$. $\mathcal{M}$ is said to be a *model* of ϕ. Alternatively we say that ϕ *holds in* $\mathcal{M}$.
3. A formula ϕ is *satisfiable* if there exists some structure $\mathcal{M}$ such that $\mathcal{M} \models \phi$, and it is *valid*, denoted by $\models \phi$, if $\mathcal{M} \models \phi$ for every structure $\mathcal{M}$. Finally, ϕ is *unsatisfiable* (or a *contradiction*) if it is not satisfiable, and refutable if it is not valid.
4. Given a set of formulas Γ, a model $\mathcal{M}$ and an assignment α, $\mathcal{M}$ is said to satisfy Γ with α, denoted by $\mathcal{M}, \alpha \models \Gamma$, if $\mathcal{M}, \alpha \models \phi$ for every $\phi \in \Gamma$. The notions of satisfiable, valid, unsatisfiable and refutable set of formulas are defined in the expected way.
5. Given a set Γ of formulas and a formula ϕ, we say that Γ *entails* ϕ (or that ϕ is a *logical consequence* of Γ), denoted by $\Gamma \models \phi$, iff for every structure $\mathcal{M}$ and assignment α, if $\mathcal{M}, \alpha \models \Gamma$ then $\mathcal{M}, \alpha \models \phi$.
6. Given two formulas ϕ and ψ, ϕ is *logically equivalent* to ψ, denoted by $\phi \equiv \psi$, iff $[\![\phi]\!]_{\mathcal{M}}(\alpha) = [\![\psi]\!]_{\mathcal{M}}(\alpha)$ for every structure $\mathcal{M}$ and assignment α, i.e. $\mathcal{M}, \alpha \models \phi$ iff $\mathcal{M}, \alpha \models \psi$.

Remark 4.21 Many logic texts define the satisfiability of a formula ϕ requiring only for some structure $\mathcal{M}$ the existence of an assignment α such that $\mathcal{M}, \alpha \models \phi$. Other texts carefully restrict the notion of satisfiability to sentences. The reader should be aware of this lack of uniformity when consulting other sources. The same applies to the entailment and equivalence relations.

When dealing with validity in first-order logic, we often need to reason about assignments. The following lemma states some simple facts about them and their interaction with substitutions.

Lemma 4.22 *Let* $\mathcal{M} = (D, I)$ *be a* $\mathcal{V}$-*structure,* t *and* u *be terms,* ϕ *a formula, and* α, α' *assignments.*

1. *If for every variable* $x \in \mathsf{Vars}(t)$, $\alpha(x) = \alpha'(x)$, *then* $[\![t]\!]_{\mathcal{M}}(\alpha) = [\![t]\!]_{\mathcal{M}}(\alpha')$
2. *If for every variable* $x \in \mathsf{FV}(\phi)$, $\alpha(x) = \alpha'(x)$, *then*

$$\mathcal{M}, \alpha \models \phi \quad \textit{iff} \quad \mathcal{M}, \alpha' \models \phi$$

3. $[\![t[u/x]]\!]_{\mathcal{M}}(\alpha) = [\![t]\!]_{\mathcal{M}}(\alpha[x \mapsto [\![u]\!]_{\mathcal{M}}(\alpha)])$

4. *If t is free for x in ϕ, then*

$$\mathcal{M}, \alpha \models \phi[t/x] \quad \textit{iff} \quad \mathcal{M}, \alpha[x \mapsto [\![\mathcal{M}]\!]_\alpha(t)] \models \phi$$

Proof (1) and (3) are proved by induction on the structure of t and are left to the reader. (2) and (4) are proved by induction on ϕ. The cases of atomic formulas are immediate consequences of (1) and (3). The cases of propositional connectives ($\neg, \wedge, \vee, \rightarrow$) follow immediately from the induction hypothesis. The cases of quantifiers are the most interesting; we detail here the case of the universal quantifier case (the case of the existential quantifier follows a similar argument).

To prove (2) when $\phi = \forall x. \phi'$, we have $\mathcal{M}, \alpha \models \forall x. \phi'$ iff, for all $a \in D$, $\mathcal{M}, \alpha[x \mapsto a] \models \phi'$. By definition of function patching, we know that $\alpha[x \mapsto a](x) = \alpha'[x \mapsto a](x)$, and since $\mathsf{FV}(\phi') \subseteq \mathsf{FV}(\phi) \cup \{x\}$ we conclude that for all $y \in \mathsf{FV}(\phi')$, $\alpha[x \mapsto a](y) = \alpha'[x \mapsto a](y)$. We are thus in the conditions to apply the induction hypothesis to conclude that, for all $a \in D$, $\mathcal{M}, \alpha'[x \mapsto a] \models \phi'$ which, by definition, is $\mathcal{M}, \alpha' \models \forall x. \phi'$.

To prove (4) when $\phi = \forall y. \phi'$, we distinguish between the cases where $x = y$ and $x \neq y$. If $x = y$, we have that $(\forall x. \phi')[t/x] = \forall x. \phi'$. Hence, $\mathcal{M}, \alpha \models (\forall x. \phi')[t/x]$ iff, for all $a \in D$, $\mathcal{M}, \alpha[x \mapsto a] \models \phi'$. On the other hand, $\mathcal{M}, \alpha[x \mapsto [\![t]\!]_\mathcal{M}(\alpha)] \models \forall x. \phi'$ iff, for all $a \in D$ $\mathcal{M}, (\alpha[x \mapsto [\![t]\!]_\mathcal{M}(\alpha)])[x \mapsto a] \models \phi'$, which by definition of function patching is $\mathcal{M}, \alpha[x \mapsto a] \models \phi'$, thus establishing the equivalence of both sides.

For the remaining case ($x \neq y$), we have that $(\forall y. \phi')[t/x]$ is $\forall y. (\phi'[t/x])$. Then $\mathcal{M}, \alpha \models (\forall y. \phi')[t/x]$ iff $\mathcal{M}, \alpha \models \forall y. (\phi'[t/x])$ iff, for all $a \in D$, $\mathcal{M}, \alpha[y \mapsto a] \models \phi'[t/x]$. By induction hypothesis, we obtain $\mathcal{M}, (\alpha[y \mapsto a])[x \mapsto [\![t]\!]_\mathcal{M}(\alpha[y \mapsto a])] \models \phi'$. Since t is free for x in ϕ, we know that $y \notin \mathsf{Vars}(t)$ and so, by (1), $[\![t]\!]_\mathcal{M}(\alpha[y \mapsto a]) = [\![t]\!]_\mathcal{M}(\alpha)$. Moreover, by definition of function patching, and since $x \neq y$, $(\alpha[y \mapsto a])[x \mapsto [\![t]\!]_\mathcal{M}(\alpha)] = (\alpha[x \mapsto [\![t]\!]_\mathcal{M}(\alpha)])[y \mapsto a]$. Hence, $\mathcal{M}, (\alpha[y \mapsto a])[x \mapsto [\![t]\!]_\mathcal{M}(\alpha[y \mapsto a])] \models \phi'$ iff $\mathcal{M}, (\alpha[x \mapsto [\![t]\!]_\mathcal{M}(\alpha)])[y \mapsto a] \models \phi'$ which, by definition, is $\mathcal{M}, \alpha[x \mapsto [\![t]\!]_\mathcal{M}(\alpha)] \models \forall y. \phi'$. $\qquad\square$

The embedding of propositional logic into first-order logic is formalised below (we use the subscripts PL and FOL to differentiate between the concepts of validity and entailment in both logics).

Proposition 4.23 *Let $\mathcal{V}$ be an arbitrary first-order vocabulary and $\lceil \cdot \rceil : \mathbf{Prop} \rightarrow \mathbf{Form}_\mathcal{V}$ be a mapping from the set of proposition symbols to first-order formulas. We let $\lceil \cdot \rceil$ also denote the extension of this map to propositional formulas (i.e. the mapping that replaces each proposition symbol in a propositional formula). For all propositional formulas A and B, we have:*

1. *For every $\mathcal{V}$-structure $\mathcal{M}$ and assignment α,*

$$\overline{\mathcal{M}_\alpha} \models_{\mathsf{PL}} A \quad \textit{iff} \quad \mathcal{M}, \alpha \models \lceil A \rceil$$

where $\overline{\mathcal{M}_\alpha}$ is the propositional model $\overline{\mathcal{M}_\alpha} = \{P \mid \mathcal{M}, \alpha \models \lceil P \rceil\}$.

2. *If* $\models_{\mathsf{PL}} A$, *then* $\models_{\mathsf{FOL}} \ulcorner A \urcorner$.
3. *If* $A \equiv_{\mathsf{PL}} B$, *then* $\ulcorner A \urcorner \equiv_{\mathsf{FOL}} \ulcorner B \urcorner$.

Proof (1) is proved by induction on the structure of A. Let $\mathcal{M}$ be a $\mathcal{V}$-structure and α an assignment. If A is atomic, the results follows by definition of $\overline{\mathcal{M}_\alpha}$. The remaining cases are also immediate. We illustrate the case when A is $B \wedge C$: $\mathcal{M}, \alpha \models \ulcorner B \wedge C \urcorner$ iff $\mathcal{M}, \alpha \models \ulcorner B \urcorner$ and $\mathcal{M}, \alpha \models \ulcorner C \urcorner$ which, by induction hypothesis, is $\overline{\mathcal{M}_\alpha} \models_{\mathsf{PL}} B$ and $\overline{\mathcal{M}_\alpha} \models_{\mathsf{PL}} C$. Finally, by definition of propositional validity, we obtain $\overline{\mathcal{M}_\alpha} \models_{\mathsf{PL}} B \wedge C$. (2) and (3) are direct consequences of (1). $\qquad\square$

The previous proposition provides a means to transfer results and methods of propositional logic to first-order logic. In particular, we obtain for free all the propositional equivalence laws given in Fig. 3.2, such as the algebraic properties of propositional connectives, De Morgan's laws, etc. Note however that validity and equivalence results cannot be transferred in the reverse direction, because of the more refined nature of validity in first-order logic. The set of "propositionally valid" first-order formulas (in the above sense) is a proper subset of the set of valid first-order formulas. We remark that it is customary to classify as a *tautology* in first-order logic every sentence that is propositionally valid. But again there exists no uniformity among authors, and sometimes the expression is used for valid first-order sentences.

As seen in the last chapter, logical equivalence can be used to conduct equational reasoning. For this we have first to assert that indeed this is an equivalence relation.

Proposition 4.24 $\equiv$ *is an equivalence relation on* $\mathbf{Form}_\mathcal{V}$.

The proof is left to the reader as Exercise 4.7. In propositional logic this result was complemented with a congruence property (Proposition 3.12) that allowed one to replace subformulas by equivalent propositions. The corresponding result here is slightly different since one needs to avoid capturing free variables.

Proposition 4.25 *Let* θ, ϕ *and* ψ *be formulas such that* ϕ *is a subformula of* θ *and* $\phi \equiv \psi$. *Then replacing some occurrence of* ϕ *by* ψ *in* θ *results in a formula that is logically equivalent to* θ, *as long as no free variables of* ψ *are captured.*

Proof By induction on the structure of θ. $\qquad\square$

As a first example of a genuine first-order equivalence (one that does not follow from a propositional equivalence) we present a distinctive feature of bound variables that says that their names are unimportant—we can freely rename them as long as the associations between bound occurrences of variables and their respective quantifiers are preserved.

Proposition 4.26 (Renaming of bound variables) *If* y *is free for* x *in* ϕ *and* $y \notin \mathsf{FV}(\phi)$, *then the following equivalences hold.*

1. $\forall x. \phi \equiv \forall y. \phi[y/x]$
2. $\exists x. \phi \equiv \exists y. \phi[y/x]$

Proof Both proofs are identical, so we focus on the first. Let $\mathcal{M} = (D, I)$ be a $\mathcal{V}$-structure and α an assignment. Then $\mathcal{M}, \alpha \models \forall y. (\phi[y/x])$ iff forall $a \in D$, $\mathcal{M}, \alpha[y \mapsto a] \models \phi[y/x]$. By Lemma 4.22(4), we have $\mathcal{M}, \alpha[y \mapsto a][x \mapsto [\![y]\!]_\mathcal{M}(\alpha[y \mapsto a])] \models \phi$ which, by definition of function patching and interpretation of terms is $\mathcal{M}, \alpha[y \mapsto a][x \mapsto a] \models \phi$. Since $y \notin FV(\phi)$, and by Lemma 4.22(2), we have $\mathcal{M}, \alpha[x \mapsto a] \models \phi$. We can then conclude $\mathcal{M}, \alpha \models \forall x. \phi$. $\qquad\square$

Note that the proviso that y be free for x in ϕ is required to prevent variable y from being captured, by some quantifier that also binds y inside ϕ. If y is not free for x in ϕ, then one should choose another variable z instead of y. One simple way to choose this variable is to observe that if $y \notin BV(\phi)$, then y is trivially free for every variable x in ϕ, so it suffices to pick a *fresh* variable, not occurring in the formula at hand. If we really want to rename x to y, there is a way to do so: starting from the innermost quantified subformulas of ϕ and moving outwards, we may use Proposition 4.26 itself, step by step, by always choosing fresh variables different from y, so that y is no longer used in the resulting formula, which is still equivalent to ϕ.

So the bottomline is that variable capture is just a bureaucratic issue that can be dealt with by simply assuming that when a substitution $\phi[t/x]$ is written, the required renamings (using fresh variables) will take place to ensure that $BV(\phi) \cap Vars(t) = \emptyset$, and thus t is free for any x in ϕ. For instance in order to apply Proposition 4.25 to replace a subformula of θ by an equivalent formula, one first uses Proposition 4.26 to perform the renamings required in θ to avoid variable capture. Henceforth we will drop the "free for" requirement when applying substitutions.

Other important class of equivalences relate quantifiers with propositional connectives. We have already mentioned that universal (resp. existential) quantifiers can be understood as generalisations of conjunction (resp. disjunction). The following lemma can be seen either as a formalisation of this intuition, or simply as a set of useful equivalences.

Lemma 4.27 *The following equivalences hold in first-order logic.*

$$\forall x. \phi \wedge \psi \equiv (\forall x. \phi) \wedge (\forall x. \psi) \qquad \exists x. \phi \vee \psi \equiv (\exists x. \phi) \vee (\exists x. \psi)$$

$$\forall x. \phi \equiv (\forall x. \phi) \wedge \phi[t/x] \qquad \exists x. \phi \equiv (\exists x. \phi) \vee \phi[t/x]$$

$$\neg \forall x. \phi \equiv \exists x. \neg \phi \qquad \neg \exists x. \phi \equiv \forall x. \neg \phi$$

Proof The proofs of all these equivalences are similar. We prove $\forall x. \phi \equiv (\forall x. \phi) \wedge \phi[t/x]$ as a representative (we invite the reader to work out the details of the remaining equivalences). Let $\mathcal{M} = (D, I)$ be a structure and α an assignment such that $\mathcal{M}, \alpha \models \forall x. \phi$, thus $\mathcal{M}, \alpha[x \mapsto a] \models \phi$ for all $a \in D$. In particular, for any term t, $\mathcal{M}, \alpha[x \mapsto [\![t]\!]_\mathcal{M}(\alpha)] \models \phi$, which, by Proposition 4.22(4), is $\mathcal{M}, \alpha \models \phi[t/x]$. Hence, we also have that $\mathcal{M}, \alpha \models (\forall x. \phi) \wedge \phi[t/x]$. $\qquad\square$

The first two equivalences establish that $\forall$ (resp. $\exists$) distributes over $\wedge$ (resp. $\vee$). We note however, that the remaining distributive laws ($\forall$ over $\vee$ and $\exists$ over $\wedge$) do not hold (see Exercise 4.10). The last two equivalences are, not surprisingly, called *De Morgan's laws* for quantifiers. They can be used to extend the *negation normal form* conversion to first-order logic, which we will do in Sect. 4.5.

Similarly to propositional logic, the first-order notion of semantic entailment is an abstract consequence relation.

Proposition 4.28 *The semantic entailment relation for first-order logic satisfies the following properties:*

1. *For all $\phi \in \Gamma$, $\Gamma \models \phi$* $\hspace{4cm}$ (*inclusion*)
2. *If $\Gamma \models \phi$, then $\Gamma, \psi \models \phi$* $\hspace{3cm}$ (*monotonicity*)
3. *If $\Gamma \models \phi$ and $\Gamma, \phi \models \psi$, then $\Gamma \models \psi$* $\hspace{2cm}$ (*cut*)

Proof Identical to the proof of Proposition 3.17. $\hspace{5cm}$ □

The following proposition states two fundamental properties on the interaction of logical connectives with semantic entailment.

Proposition 4.29 *Let ϕ and ψ be first-order formulas and Γ a set of formulas.*

1. *$\Gamma \models \phi \rightarrow \psi$ iff $\Gamma, \phi \models \psi$*
2. *If x does not occur free in Γ, then $\Gamma \models \forall x . \phi$ iff $\Gamma \models \phi$.*

Proof

1. Let $\mathcal{M}$ be a structure and α an assignment. To prove the left-to-right implication, assume $\mathcal{M}, \alpha \models \Gamma, \phi$. Then, $\mathcal{M}, \alpha \models \phi \rightarrow \psi$ (since $\Gamma \models \phi \rightarrow \psi$). But then, necessarily $\mathcal{M}, \alpha \models \psi$. For proving the converse, assume $\mathcal{M}, \alpha \models \Gamma$. We consider the cases whether $\mathcal{M}, \alpha \models \phi$ or $\mathcal{M}, \alpha \not\models \phi$. When $\mathcal{M}, \alpha \models \phi$, we have that $\mathcal{M}, \alpha \models \Gamma, \phi$. Hence, $\mathcal{M}, \alpha \models \psi$ (since $\Gamma, \phi \models \psi$). But then $\mathcal{M}, \alpha \models \phi \rightarrow \psi$. When $\mathcal{M}, \alpha \not\models \phi$, one immediately gets that $\mathcal{M}, \alpha \models \phi \rightarrow \psi$.
2. Let $\mathcal{M} = (D, I)$ be a structure and α an assignment. To prove the left-to-right implication, assume $\mathcal{M}, \alpha \models \Gamma$. Since $\Gamma \models \forall x . \phi$, we have $\mathcal{M}, \alpha \models \forall x . \phi$. Then $\mathcal{M}, \alpha[x \mapsto a] \models \phi$, for all $a \in D$. In particular, $\mathcal{M}, \alpha[x \mapsto \alpha(x)] \models \phi$ which is $\mathcal{M}, \alpha \models \phi$. To prove the converse, assume $\mathcal{M}, \alpha \models \Gamma$. Note that, since x does not occur free in Γ, we have that $\mathcal{M}, \alpha[x \mapsto a] \models \Gamma$, for any $a \in D$ (by Proposition 4.22(2)). Since $\Gamma \models \phi$, we also have that $\mathcal{M}, \alpha[x \mapsto a] \models \phi$. Hence we conclude that $\mathcal{M}, \alpha \models \forall x . \phi$. $\hspace{2cm}$ □

Using inter-definability of connectives, these two fundamental equivalences can be extended to the remaining connectives. In fact, whenever $\phi \equiv \psi$, we have as direct consequences of the definitions of equivalence and entailment:

$$\Gamma \models \phi \quad \text{iff} \quad \Gamma \models \psi$$

$$\Gamma, \phi \models \theta \quad \text{iff} \quad \Gamma, \psi \models \theta$$

Proposition 4.30

1. $\Gamma \models \neg\phi$ iff $\Gamma, \phi \models \perp$
2. $\Gamma \models \phi$ iff $\Gamma, \neg\phi \models \perp$
3. $\Gamma \models \phi \wedge \psi$ iff $\Gamma \models \phi$ and $\Gamma \models \psi$
4. $\Gamma \models \phi \vee \psi$ iff $\Gamma \models \phi$ or $\Gamma \models \psi$
5. If x does not occur free in Γ or ψ, then $\Gamma, \exists x . \phi \models \psi$ iff $\Gamma, \phi \models \psi$.
6. If $\Gamma \models \forall x . \phi$, then $\Gamma \models \phi[t/x]$.
7. If $\Gamma \models \phi[t/x]$, then $\Gamma \models \exists x . \phi$.

Proof

(1), (2), (3) and (4) are left as Exercise 4.12

(5) $\Gamma, \exists x . \phi \models \psi$
     iff   (*since* $\exists x . \phi \equiv \neg\forall x . \neg\phi$)
                $\Gamma, \neg\forall x . \neg\phi \models \psi$
     iff   (*by* (2))
                $\Gamma, \neg\forall x . \neg\phi, \neg\psi \models \perp$
     iff   (*by* (2))
                $\Gamma, \neg\psi \models \forall x . \neg\phi$
     iff   (*by Proposition* 4.29(2)—*since* x *is not free in* Γ *or* ψ)
                $\Gamma, \neg\psi \models \neg\phi$
     iff   (*by* (1))
                $\Gamma, \neg\psi, \phi \models \perp$
     iff   (*by* (2))
                $\Gamma, \phi \models \psi$

(6) $\Gamma \models \forall x . \phi$
     iff   (*by Proposition* 4.27)
                $\Gamma \models \forall x . \phi \wedge \psi[t/x]$
     then   (*by* (3))
                $\Gamma \models \phi[t/x]$

(7) $\Gamma \models \phi[t/x]$
     then   (*by* (4))
                $\Gamma \models \exists x . \phi \vee \phi[t/x]$
     iff   (*by Proposition* 4.27)
                $\Gamma \models \exists x . \phi$ $\square$

Notice that the previous propositions include all the entailment-related results shown for propositional logic, in particular those of Proposition 3.18, which relate propositional connectives with entailment.

4.3 Proof System

Following the approach taken for propositional logic, we present here the natural deduction proof system for classical first-order logic in sequent style. Derivations in classical first-order logic will be similar to derivations in classical propositional logic, except that we will have new proof rules for dealing with the quantifiers.

$$(Ax) \frac{}{\Gamma, \phi \vdash \phi} \qquad (I_\rightarrow) \frac{\Gamma, \phi \vdash \psi}{\Gamma \vdash \phi \rightarrow \psi} \qquad (E_\rightarrow) \frac{\Gamma \vdash \phi \qquad \Gamma \vdash \phi \rightarrow \psi}{\Gamma \vdash \psi}$$

$$(I_\neg) \frac{\Gamma, \phi \vdash \bot}{\Gamma \vdash \neg\phi} \qquad (E_\neg) \frac{\Gamma \vdash \phi \qquad \Gamma \vdash \neg\phi}{\Gamma \vdash \psi} \qquad (RAA) \frac{\Gamma, \neg\phi \vdash \bot}{\Gamma \vdash \phi}$$

$$(I_\wedge) \frac{\Gamma \vdash \phi \qquad \Gamma \vdash \psi}{\Gamma \vdash \phi \wedge \psi} \qquad (E_{\wedge 1}) \frac{\Gamma \vdash \phi \wedge \psi}{\Gamma \vdash \phi} \qquad (E_{\wedge 2}) \frac{\Gamma \vdash \phi \wedge \psi}{\Gamma \vdash \psi}$$

$$(I_{\vee 1}) \frac{\Gamma \vdash \phi}{\Gamma \vdash \phi \vee \psi} \qquad (I_{\vee 2}) \frac{\Gamma \vdash \psi}{\Gamma \vdash \phi \vee \psi} \qquad (E_\vee) \frac{\Gamma \vdash \phi \vee \psi \qquad \Gamma, \phi \vdash \theta \qquad \Gamma, \psi \vdash \theta}{\Gamma \vdash \theta}$$

$$(I_\forall) \frac{\Gamma \vdash \phi[y/x]}{\Gamma \vdash \forall x. \phi} \; (a) \qquad (E_\forall) \frac{\Gamma \vdash \forall x. \phi}{\Gamma \vdash \phi[t/x]}$$

(a) y must not occur free in either Γ or ϕ

$$(I_\exists) \frac{\Gamma \vdash \phi[t/x]}{\Gamma \vdash \exists x. \phi} \qquad (E_\exists) \frac{\Gamma \vdash \exists x. \phi \qquad \Gamma, \phi[y/x] \vdash \theta}{\Gamma \vdash \theta} \; (b)$$

(b) y must not occur free in either Γ, ϕ or θ

Fig. 4.1 System $\mathcal{N}_{\mathsf{FOL}}$ for classical first-order logic

More precisely, we overload the proof rules of propositional logic, changing the Roman letters to Greek letters, and we add introduction and elimination rules for the quantifiers. This means that the proofs developed for propositional logic in Chap. 3 still hold in this proof system.

As before, an inference rule consists of zero or more premises and a single conclusion, all of which are sequents, i.e., judgments of the form $\Gamma \vdash \phi$ denoting that ϕ can be formally deduced from the set of assumptions Γ. An instance of an inference rule is obtained by replacing all occurrences of each meta-variable by a phrase in its range. In some rules, there may be side conditions that must be satisfied by this replacement. Also, there may be syntactic operations (such as substitutions) that have to be carried out after the replacement.

Definition 4.31 The proof system $\mathcal{N}_{\mathsf{FOL}}$ of natural deduction for first-order logic is defined by the rules presented in Fig. 4.1. A derivation (or proof) in $\mathcal{N}_{\mathsf{FOL}}$ is inductively defined in the same way given in Definition 3.19 for propositional logic.

We say that $\Gamma \vdash \phi$ is *derivable* in this system if it is the conclusion of some derivation tree. The *deduction relation* $\vdash \in \mathcal{P}(\mathbf{Form}_\mathcal{V}) \times \mathbf{Form}_\mathcal{V}$ is the relation induced by derivability in $\mathcal{N}_{\mathsf{FOL}}$. If $\vdash \phi$ is derivable, we say that ϕ is a *theorem*.

We remark that some rules now have *side conditions* that are expected to be respected by all instances used in a derivation. Let us give a brief explanation of the proof rules for quantifiers.

- Rule ($I_\forall$) tells us that if $\phi[y/x]$ can be deduced from Γ for a variable y that does not occur free in either Γ or ϕ, then $\forall x.\,\phi$ can also be deduced from Γ because y is fresh. The side condition (a) stating that y must be not free in ϕ or in any formula of Γ is crucial for the soundness of this rule. As y is a fresh variable we can think of it as an indeterminate term, which justifies that $\forall x.\,\phi$ can be deduced from Γ.
- Rule ($E_\forall$) says that if $\forall x.\phi$ can be deduced from Γ then the x in ϕ can be replaced by any term t assuming that t is free for x in ϕ (this is implicit in the notation). It is easy to understand that this rule is sound: if ϕ is true for all x, then it must be true for any particular term t.
- Rule ($I_\exists$) tells us that if it can be deduced from Γ that $\phi[t/x]$ for some term t which is free for x in ϕ (this proviso is implicit in the notation), then $\exists x.\,\phi$ can also be deduced from Γ.
- The second premise of rule ($E_\exists$) tells us that θ can be deduced if, additionally to Γ, ϕ holds for an indeterminate term. But the first premise states that such a term exists, thus θ can be deduced from Γ with no further assumptions.

Note the similarity between these rules and the introduction and elimination rules for $\wedge$ and $\vee$, following the reading of universal (resp. existential) formulas as generalised conjunctions (resp. disjunctions).

We will now illustrate the construction of first-order derivations, in particular the use of the quantifier rules, using different presentation styles. Note that no particular vocabulary needs to be assumed in these derivations; predicates and terms may be represented by meta-variables.

Example 4.32 A derivation for the sequent $P(t), \forall x.\, P(x) \to \neg Q(x) \vdash \neg Q(t)$ is

$$
\cfrac{\cfrac{\text{(Ax)} \cfrac{}{P(t), \forall x.\, P(x) \to \neg Q(x) \vdash \forall x.\, P(x) \to \neg Q(x)}}{\text{(E}_\forall)\ \ \cfrac{}{P(t), \forall x.\, P(x) \to \neg Q(x) \vdash P(t) \to \neg Q(t)}} \qquad \text{(Ax)}\ \cfrac{}{P(t), \forall x.\, P(x) \to \neg Q(x) \vdash P(t)}}{\text{(E}_\to)\ \ \ P(t), \forall x.\, P(x) \to \neg Q(x) \vdash \neg Q(t)}
$$

As we did for propositional logic, we adopt alternative formats for presenting bigger proof trees:

- for the development of derivations in the backward direction,

$P(t), \forall x.\, P(x) \to \neg Q(x) \vdash \neg Q(t),$	($E_\to$)
$\quad$ **1.** $\ P(t), \forall x.\, P(x) \to \neg Q(x) \vdash P(t) \to \neg Q(t),$	($E_\forall$)
$\qquad$ **1.** $\ P(t), \forall x.\, P(x) \to \neg Q(x) \vdash \forall x.\, P(x) \to \neg Q(x),$	(Ax)
$\quad$ **2.** $\ P(t), \forall x.\, P(x) \to \neg Q(x) \vdash P(t),$	(Ax)

- for the development of derivations in the forward direction,

	Judgments	Justification
1.	$P(t), \forall x.\, P(x) \to \neg Q(x) \vdash P(t)$	(Ax)
2.	$P(t), \forall x.\, P(x) \to \neg Q(x) \vdash \forall x.\, P(x) \to \neg Q(x)$	(Ax)
3.	$P(t), \forall x.\, P(x) \to \neg Q(x) \vdash P(t) \to \neg Q(t)$	($E_\forall$) 2
4.	$P(t), \forall x.\, P(x) \to \neg Q(x) \vdash \neg Q(t)$	($E_\to$) 1, 3

Example 4.33 Let us show that $(\exists x. \neg\psi) \to \neg\forall x. \psi$ is a theorem.

$$\vdash (\exists x. \neg\psi) \to \neg\forall x. \psi, \qquad (I_\to)$$

$$\textbf{1.}\ \ \exists x. \neg\psi \vdash \neg\forall x. \psi, \qquad (I_\neg)$$

$$\textbf{1.}\ \ \exists x. \neg\psi, \forall x. \psi \vdash \bot, \qquad (E_\exists)$$

$$\textbf{1.}\ \ \exists x. \neg\psi, \forall x. \psi \vdash \exists x. \neg\psi, \qquad (Ax)$$

$$\textbf{2.}\ \ \exists x. \neg\psi, \forall x. \psi, \neg\psi[x_0/x] \vdash \bot, \qquad (E_\neg)$$

$$\textbf{1.}\ \ \exists x. \neg\psi, \forall x. \psi, \neg\psi[x_0/x] \vdash \psi[x_0/x], \qquad (E_\forall)$$

$$\textbf{1.}\ \ \exists x. \neg\psi, \forall x. \psi, \neg\psi[x_0/x] \vdash \forall x. \psi, \qquad (Ax)$$

$$\textbf{2.}\ \ \exists x. \neg\psi, \forall x. \psi, \neg\psi[x_0/x] \vdash \neg\psi[x_0/x], \qquad (Ax)$$

Note that when the rule $(E_\exists)$ is applied a fresh variable x_0 is introduced. The side condition imposes that x_0 must not occur free either in $\exists x. \neg\psi$ or in $\forall x. \psi$.

In the previous example it was necessary to explicitly write the substitutions, because we could not evaluate them in generic formulas like ψ. This may result in the notation overhead rapidly becoming overwhelming. To obviate this, the convention of writing $\phi(x_1, \ldots, x_n)$ to denote a formula having free variables $x_1, \ldots, x_n$ is often adopted. Additionally $\phi(t_1, \ldots, t_n)$ is written to denote the formula obtained by replacing each free occurrence of x_i in ϕ by the term t_i. Note that when we write $\phi(x_1, \ldots, x_n)$, it is not meant that $x_1, \ldots, x_n$ are the *only* variables of ϕ.

The following derivation adopts this convention to establish the converse implication.

Example 4.34 The following derivation shows that $(\neg\forall x. \psi(x)) \to \exists x. \neg\psi(x)$ is a theorem.

$$\vdash (\neg\forall x. \psi(x)) \to \exists x. \neg\psi(x), \qquad (I_\to)$$

$$\textbf{1.}\ \ \neg\forall x. \psi(x) \vdash \exists x. \neg\psi(x), \qquad (RAA)$$

$$\textbf{1.}\ \ \neg\forall x. \psi(x), \neg\exists x. \neg\psi(x) \vdash \bot, \qquad (E_\neg)$$

$$\textbf{1.}\ \ \neg\forall x. \psi(x), \neg\exists x. \neg\psi(x) \vdash \neg\forall x. \psi(x), \qquad (Ax)$$

$$\textbf{2.}\ \ \neg\forall x. \psi(x), \neg\exists x. \neg\psi(x) \vdash \forall x. \psi(x), \qquad (I_\forall)$$

$$\textbf{1.}\ \ \neg\forall x. \psi(x), \neg\exists x. \neg\psi(x) \vdash \psi(x_0), \qquad (RAA)$$

$$\textbf{1.}\ \ \neg\forall x. \psi(x), \neg\exists x. \neg\psi(x), \neg\psi(x_0) \vdash \bot, \qquad (E_\neg)$$

$$\textbf{1.}\ \ \neg\forall x. \psi(x), \neg\exists x. \neg\psi(x), \neg\psi(x_0) \vdash \neg\exists x. \neg\psi(x), \qquad (Ax)$$

$$\textbf{2.}\ \ \neg\forall x. \psi(x), \neg\exists x. \neg\psi(x), \neg\psi(x_0) \vdash \exists x. \neg\psi(x), \qquad (I_\exists)$$

$$\textbf{1.}\ \ \neg\forall x. \psi(x), \neg\exists x. \neg\psi(x), \neg\psi(x_0) \vdash \neg\psi(x_0), \qquad (Ax)$$

Note that the fresh variable x_0 is introduced when rule $(I_\forall)$ is applied. The side condition imposes that x_0 must not occur free in $\neg\forall x. \psi(x)$, $\neg\exists x. \neg\psi(x)$, or $\psi(x)$. Later when $(I_\exists)$ is used any term can be chosen to replace x in $\neg\psi(x)$, so we choose x_0 in order to close the derivation using (Ax).

This last derivation is not very intuitive because it uses the rule (RAA), which means that some steps are proved by contradiction. This derivation is not constructive and is thus only valid classically; $(\neg\forall x. \psi(x)) \to \exists x. \neg\psi(x)$ is not a theorem in intuicionistic logic. The next two examples both present constructive derivations.

Example 4.35 The following is a proof tree for the sequent $(\forall x . \phi(x)) \vee (\forall x . \psi(x))$ $\vdash \forall x . \phi(x) \vee \psi(x)$.

$$
\begin{array}{ll}
(\forall x . \phi(x)) \vee (\forall x . \psi(x)) \vdash \forall x . \phi(x) \vee \psi(x), & (E_\vee) \\
\quad\textbf{1.}\;\; (\forall x . \phi(x)) \vee (\forall x . \psi(x)) \vdash (\forall x . \phi(x)) \vee (\forall x . \psi(x)), & (Ax) \\
\quad\textbf{2.}\;\; (\forall x . \phi(x)) \vee (\forall x . \psi(x)), \forall x . \phi(x) \vdash \forall x . \phi(x) \vee \psi(x), & (I_\forall) \\
\quad\quad\textbf{1.}\;\; (\forall x . \phi(x)) \vee (\forall x . \psi(x)), \forall x . \phi(x) \vdash \phi(x_0) \vee \psi(x_0), & (I_{\vee 1}) \\
\quad\quad\quad\textbf{1.}\;\; (\forall x . \phi(x)) \vee (\forall x . \psi(x)), \forall x . \phi(x) \vdash \phi(x_0), & (E_\forall) \\
\quad\quad\quad\quad\textbf{1.}\;\; (\forall x . \phi(x)) \vee (\forall x . \psi(x)), \forall x . \phi(x) \vdash \forall x . \phi(x), & (Ax) \\
\quad\textbf{3.}\;\; (\forall x . \phi(x)) \vee (\forall x . \psi(x)), \forall x . \psi(x) \vdash \forall x . \phi(x) \vee \psi(x), & (I_\forall) \\
\quad\quad\textbf{1.}\;\; (\forall x . \phi(x)) \vee (\forall x . \psi(x)), \forall x . \psi(x) \vdash \phi(x_0) \vee \psi(x_0), & (I_{\vee 2}) \\
\quad\quad\quad\textbf{1.}\;\; (\forall x . \phi(x)) \vee (\forall x . \psi(x)), \forall x . \psi(x) \vdash \psi(x_0), & (E_\forall) \\
\quad\quad\quad\quad\textbf{1.}\;\; (\forall x . \phi(x)) \vee (\forall x . \psi(x)), \forall x . \psi(x) \vdash \forall x . \psi(x), & (Ax)
\end{array}
$$

For each use of $(E_\exists)$ a fresh variable x_0 is introduced. When the rule $(I_\forall)$ is applied later, we choose x_0 to replace x in order to close the proof with (Ax).

Example 4.36 The sequent $\exists x . \exists y . \phi(x, y) \vdash \exists y . \exists x . \phi(x, y)$ can be derived as follows.

$$
\begin{array}{ll}
\exists x . \exists y . \phi(x, y) \vdash \exists y . \exists x . \phi(x, y), & (E_\exists) \\
\quad\textbf{1.}\;\; \exists x . \exists y . \phi(x, y) \vdash \exists x . \exists y . \phi(x, y), & (Ax) \\
\quad\textbf{2.}\;\; \exists x . \exists y . \phi(x, y), \exists y . \phi(x_0, y) \vdash \exists y . \exists x . \phi(x, y), & (E_\exists) \\
\quad\quad\textbf{1.}\;\; \exists x . \exists y . \phi(x, y), \exists y . \phi(x_0, y) \vdash \exists y . \phi(x_0, y), & (Ax) \\
\quad\quad\textbf{2.}\;\; \exists x . \exists y . \phi(x, y), \exists y . \phi(x_0, y), \phi(x_0, y_0) \vdash \exists y . \exists x . \phi(x, y), & (I_\exists) \\
\quad\quad\quad\textbf{1.}\;\; \exists x . \exists y . \phi(x, y), \exists y . \phi(x_0, y), \phi(x_0, y_0) \vdash \exists x . \phi(x, y_0), & (I_\exists) \\
\quad\quad\quad\quad\textbf{1.}\;\; \exists x . \exists y . \phi(x, y), \exists y . \phi(x_0, y), \phi(x_0, y_0) \vdash \phi(x_0, y_0), & (Ax)
\end{array}
$$

For each instance of rule $(E_\exists)$ we choose fresh variables x_0 and y_0, respectively, and later apply $(I_\exists)$ choosing y_0 and x_0 to replace y and x, in order to close the proof by (Ax).

As was the case in $\mathcal{N}_{\mathsf{PL}}$, the deduction relation for $\mathcal{N}_{\mathsf{FOL}}$ is an abstract consequence relation.

Proposition 4.37 *The deduction relation for $\mathcal{N}_{\mathsf{FOL}}$ satisfies the following properties*:

1. *For all $\phi \in \Gamma$, $\Gamma \vdash \phi$.*   *(inclusion)*
2. *If $\Gamma \vdash \phi$, then $\Gamma, \psi \vdash \phi$.*   *(monotonicity)*
3. *If $\Gamma \vdash \phi$ and $\Gamma, \phi \vdash \psi$, then $\Gamma \vdash \psi$.*   *(cut)*

Proof The proof of inclusion and cut are exactly the same as in the proof of Proposition 3.22. For the proof of monotonicity, let $\mathcal{D}$ be the derivation tree with conclusion $\Gamma \vdash \phi$. The simple inductive reasoning used in the proof of Proposition 3.22 cannot be reproduced here carelessly, since we need to ensure that $\mathsf{FV}(\psi)$ do not disturb any of the side-conditions of $\mathcal{D}$.

For that, we introduce the concept of *proper variable* of a derivation $\mathcal{D}$ as being a variable that appears in a side-condition of $\mathcal{D}$. Note that a variable constrained by a side-condition of a rule never appears as a free variable in the rule's conclusion, and hence proper variables do not occur free in the conclusion of $\mathcal{D}$. Clearly, a naive insertion of ψ in the assumption set of the sequents in $\mathcal{D}$ will only disturb the side-conditions of $\mathcal{D}$ if any of its proper variables is also a free variable of ψ. To avoid that, we define $\mathcal{D}[y/x]$ as being the result of applying the substitution $[y/x]$ to every formula of $\mathcal{D}$, where y is a variable distinct from any variable that appears free in (any formula in) $\mathcal{D}$. Note that this construction always produces a derivation tree (i.e. all side conditions are respected), because if x appears in $\mathcal{D}$ constrained by a side-condition, the new variable y must satisfy that same condition.

Returning to our proof, let $\{y_1, \ldots, y_n\}$ be the (possibly empty) set of proper variables of $\mathcal{D}$ that are also free variables of ψ. We iteratively apply to $\mathcal{D}$ the substitutions $[z_i/y_i]$, where z_i is a new fresh variable. This procedure results in a new (well-formed) derivation $\mathcal{D}'$, also with conclusion $\Gamma \vdash \phi$, and with all proper variables distinct from the free variables of ψ. We can then proceed by adding ψ to all assumption sets of $\mathcal{D}'$ to obtain a valid derivation of the sequent $\Gamma, \psi \vdash \phi$, in the same way as in the proof of Proposition 3.22.   □

The notions of *admissible* and *derivable* rule are, as in system $\mathcal{N}_{\mathsf{PL}}$, powerful tools that allow us to extend the system for practical use. Of course, the admissibility of the *weakening rule* is as in $\mathcal{N}_{\mathsf{PL}}$ a direct consequence of the monotonicity property of the deduction relation.

Proposition 4.38 (Weakening) *The following rule, named* weakening, *is admissible in* $\mathcal{N}_{\mathsf{FOL}}$

$$(\mathrm{W}) \; \frac{\Gamma \vdash \phi}{\Gamma, \psi \vdash \phi}$$

Proof Similar to the proof of Proposition 3.24, using Proposition 4.37 (instead of 3.22).   □

Since all the admissible rules discussed in Chap. 3 are derivable in the system $\mathcal{N}_{\mathsf{PL}} \cup \{W\}$, we conclude that they are also admissible in $\mathcal{N}_{\mathsf{FOL}}$.

4.4 Soundness and Completeness

The definition of the proof system is given from a completely syntactic point of view, but the derivable syntactic objects possess the desired semantic properties. In other words the calculus is sound, as was the case for propositional logic.

Theorem 4.39 (Soundness) *For every set of formulas Γ and formula ϕ, if $\Gamma \vdash \phi$ then $\Gamma \models \phi$.*

Proof By induction on the derivation of $\Gamma \vdash \phi$. We only treat here the cases of the quantifiers. The remaining cases are proved in a similar way to what was done for propositional logic (recall that we have established all the supporting lemmas).

— Assume the last step is

$$(\text{I}_\forall) \ \frac{\Gamma \vdash \phi[y/x]}{\Gamma \vdash \forall x . \phi} \ , \qquad \text{where } y \text{ must not occur free in } \Gamma \text{ or } \phi$$

By induction hypothesis we have $\Gamma \models \phi[y/x]$. Since y does not occur free in Γ, by Proposition 4.29(2) we have $\Gamma \models \forall x . \phi$.

— Assume the last step is

$$(\text{E}_\forall) \ \frac{\Gamma \vdash \forall x . \phi}{\Gamma \vdash \phi[t/x]}$$

By induction hypothesis $\Gamma \models \forall x . \phi$. Then, by Proposition 4.30(6), we have $\Gamma \models \phi[t/x]$.

— Assume the last step is

$$(\text{I}_\exists) \ \frac{\Gamma \vdash \phi[t/x]}{\Gamma \vdash \exists x . \phi}$$

By induction hypothesis $\Gamma \models \phi[t/x]$. Then, by Proposition 4.30(7), we have $\Gamma \models \exists x . \phi$

— Assume the last step is

$$(\text{E}_\exists) \ \frac{\Gamma \vdash \exists x . \phi \qquad \Gamma , \phi[y/x] \vdash \theta}{\Gamma \vdash \theta} \ , \qquad \text{where } y \text{ must not occur free in } \Gamma , \phi \text{ or } \theta$$

By induction hypothesis we have $\Gamma \models \exists x . \phi$ (IH1) and $\Gamma , \phi[y/x] \models \theta$ (IH2), where y does not occur free in Γ, ϕ or θ. From the IH2 and Proposition 4.30(5) we conclude $\Gamma , \exists x . \phi \models \theta$. Using cut (Proposition 4.28(3)) with IH1 we conclude $\Gamma \models \theta$. □

Conversely it can be proved that the calculus is complete, i.e. it derives every sequent corresponding to a semantic entailment. The proof of completeness of system $\mathcal{N}_{\text{FOL}}$ follows the same pattern as for $\mathcal{N}_{\text{PL}}$, but requires a significant amount of additional technicalities that make it significantly more complicated. For this reason, we omit it here and refer the interested reader to a standard text like [21].

Theorem 4.40 (Completeness) *For every set of formulas Γ and formula ϕ, if $\Gamma \models \phi$ then $\Gamma \vdash \phi$.*

Theorem 4.41 (Compactness) *A (possibly infinite) set of sentences Γ is satisfiable iff every finite subset of Γ is satisfiable.*

Proof Similar to the proof of Proposition 3.32. □

4.5 Validity Checking

Unsurprisingly, the problem of determining whether an arbitrary first-order sentence is valid is significantly harder than for the propositional case. In fact, it is impossible to solve this problem in its full generality: given an arbitrary first-order formula it is impossible to devise a decision procedure that checks if the formula is valid.

Proposition 4.42 *The validity problem for first-order logic is undecidable.*

Proof A detailed proof of this result demands a considerable amount of auxiliary notions, so we will provide only a brief sketch of the argument (the interested reader can find a detailed proof in the literature, for instance [11, Thm. 6.20]).

To show that there exists no decision procedure to answer the validity problem, we assume that such a decision procedure exists, and show that a contradiction is reached. For that, we encode a known undecidable problem (say, the halting problem) as a validity problem in first-order logic. Concretely, one provides a construction that, given an instance of the problem, returns a first-order formula whose validity answers the encoded problem on that particular instance. This, combined with a decision procedure for validity of first-order logic, would provide a decision procedure for the undecidable problem, which would be a contradiction. So we conclude that no decision procedure for validity in first-order logic exists.   □

It is worth to mention that this negative result (undecidability) is a direct consequence of a positive feature of first-order logic—its expressive power. The result does not preclude however restricted instances of the general problem from being solvable. We will see that the problem of validity-checking of first-order formulas can, to some extent, be reduced to the propositional case, allowing to transfer methods and strategies presented in Sect. 3.5. This requires the introduction of specific notions of normal forms to restrict the use of quantifiers, but let us first extend our study of negation normal forms from propositional logic to first-order logic.

4.5.1 Negation and Prenex Normal Forms

Every first-order formula has an equivalent formula in negation normal form. Of course, computing it takes quantifiers in account.

Definition 4.43 A first-order formula is in *negation normal form* (NNF) if the implication connective is not used in it, and negation is only applied to atomic formulas.

Proposition 4.44 *Every first-order formula ϕ is equivalent to a NNF formula ψ.*

Proof Let ϕ be an arbitrary first-order formula. We compute its NNF by repeatedly and exhaustively rewriting using the following equivalences (oriented from left to right):

$$\phi \to \psi \equiv \neg\phi \lor \psi$$

$$\neg\neg\phi \equiv \phi$$

$$\neg(\phi \land \psi) \equiv \neg\phi \lor \neg\psi$$

$$\neg(\phi \lor \psi) \equiv \neg\phi \land \neg\psi$$

$$\neg\forall x.\,\phi \equiv \exists x.\,\neg\phi$$

$$\neg\exists x.\,\phi \equiv \forall x.\,\neg\phi$$

These are the same equivalences used in the propositional NNF transformation, augmented with two new rules for handling quantifiers. As in the propositional case, the resulting formula is necessarily in NNF since otherwise at least one of the equivalences would be applicable. The discussion concerning termination and the size of the resulting formula in the propositional case applies equally here. Proposition 4.24 and Proposition 4.25 justify the equivalence between the original formula and its NNF. □

Example 4.45 Let us compute the NNF of $\forall x.\,(\forall y.\,P(x,y) \lor Q(x)) \to \exists z.\,P(x,z)$.

$$\forall x.\,(\forall y.\,P(x,y) \lor Q(x)) \to \exists z.\,P(x,z) \quad\equiv$$
$$\forall x.\,\neg(\forall y.\,P(x,y) \lor Q(x)) \lor \exists z.\,P(x,z) \quad\equiv$$
$$\forall x.\,\exists y.\,(\neg P(x,y) \land \neg Q(x)) \lor \exists z.\,P(x,z)$$

The conversion to NNF pushes negation towards the atoms. In first-order formulas we can move quantifiers in the opposite direction.

Lemma 4.46 *If x does not occur free in ψ, then the following equivalences hold.*

$(\forall x.\,\phi) \land \psi \equiv \forall x.\,\phi \land \psi$	$\psi \land (\forall x.\,\phi) \equiv \forall x.\,\psi \land \phi$
$(\forall x.\,\phi) \lor \psi \equiv \forall x.\,\phi \lor \psi$	$\psi \lor (\forall x.\,\phi) \equiv \forall x.\,\psi \lor \phi$
$(\exists x.\,\phi) \land \psi \equiv \exists x.\,\phi \land \psi$	$\psi \land (\exists x.\,\phi) \equiv \exists x.\,\psi \land \phi$
$(\exists x.\,\phi) \lor \psi \equiv \exists x.\,\phi \lor \psi$	$\psi \lor (\exists x.\,\phi) \equiv \exists x.\,\psi \lor \phi$

Proof The equivalences in the right column are obtained from the equivalences in the left column using commutativity of either $\land$ or $\lor$. Those in the left column are all similar; we prove the first as a representative case. Let $\mathcal{M} = (D, I)$ be a structure and α an assignment such that $\mathcal{M}, \alpha \models \forall x.\,\phi \land \psi$. By definition, $\mathcal{M}, \alpha[x \mapsto a] \models \phi$ and $\mathcal{M}, \alpha[x \mapsto a] \models \psi$, for all $a \in D$. But since $x \notin \mathsf{FV}(\psi)$, by Proposition 4.22(2) $\mathcal{M}, \alpha \models \psi$. Again by definition we obtain $\mathcal{M}, \alpha \models (\forall x.\,\phi) \land \psi$. □

Definition 4.47 A formula is in *prenex form* if it is of the form $Q_1x_1.Q_2x_2....$ $Q_nx_n.\psi$ where each Q_i is a quantifier (either $\forall$ or $\exists$) and ψ is a quantifier-free formula.

Proposition 4.48 *For any formula of first-order logic, there exists an equivalent formula in prenex form.*

Proof Let ϕ be an arbitrary first-order formula and ψ its NNF. In ψ we rewrite repeatedly and exhaustively the equivalences of Lemma 4.46 (oriented from left to right), possibly preceding each rewrite step by a renaming of bound variables to meet the conditions of the lemma. This process clearly terminates and the result is necessarily in prenex form (because the NNF translation has removed all implications and pushed negations to the atomic formulas). $\square$

We remark that a translation to prenex form could be defined without using the NNF translation as a first step. In this case, one should consider additional equivalences for handling the implication connective (see Exercise 4.11) and De Morgan's quantifier laws for handling negation. Moreover, when implementing the translation, one would have to choose a strategy for applying these equivalences—different choices would lead to different, but equivalent, results.

It is also worth considering equivalences that may produce shorter normal forms in specific patterns, such as the distributive laws of $\forall$ and $\exists$ over $\land$ and $\lor$ (Lemma 4.27), whose application will reduce the number of quantifiers in the result.

4.5.2 Herbrand/Skolem Normal Forms and Semi-Decidability

Let us now focus on what is perhaps the most distinctive feature of first-order logic: quantifiers. We have already seen that quantifiers in first-order formulas can always be moved to the outermost position in prenex normal forms (Proposition 4.48). The following constructions allow us to further restrict the roles of these quantifiers in first-order formulas.

Definition 4.49 (Herbrand and Skolem Forms) Let ϕ be a first-order formula in prenex normal form. The *Herbrandization* of ϕ (written ϕ^H) is an existential formula obtained from ϕ by repeatedly and exhaustively applying the following transformation, that removes the first universal quantification occurring in the prenex formula:

$$\exists x_1, \ldots, x_n . \forall y . \psi \rightsquigarrow \exists x_1, \ldots, x_n . \psi[f(x_1, \ldots, x_n)/y]$$

with f a fresh function symbol with arity n (i.e. f does not occur in ψ).

Dually, the *Skolemization* of ϕ (written ϕ^S) is a universal formula obtained from ϕ by repeatedly applying the transformation:

$$\forall x_1, \ldots, x_n . \exists y . \psi \rightsquigarrow \forall x_1, \ldots, x_n . \psi[f(x_1, \ldots, x_n)/y]$$

again, f is a fresh function symbol with arity n.

Herbrand normal form (resp. *Skolem normal form*) formulas are those obtained by the process of Herbrandization (resp. Skolemization).

It is convenient to write Herbrand and Skolem formulas using vector notation for the sequence of quantified variables, $\exists \overline{x}. \ \psi$ and $\forall \overline{x}. \ \psi$ (with ψ quantifier free), respectively. Note that Herbrandization and Skolemization change the underlying vocabulary $\mathcal{V}$, since they rely on adding fresh constants and/or function symbols. These additional symbols are often called *Herbrand functions* and *Skolem functions* respectively. This observation alone suffices to show that a formula ϕ and its Herbrandization/Skolemization are not logically equivalent. The following result clarifies the relation between a formula and its Herbrandization/Skolemization.

Proposition 4.50 *Let ϕ be a first-order formula in prenex normal form. ϕ is valid iff its Herbrandization ϕ^H is valid. Dually, ϕ is unsatisfiable iff its Skolemization ϕ^S is unsatisfiable.*

Proof We will prove the Skolemization case (the Herbrandization case is similar and left to the reader as an exercise).

The proof is by induction on the number of existential variables removed in the process. The base case is trivial since the formula is already in Skolem form. For the induction step, let us assume that $\forall \overline{x}. \ \exists y. \ \psi$ is satisfiable, i.e., there exists a model $\mathcal{M} = (D, I)$ such that $\mathcal{M} \models \forall \overline{x}. \ \exists y. \ \psi$. By definition, this means that for every n-tuple $\overline{a} = (a_1, \ldots, a_n) \in D^n$, there exists $b \in D$ such that $\mathcal{M} \models \psi[\overline{a}/\overline{x}, b/y]$. Consider the indexed family of sets $\{B_{\overline{a}}\}_{\overline{a} \in D^n}$, where $B_{\overline{a}}$ is the (necessarily nonempty) set $\{b \in D \mid \mathcal{M} \models \psi[\overline{a}/\overline{x}, b/y]\}$, and let $f^\star$ be a choice function for this family of sets (i.e., for every $\overline{a} \in D^n$, $f^\star(\overline{a})$ selects an element from $B_{\overline{a}}$). We construct a new interpretation $\mathcal{M}' = (D, I')$ where I' expands I with $I'(f) = f^\star$. Clearly $\mathcal{M}' \models \forall \overline{x}. \ \psi[f(\overline{x})/y]$, witnessing the satisfiability of $\forall \overline{x}. \ \psi[f(\overline{x})/y]$.

For the reverse implication, consider a model $\mathcal{M}' = (D, I')$ witnessing the satisfiability of $\forall \overline{x}. \ \psi[f(\overline{x})/y]$. We construct the model $\mathcal{M} = (D, I)$, where I restricts I' for the vocabulary of symbols not equal to f. Clearly, $\mathcal{M} \models \forall \overline{x}. \ \exists y. \ \psi$, since for every n-tuple $\overline{a} \in D^n$, $I'(f)(\overline{a})$ provides a specific choice for $b \in D$ such that $\mathcal{M} \models \psi[\overline{a}/\overline{x}, b/y]$. $\qquad \square$

Remark 4.51 The requirement that the original formula be in prenex normal form is too strong: Skolemization/Herbrandization can be applied to formulas in negative normal form. In fact, this relaxation results in potentially simpler Skolem normal forms (with fewer Skolem functions, or functions with lower arity).

Herbrand (resp. Skolem) normal form is important since it allows to reduce validity (resp. unsatisfiability) of a first-order formula to the existence of suitable instances of a quantifier-free formula, considering models constructed directly from the syntax of terms.

Definition 4.52 (Herbrand Interpretation) Let $\mathcal{V}$ be a first-order vocabulary and assume $\mathcal{V}$ has at least one constant symbol (otherwise, we explicitly expand the vocabulary with such a symbol). A *Herbrand Interpretation* $\mathcal{H} = (D_{\mathcal{H}}, I_{\mathcal{H}})$ is a $\mathcal{V}$-structure specified by a set of closed atomic predicates (i.e. atomic predicates applied to ground terms), also denoted by $\mathcal{H}$. The interpretation structure is given as follows:

- Interpretation domain: $D_{\mathcal{H}}$ is the set of ground terms for the vocabulary $\mathcal{V}$. It is called the *Herbrand universe* for $\mathcal{V}$;
- Interpretation of constants: for every $c \in \mathcal{V}$, $I_{\mathcal{H}}(c) = c$;
- Interpretation of functions: for every $f \in \mathcal{V}$ with $\mathrm{ar}(f) = n$, $I_{\mathcal{H}}(f)$ consists of the n-ary function that, given ground terms $t_1, \ldots, t_n$, returns the ground term $f(t_1, \ldots, t_n)$;
- Interpretation of predicates: for every $P \in \mathcal{V}$ with $\mathrm{ar}(P) = n$, $I_{\mathcal{H}}(P)$ is the n-ary relation $\{(t_1, \ldots, t_n) \mid P(t_1, \ldots, t_n) \in \mathcal{H}\}$.

The requirement on $\mathcal{V}$ ensures that the interpretation domain $D_{\mathcal{H}}$ is not empty. We remark that assignments in Herbrand interpretations are just ground substitutions, thus an existential formula $\exists x. \psi$ (resp. universal formula $\forall x. \psi$), with ψ quantifier-free, is valid in a Herbrand interpretation $\mathcal{H}$ iff there exists a ground instance $\psi\sigma$ of ψ such that $\mathcal{H} \models \psi\sigma$ (resp. $\mathcal{H} \models \psi\sigma$ for all ground instances $\psi\sigma$ of ψ).

Lemma 4.53   *An existential formula ϕ is valid iff for every Herbrand model $\mathcal{H}$, $\mathcal{H} \models \phi$. Dually, a universal formula ϕ is unsatisfiable iff there exists no Herbrand model $\mathcal{H}$ such that $\mathcal{H} \models \phi$.*

Proof We prove the existential case. The dual case is obtained by negation.

The left-to-right implication is immediate. For the reverse implication, let ϕ be $\exists x. \psi$, with ψ a quantifier-free formula. Given an arbitrary interpretation $\mathcal{M} = (D, I)$, construct the Herbrand interpretation $\mathcal{H} = \{P(t_1, \ldots, t_n) \mid t_1, \ldots, t_n$ are ground terms, and $\mathcal{M} \models P(t_1, \ldots, t_n)\}$. Since $\mathcal{H} \models \exists x. \psi$, there exists a ground-instance $\psi\sigma$ that holds in $\mathcal{H}$ (assignments in $\mathcal{H}$ are just ground substitutions), and hence $\mathcal{M} \models \psi\sigma$ also. But then, by Proposition 4.30(7), $\mathcal{M} \models \exists x. \psi$. $\square$

The class of Herbrand models are then sufficient for checking the validity of formulas. Note that Herbrand models resemble propositional models, with ground atomic predicates playing the role of propositional symbols. In fact for ground formulas, the definition of validity perfectly matches validity of propositional formulas. Hence it makes sense to say of a valid ground formula that it is "propositionally valid".

Theorem 4.54 (Herbrand's Theorem) *An existential first-order formula $\exists x. \psi$ (with ψ quantifier-free) is valid iff there exists an integer k and ground instances $\psi\sigma_1, \ldots, \psi\sigma_k$ such that $\psi\sigma_1 \vee \cdots \vee \psi\sigma_k$ is propositionally valid.*

Dually, a universal formula $\forall x. \psi$ (with ψ quantifier-free) is unsatisfiable iff there exists an integer k and closed instances $\psi\sigma_1, \ldots, \psi\sigma_k$ such that $\psi\sigma_1 \wedge \cdots \wedge \psi\sigma_k$ is propositionally unsatisfiable.

Proof We prove the existential case (again, the dual case follows by negation). By Lemma 4.53, $\exists \overline{x}. \psi$ is valid iff every Herbrand interpretation validates it, that is, iff for each Herbrand interpretation $\mathcal{H}$ there exists a ground instance $\psi \sigma$ of ψ such that $\mathcal{H} \models \psi \sigma$, which in turn implies that $\mathcal{H} \not\models \neg \psi \sigma$. It follows that the set $\Gamma = \{\neg \psi \sigma \mid \sigma$ is a ground substitution$\}$ is propositionally unsatisfiable in the class of Herbrand models. Hence, by propositional compactness (Theorem 3.32), there exists a finite subset $\Gamma' \subset \Gamma$ that is already unsatisfiable. Let k be the cardinality of Γ', i.e. $\Gamma' = \{\neg \psi \sigma_1, \ldots, \neg \psi \sigma_k\}$. Unsatisfiability of Γ' implies that $\neg \psi \sigma_1 \wedge \cdots \wedge \neg \psi \sigma_k$ is also propositionally unsatisfiable, which in turn implies that $\psi \sigma_1 \vee \cdots \vee \psi \sigma_k$ is valid in every Herbrand model. The reverse implication is trivial. □

Herbrand's theorem is remarkable since it effectively reduces the problem of first-order validity to propositional validity of ground instances. Of course, the set of ground instances of a first-order formula is in general infinite, which prevents transferring decidability results from propositional logic. Nevertheless, it allows us to conclude that the set of valid first-order formulas is recursively enumerable, which makes the validity problem of first-order logic semi-decidable.

Theorem 4.55 (Semi-Decidability) *The problem of validity of first-order formulas is semi-decidable, i.e. there exists a procedure that, given a first-order formula, answers "yes" iff the formula is valid (but might not terminate if the formula is not valid).*

Proof Let ϕ be a first-order formula. First, we construct the Herbrand normal form ϕ^H of ϕ. Lemma 4.50 guarantees that if ϕ is valid, ϕ^H is also valid. By Herbrand's theorem, there exist k ground instances whose disjunction is propositionally valid. We do not know in advance what the value of k is, but we can enumerate all the sequences of ground instances and check, for each of these sequences, if the disjunction of the formulas in it is propositionally valid. If the disjunction is valid, the procedure halts returning "yes". If not, it continues checking the next sequence in the enumeration. If the original formula is indeed valid, the procedure will necessarily terminate (it will reach the position in the enumeration containing the sequence of the required ground instances). □

In theory it is then always possible to confirm that a first-order theorem is indeed valid, and analogously that an unsatisfiable formula is indeed unsatisfiable. Of course, directly encoding a strategy for checking the validity of first-order formulas will necessarily face efficiency and feasibility problems.

Note also that the procedure described above uses propositional validity checking as a subroutine, a feature shared by most modern techniques for validity and satisfiability checking. This fact has promoted significant advances in the field of propositional satisfiability. For example the DPLL procedure mentioned in the last chapter was proposed in the context of solving validity checking for first-order formulas.

Remark 4.56 The search through the space of ground instances, as prescribed in the proof of Theorem 4.55, is in some verification methods replaced by the capability of equating first-order terms—the so called *unification* of terms. Unification is an important subject with a wide range of applications in automated verification. One such application is *first-order resolution* where, roughly speaking, clauses of a quantifier-free CNF are combined to obtain a new clause, that can be added to the CNF without disturbing its validity status. If the clause obtained is the empty clause ($\perp$), this establishes that the formula is unsatisfiable.

The interested reader can find information on these matters in any textbook on logic, see Sect. 4.8.

4.5.3   Decidable Fragments

An interesting refinement of Theorem 4.55 is to investigate fragments in which bounds can be established for searching the ground instance space. Clearly if the set of ground terms is finite the set of ground instances of the formula under scrutiny will be finite as well. This immediately leads to a bound on the number of instances whose search is required by Herbrand's theorem, which will make possible to decide if the formula is valid or otherwise. So the question is under what conditions can a bound be known for the number of ground instances of a formula ϕ.

First note that ϕ should not use any function symbols, since this would immediately result in an infinite set of ground terms. Moreover, such a function symbol should also not be introduced by the Herbrandization process, the first step of the procedure outlined in the proof of Theorem 4.55. This will be the case if we restrict our attention to formulas whose prenex normal form has the shape

$$\forall \overline{x}.\, \exists \overline{y}.\, \psi$$

with ψ quantifier-free. This fragment of formulas is normally known as the *AE fragment*, owing its name to the alternation of quantifiers allowed (*A* refers to the universal quantifier and *E* to existential quantifier). This discussion shows that the validity problem for the AE fragment is decidable.

The class of formulas can be further enlarged by observing that a formula not in AE may be equivalent to one in AE. Of course, such an observation should be complemented with a simple, mechanisable means to test the equivalence implicit in the characterisation (after all, validity checking can itself be expressed as a test of equivalence with $\top$).

Let us illustrate this with a small example: consider the formula

$$\exists x.\, \forall y.\, P(x) \vee Q(y)$$

This formula does not belong to AE. Indeed, its Herbrand normal form is $\exists x.\, P(x) \vee Q(f(x))$, which has an infinite set of ground instances. But we observe that, applying the prenex transformation in the reverse direction (Lemma 4.46), we obtain $(\exists x.\, P(x)) \vee (\forall y.\, Q(y))$. Now, applying again the prenex transformation, but choosing y to be moved first, we reach $\forall y.\, \exists x.\, P(x) \vee Q(y)$, which belongs to AE.

This example illustrates the idea behind a technique known as *miniscope*, which consists in pushing existential quantifiers inside the formula (minimizing their scopes). The technique can be used to show that the class of *monadic formulas*, which are formulas containing only unary predicates, can be transformed into formulas of the AE fragment. Monadic formulas thus constitute a decidable fragment of first-order logic.

4.6 Variations and Extensions

Equality of terms can be treated in first-order logic by introducing a binary equality predicate, treated in the same way as any other predicate. A theory for reasoning with equality in this way will be studied in Sect. 4.7. In this section we cover an alternative approach, which consists in having equality as a logical symbol whose interpretation is fixed. We then introduce *many-sorted first-order logic*, an extension of first-order logic that attaches sorts to terms of the language (we will see that this has no impact on the expressive power). Many-sorted first-order logic is particularly important to us because it will be used in Chap. 5 of this book. Finally, we briefly mention second-order logic, as an example of a logic that is more expressive than first-order logic.

4.6.1 First-Order Logic with Equality

Equality is a very special relation in mathematical discourse. In our presentation of first-order logic the equality relation has no special status. An alternative approach, usually called *first-order logic with equality*, considers equality as a logical symbol with a fixed interpretation. In this approach the equality symbol ($=$) is interpreted as the equality relation in the domain of interpretation. So we have, for an assignment $\alpha \in \Sigma_D$ and a structure $\mathcal{M}$, that

$$[\![t_1 = t_2]\!]_{\mathcal{M}}(\alpha) = \mathbf{T} \quad \text{iff} \quad [\![t_1]\!]_{\mathcal{M}}(\alpha) \text{ and } [\![t_2]\!]_{\mathcal{M}}(\alpha) \text{ are the same element of } D$$

In terms of proof theory, rules about equality are added to the proof system. The rules are usually an axiom for reflexivity and a rule for substitution.

$$\text{(refl)} \; \frac{}{\Gamma \vdash t = t} \qquad \text{(subs)} \; \frac{\Gamma \vdash t_1 = t_2 \qquad \Gamma \vdash \phi[t_1/x]}{\Gamma \vdash \phi[t_2/x]}$$

The rules (refl) and (subs) correspond, respectively, to introduction an elimination rules for equality. The (refl) rule tells us that any term has to be equal to itself, and is evidently sound. (subs) allows us to substitute equals for equals in a formula.

Many other properties of equality follow from the rules above. For instance, the rules for symmetry and transitivity of equality are derivable

$$\text{(sym)} \; \frac{\Gamma \vdash t_1 = t_2}{\Gamma \vdash t_2 = t_1} \qquad \text{(trans)} \; \frac{\Gamma \vdash t_1 = t_2 \qquad \Gamma \vdash t_2 = t_3}{\Gamma \vdash t_1 = t_3}$$

Rule (sym) is derived as follows.

$$\text{(subs)} \ \dfrac{\Gamma \vdash t_1 = t_2 \qquad \text{(refl)} \ \dfrac{}{\Gamma \vdash t_1 = t_1}}{\Gamma \vdash t_2 = t_1}$$

Rule (trans) follows directly by applying (subs) to the premises.

To understand the significant difference between having equality with the status of any other predicate, or with a fixed interpretation as in first-order logic with equality, consider the formula $\exists x_1, x_2.\forall y. y = x_1 \vee y = x_2$. With a fixed interpretation of equality, the validity of this formula implies that the cardinality of the interpretation domain is at most two—the quantifiers can actually be used to fix the cardinality of the domain, which is not otherwise possible in first-order logic. As a second example, the formula write $\exists x_1, x_2.\neg(x_1 = x_2)$ implies that there exist at least two distinct elements in the domain, thus its cardinality must be at least two.

4.6.2 Many-Sorted First-Order Logic

A natural extension of first-order logic that can be considered is the one that results from allowing different domains of elements to coexist in the framework. This allows distinct "sorts" or types of objects to be distinguished at the syntactical level, constraining how operations and predicates interact with these different sorts. For instance, a function may have an argument of one sort, and produce a result of some other sort. Having full support for different sorts of objects in the language allows for cleaner and more natural encodings of whatever we are interested in modeling and reasoning about.

Many-sorted first-order logic extends first-order logic by attaching sorts from a given set to terms of the language. We will briefly sketch here how the basic definitions are adapted to the many-sorted scenario, omitting proofs of the results. The interested reader can find them, for instance, in [9].

Definition 4.57 A *many-sorted vocabulary* $\mathcal{V}$ is a triple $\mathcal{V} = (\mathcal{S}, \mathcal{F}, \text{rank})$ where

— $\mathcal{S}$ is a nonempty enumerable set of *sorts*. We let $\mathcal{S}_b = \mathcal{S} \cup \{\text{bool}\}$.
— $\mathcal{F}$ is an $\mathcal{S}_b$-indexed set of enumerable sets $\mathcal{F} = \{\mathcal{F}_s\}_{s \in \mathcal{S}_b}$, where $\mathcal{F}_s$ is the set of function symbols whose codomain has sort s. Moreover, we demand that $\{\bot, \neg, \wedge, \vee, \rightarrow\} \cap \mathcal{F}_{\text{bool}} = \emptyset$.
— rank is an $\mathcal{S}_b$-indexed set of mappings $\text{rank} = \{\text{rank}_s : \mathcal{F}_s \rightarrow \mathcal{S}^\star\}_{s \in \mathcal{S}_b}$, where $\mathcal{S}^\star$ is the set of (possibly empty) sequences of sorts, and $\text{rank}_s : \mathcal{F}_s \rightarrow \mathcal{S}^\star$ assigns to each function symbol the sorts of its arguments.

We write $f^s \in \mathcal{F}$ as an abbreviation for $f \in \mathcal{F}_s$, for some $s \in \mathcal{S}_b$. Likewise, we write $\text{rank}(f^s)$ as an abbreviation for $\text{rank}_s(f)$ for some $s \in \mathcal{S}_b$ such that $f \in \mathcal{F}_s$. Elements of f^s of $\mathcal{F}$ for which $\text{rank}(f^s)$ is the empty sequence are called *constants of sort s*, and elements of $\mathcal{F}_{\text{bool}}$ are called *predicate symbols*.

Note that in many-sorted vocabularies constant and predicate symbols are just special cases of function symbols.

Definition 4.58 Let $V = (S, \mathcal{F}, \text{rank})$ be a many-sorted vocabulary. For each sort $s \in S$, we consider an infinite enumerable set $\mathcal{X}_s$ of *variables of sort s*. $\mathcal{X}$ is the S-indexed set of sets $\mathcal{X} = \{\mathcal{X}_s\}_{s \in S}$. Again, we write $x^s \in \mathcal{X}$ as an abbreviation for $x \in \mathcal{X}_s$, for some $s \in S$.

Terms and *formulas* are defined by the grammar

$$\begin{aligned}
S &\ni s \\
\mathcal{X} &\ni x^s \\
\textbf{Term}^s \ni t^s &::= x^s \mid f^s(t_1^{s_1}, \ldots, t_n^{s_n}), \quad \text{for } f^s \in \mathcal{F} \\
&\qquad\qquad\qquad\qquad\qquad \text{and rank}(f^s) = \langle s_1, \ldots, s_n \rangle \\
\textbf{Form} \ni \phi &::= t^{\text{bool}} \mid \bot \mid (\neg\phi) \mid (\phi \wedge \phi) \mid (\phi \vee \phi) \mid (\phi \rightarrow \phi) \mid \\
&\qquad (\forall x^s . \phi) \mid (\exists x^s . \phi)
\end{aligned}$$

The notions of free and bound variables and substitution are adapted accordingly from their unsorted counterparts.

Definition 4.59 Let $V = (S, \mathcal{F}, \text{rank})$ be a many-sorted vocabulary. A *V-structure* $\mathcal{M}$ is a pair $\mathcal{M} = (D, I)$ where D is an S-indexed set of nonempty sets. D_s is the interpretation domain for sort s and we further fix the interpretation domain for sort bool as $D_{\text{bool}} = \{\mathbf{F}, \mathbf{T}\}$. I is an *interpretation function* that assigns meaning to the symbols of V as follows: for each $f^s \in \mathcal{F}$ with $\text{rank}(f^s) = \langle s_1, \ldots, s_n \rangle$, $I(f^s) : D_{s_1} \times \cdots \times D_{s_n} \rightarrow D_s$ is an n-ary function.

An *assignment* α is an S-indexed set of mappings $\alpha = \{\alpha_s\}_{s \in S}$, where $\alpha_s : \mathcal{X}_s \rightarrow D_s$. We write $\alpha(x^s)$ as an abbreviation for $\alpha_s(x)$, for some $s \in S$ with $x \in \mathcal{X}_s$. The set of all assignments into an interpretation domain D is denoted by Σ_D.

Definition 4.60 Let $V = (S, \mathcal{F}, \text{rank})$ be a many-sorted vocabulary, $\mathcal{M}$ a V-structure, and α an assignment. The *interpretation of terms of sort $s \in S$ with respect to $\mathcal{M}$ and α* is inductively defined by:

$$[\![x^s]\!]_{\mathcal{M}}(\alpha) = \alpha(x^s)$$

$$[\![f^s(t_1^{s_1}, \ldots, t_n^{s_n})]\!]_{\mathcal{M}}(\alpha) = I(f^s)([\![t_1^{s_1}]\!]_{\mathcal{M}}(\alpha), \ldots, [\![t_n^{s_n}]\!]_{\mathcal{M}}(\alpha))$$

The *interpretation of formulas with respect to $\mathcal{M}$ and α* is defined by:

$$[\![\bot]\!]_{\mathcal{M}}(\alpha) = \mathbf{F}$$

$$[\![f^{\text{bool}}(t_1^{s_1}, \ldots, t_n^{s_n})]\!]_{\mathcal{M}}(\alpha) = \mathbf{T} \quad \text{iff} \quad I(f^{\text{bool}})([\![t_1^{s_1}]\!]_{\mathcal{M}}(\alpha), \ldots, [\![t_n^{s_n}]\!]_{\mathcal{M}}(\alpha)) = \mathbf{T}$$

$$[\![\neg\phi]\!]_{\mathcal{M}}(\alpha) = \mathbf{T} \quad \text{iff} \quad [\![\phi]\!]_{\mathcal{M}}(\alpha) = \mathbf{F}$$

$$[\![\phi \wedge \psi]\!]_{\mathcal{M}}(\alpha) = \mathbf{T} \quad \text{iff} \quad [\![\phi]\!]_{\mathcal{M}}(\alpha) = \mathbf{T} \text{ and } [\![\psi]\!]_{\mathcal{M}}(\alpha) = \mathbf{T}$$

$$[\![\phi \vee \psi]\!]_{\mathcal{M}}(\alpha) = \mathbf{T} \quad \text{iff} \quad [\![\phi]\!]_{\mathcal{M}}(\alpha) = \mathbf{T} \text{ or } [\![\psi]\!]_{\mathcal{M}}(\alpha) = \mathbf{T}$$

$$[\![\phi \rightarrow \psi]\!]_{\mathcal{M}}(\alpha) = \mathbf{T} \quad \text{iff} \quad [\![\phi]\!]_{\mathcal{M}}(\alpha) = \mathbf{F} \text{ or } [\![\psi]\!]_{\mathcal{M}}(\alpha) = \mathbf{T}$$

$$[\![\forall x^s.\phi]\!]_{\mathcal{M}}(\alpha) = \mathbf{T} \quad \text{iff} \quad [\![\phi]\!]_{\mathcal{M}}(\alpha[x^s \mapsto a^s]) = \mathbf{T} \text{ for all } a^s \in D_s$$

$$[\![\exists x^s.\phi]\!]_{\mathcal{M}}(\alpha) = \mathbf{T} \quad \text{iff} \quad [\![\phi]\!]_{\mathcal{M}}(\alpha[x^s \mapsto a^s]) = \mathbf{T} \text{ for some } a^s \in D_s$$

The notions of validity, satisfiability, entailment, and logical equivalence are all defined as for the unsorted case.

We conclude this section with a translation from many-sorted first-order logic to (unsorted) first-order logic. The translation shows that many-sorted logic has the same expressive power as first-order logic. On the other hand, it allows for all the meta-theoretic results to be transferred from the unsorted to the many-sorted version.

We fix a many-sorted vocabulary $\mathcal{V} = (\mathcal{S}, \mathcal{F}, \text{rank})$, and use the notation $\lceil \cdot \rceil$ for the translation of each many-sorted concept into its unsorted analogue. For each sort $s \in \mathcal{S}$, we consider an additional predicate symbol P_s on the unsorted vocabulary that will identify the "elements of sort s" in the interpretation domain. We then construct formally the unsorted vocabulary $\lceil \mathcal{V} \rceil = (\lceil C \rceil, \lceil \mathcal{F} \rceil, \lceil \mathcal{P} \rceil)$:

$$\lceil C \rceil = \bigcup_{s \in \mathcal{S}} \{f \in \mathcal{F}_s \mid \text{rank}(f) = \epsilon\}$$

$$\lceil \mathcal{F} \rceil = \bigcup_{s \in \mathcal{S}} \{f \in \mathcal{F}_s \mid \text{rank}(f) \neq \epsilon\}$$

$$\lceil \mathcal{P} \rceil = \mathcal{F}_{\text{bool}} \cup \{P_s \mid s \in \mathcal{S}\}$$

where ϵ denotes the empty sequence. For the set of variables, we take $\lceil \mathcal{X} \rceil = \bigcup_{s \in \mathcal{S}} \mathcal{X}_s$. Clearly, any sorted term t^s is also a term of the unsorted vocabulary, thus the translation function for terms is the identity $\lceil t \rceil = t$. The translation of formulas is adjusted in order to pass information of sorts to the unsorted world. We translate quantifiers as follows:

$$\lceil \forall x^s.\phi \rceil = \forall x. P_s(x) \rightarrow \lceil \phi \rceil$$

$$\lceil \exists x^s.\phi \rceil = \exists x. P_s(x) \rightarrow \lceil \phi \rceil$$

and the remaining cases just follow the structure of the formulas, for instance $\lceil \phi \wedge \psi \rceil = \lceil \phi \rceil \wedge \lceil \psi \rceil$. In particular, when ϕ is quantifier free, $\lceil \phi \rceil = \phi$.

Finally we address models. Each many-sorted $\mathcal{V}$-structure $\mathcal{M} = (D, I)$ is translated into an unsorted $\lceil \mathcal{V} \rceil$-structure $\lceil \mathcal{M} \rceil = (\lceil D \rceil, \lceil I \rceil)$:

$$\lceil D \rceil = \bigcup_{s \in \mathcal{S}} D_s$$

$$\lceil I \rceil(c) = I(c)$$

$$\lceil I \rceil(f) = \text{an arbitrary extension of } I(f)$$

$$\lceil I \rceil(P) = \begin{cases} D_s & \text{if } P \text{ is a sort predicate } P_s \\ I(P) & \text{otherwise} \end{cases}$$

where $c \in \lceil \mathcal{C} \rceil$, $f \in \lceil \mathcal{F} \rceil$, and $P \in \lceil \mathcal{P} \rceil$.

Strictly speaking this construction does not fully characterise a model of the unsorted vocabulary, due to the choice of arbitrary extensions of function interpretations. This turns out to be irrelevant since the following result holds for any model fulfilling this specification.

Lemma 4.61 *Let $\mathcal{V}$ be a many-sorted first-order vocabulary, and ϕ and $\mathcal{M}$ be respectively a $\mathcal{V}$-formula and a $\mathcal{V}$-structure. Then*

$$\mathcal{M} \models \phi \quad \text{iff} \quad \lceil \mathcal{M} \rceil \models \lceil \phi \rceil$$

Hence $\Gamma \models \phi$ implies $\lceil \Gamma \rceil \models \lceil \phi \rceil$. For the converse implication, we need to impose certain constraints on the unsorted $\lceil \mathcal{V} \rceil$-structure in order to ensure its well-sortedness. Consider the following set of formulas Φ:

- $\exists x . P_s(x)$, for every $s \in \mathbf{Sort}$
- $P_s(c)$, for every $c \in \mathcal{F}_s$ with $\text{rank}_s(c) = \epsilon$
- $\forall x_1, \ldots, x_n . (P_{s_1}(x_1) \wedge \cdots \wedge P_{s_n}(x_n) \rightarrow P_s(f(x_1, \ldots, x_n)))$, for every $f \in \mathcal{F}_s$ with $\text{ar}(f) = \langle s_1, \ldots, s_n \rangle$

Observe that for every $\mathcal{V}$-structure $\mathcal{M}$, $\lceil \mathcal{M} \rceil \models \Phi$. Moreover, every $\lceil \mathcal{V} \rceil$-structure $\lceil \mathcal{M} \rceil$ such that $\lceil \mathcal{M} \rceil \models \Phi$ can be turned into a many-sorted $\mathcal{V}$-structure (the validity of Φ ensures that $\lceil \cdot \rceil$ can be inverted). Combining these results leads to the following correspondence theorem.

Theorem 4.62 *Let $\mathcal{V}$ be a many-sorted first-order vocabulary, Γ be a set of $\mathcal{V}$-formulas, and ϕ a $\mathcal{V}$-formula. Then*

$$\Gamma \models \phi \quad \text{iff} \quad \lceil \Gamma \rceil, \Phi \models \lceil \phi \rceil$$

This proposition can be used to transfer the meta-theoretical results of first-order logic to the many-sorted case. Of course, the concepts used in those results need to be mapped to the many-sorted world—for instance the many-sorted analogues of the soundness and completeness theorems of Sect. 4.4 will refer to a suitable many-sorted version of the proof-system. Proof theory for many-sorted first-order logic is addressed for instance in [10].

4.6.3 Second-Order Logic

In first-order logic the quantifiers range over the elements of the interpretation domain under consideration. There are however situations in which one may be interested in quantifying over arbitrary properties (like subsets, or more generally, relations) on the interpretation domain. When first-order logic is extended with such a capability, we obtain what is called *second-order logic*. However, and in clear contrast with the previously mentioned extensions of first-order logic, the theoretical framework must now be considerably disturbed. On the one hand, the corresponding expressive power is greatly improved: it becomes possible to express properties that are not expressible in first-order logic. On the other hand, most of the main meta-theoretical results of first-order logic fail to hold in this extended system.

Conceptually, quantification over subsets of the interpretation domain is not different from ordinary first-order quantification. Consider the first-order formula $\forall x.\ P(x) \to P(x)$. It is valid because, no matter what interpretation domain element is assigned to x, the resultant formula is always valid. But the validity of the formula $P(x) \to P(x)$ is also independent from the interpretation of the predicate P, and it is thus conceivable to write something like $\forall P.\ \forall x.\ P(x) \to P(x)$. Second-order logic extends the language of logic in order to allow expressing such quantifications over (the possible meanings of) predicates and functions.

In concrete terms, this amounts to introducing in the syntax new classes of variables for functions and predicates of every possible arity. That is, for every positive integer n we consider:

- a countable set of function variables with arity n: $F_1^n, F_2^n, \ldots$
- a countable set of predicate variables with arity n: $X_1^n, X_2^n, \ldots$

We will omit the superscript whenever the arity of the variable is clear from the context. The term and formula formation rules are extended accordingly, i.e. when $t_1, \ldots, t_n$ are terms, $F(t_1, \ldots, t_n)$ is also a term and $X(t_1, \ldots, t_n)$ is a formula. Moreover, we consider the following additional cases for the syntax of formulas:

$$\forall F.\ \phi \qquad \exists F.\ \phi$$
$$\forall X.\ \phi \qquad \exists X.\ \phi$$

Assignments are extended to map these variables respectively to n-ary functions ($D^n \to D$) and n-ary relations ($R \subseteq D^n$). Finally, the meaning of quantified formulas with respect to an assignment α is extended as follows

$$[\![\forall F^n.\ \phi]\!]_{\mathcal{M}}(\alpha) = \mathbf{T} \quad \text{iff} \quad [\![\phi]\!]_{\mathcal{M}}(\alpha[F^n \mapsto f]) = \mathbf{T} \text{ for all } f \in D^n \to D$$

$$[\![\exists F^n.\ \phi]\!]_{\mathcal{M}}(\alpha) = \mathbf{T} \quad \text{iff} \quad [\![\phi]\!]_{\mathcal{M}}(\alpha[F^n \mapsto f]) = \mathbf{T} \text{ for some } f \in D^n \to D$$

$$[\![\forall X^n.\ \phi]\!]_{\mathcal{M}}(\alpha) = \mathbf{T} \quad \text{iff} \quad [\![\phi]\!]_{\mathcal{M}}(\alpha[X^n \mapsto R]) = \mathbf{T} \text{ for all } R \subseteq D^n$$

$$[\![\exists X^n.\ \phi]\!]_{\mathcal{M}}(\alpha) = \mathbf{T} \quad \text{iff} \quad [\![\phi]\!]_{\mathcal{M}}(\alpha[X^n \mapsto R]) = \mathbf{T} \text{ for some } R \subseteq D^n$$

The following example illustrates both the increased expressive power of second-order logic and how it fails to meet theoretical results that hold for first-order logic.

Example 4.63 Consider the following family of first-order sentences:

$$\phi_n = \exists x_1, \ldots, x_n.\ \bigwedge_{i,j \in \{1,\ldots,n\} \text{ and } i \neq j} x_i \neq x_j$$

If a model $\mathcal{M}$ validates ϕ_n, its interpretation domain has at least cardinality n.[1] This is clear when we look at concrete instances, e.g., $\phi_2 = \exists x_1, x_2. \, x_1 \neq x_2$. Hence, the (infinite) set of first-order sentences $\{\phi_2, \phi_3, \ldots\}$ characterises models with an infinite interpretation domain.

In second-order logic it is possible to characterise such a class of models with a single formula, which basically asserts that there exists an irreflexive and transitive relation whose domain is the interpretation domain itself:

$$\phi_\infty = \exists X^2. \, (\forall x_1, x_2, x_3. \, X(x_1, x_2) \wedge X(x_2, x_3) \rightarrow X(x_1, x_3))$$

$$\wedge \, (\forall x. \, \neg X(x, x)) \wedge (\forall x. \, \exists y. \, X(x, y))$$

The same effect is not possible in first-order logic: the ability to express such a formula is incompatible with an important result of first-order logic, the compactness theorem. In fact, consider the set of sentences $\Gamma = \{\neg \phi_\infty, \phi_2, \phi_3, \ldots\}$. Observe that if $\mathcal{M} \models \Gamma$, then the domain of interpretation of $\mathcal{M}$ cannot be finite (if its cardinality were n it would not satisfy ϕ_{n+1}), nor infinite (since it would not satisfy $\neg \phi_\infty$). Hence we conclude that $\Gamma \models \bot$. But clearly any finite subset $\Gamma' \subset \Gamma$ is satisfiable (we just need to choose an interpretation domain with cardinality n, with n the biggest natural number such $\phi_n \in \Gamma'$), thus contradicting the compactness property.

Other meta-theoretical results of first-order logic fail in the context of second-order logic, such as the existence of a sound and complete proof system.[2]

4.7 First-Order Theories

When judging the validity of first-order formulas we are typically interested in a particular domain of discourse, which in addition to a specific underlying vocabulary includes also properties that one expects to hold. Stated differently, we are often interested in moving away from pure logical validity (i.e. validity in all models) towards a more refined notion of validity restricted to a specific class of models.

In the next chapters of this book we will make use of first-order reasoning for verifying the correctness of programs of imperative programming languages. In the verification context, one typically needs to reason about the data types that are present in the language, which is precisely a situation as we have just described: we well be interested in a notion of validity restricted to models that capture the properties of the data structures present in the language.

---

[1] This argument assumes a fixed interpretation of equality, but can easily be rephrased without that assumption.

[2] This is an immediate corollary of Gödel's incompleteness theorem, see Sect. 4.7.2. Note however that there exists an alternative semantics of second-order logic, Henkin models, that support complete proof systems and results such as compactness. This *general semantics*, as opposed to the *standard semantics* that we have been considering, essentially reduce second-order logic to many-sorted first-order logic [9].

Of course, a natural way for specifying such a class of models is by providing a *set of axioms* (sentences that are expected to hold in them). In this section we discuss *first-order theories*, which provide a basis for the kind of reasoning just described. We will in particular discuss the theories that are more useful in the context of program verification.

Definition 4.64 Let V be a vocabulary of a first-order language.

- A first-order *theory* T is a set of V-sentences that is closed under derivability (i.e., $T \vdash \phi$ implies $\phi \in T$). A T-*structure* is a V-structure that validates every formula of T.
- A formula ϕ is T-*valid* (resp. T-*satisfiable*) if every (resp. some) T-structure validates ϕ. $T \models \phi$ denotes the fact that ϕ is T-valid. Other concepts regarding validity of first-order formulas are carried over to theories in the obvious way.
- A first-order theory T is said to be a *consistent* theory if at least one T-structure exists. A consistent theory can also be characterised as a theory such that, for every V-sentence ψ, $\{\psi, \neg\psi\} \not\subseteq T$. T is said to be a *complete theory* if, for every V-sentence ϕ, either $T \models \phi$ or $T \models \neg\phi$. T is said to be a *decidable* theory if there exists a decision procedure for checking T-validity.
- Let K be a class of V-structures. The *theory of* K, denoted by $\text{Th}(K)$, is the set of sentences valid in all members of K, i.e., $\text{Th}(K) = \{\psi \mid \mathcal{M} \models \psi$, for all $\mathcal{M} \in K\}$. Conversely, given a set of V-sentences Γ, the class of *models for* Γ is defined as $\text{Mod}(\Gamma) = \{\mathcal{M} \mid$ for all $\phi \in \Gamma, \mathcal{M} \models \phi\}$.
- A subset $\mathcal{A} \subseteq T$ is called an *axiom set* for the theory T when T is the deductive closure of $\mathcal{A}$, i.e. $\psi \in T$ iff $\mathcal{A} \vdash \psi$. A theory T is *finitely* (resp. *recursively*) *axiomatisable* if it possesses a finite (resp. recursive) set of axioms.

Whenever a theory T is axiomatisable (by a finite or recursive set of axioms $\mathcal{A}$), it makes sense to extend the first-order logic proof system $\mathcal{N}_{\text{FOL}}$ with an axiom-schema:

$$\frac{}{\Gamma \vdash \phi} \quad \text{if } \phi \in \mathcal{A}$$

Observe that the requirement that $\mathcal{A}$ be a recursive set is crucial to ensure that the applicability of these axioms can effectively be checked. Moreover, if a theory T has a recursive set of axioms, the theory itself is recursively enumerable (hence, the T-validity problem is semi-decidable).

If T is a complete theory, then any T-structure validates exactly the same set of T-sentences (the theory itself). For a given V-structure $\mathcal{M}$, the theory $\text{Th}(\mathcal{M})$ (of a single-element class of V-structures) is complete. These semantically defined theories are useful when one is interested in reasoning in some specific mathematical domain such as the natural numbers, rational numbers, etc. However, we remark that such theory may lack an axiomatisation, which seriously compromises its use in purely deductive reasoning.

On the other hand, if a theory is complete and has a recursive set of axioms, it can be shown to be *decidable* (intuitively, for checking T-validity of ϕ, we simultaneously look for ϕ and $\neg\phi$ in an enumeration for T). As an example of a non-complete theory, consider the theory axiomatised by the empty set of axioms—any satisfiable sentence ϕ that is not valid witnesses the incompleteness of the theory (since neither $\models \phi$ nor $\models \neg\phi$).

The decidability criterion for $\mathcal{T}$-validity is crucial for mechanised reasoning in the theory $\mathcal{T}$. It may be necessary to restrict the class of formulas under consideration to a suitable *fragment*; the $\mathcal{T}$-validity problem in a fragment refers to the decision about whether or not $\phi \in \mathcal{T}$ when ϕ belongs to the fragment under consideration.

One particular fragment that will be mentioned throughout this section is the fragment consisting of universal formulas, often referred to as the *quantifier-free fragment*, alluding to the fact that a formula with free variables is valid iff its universal closure is (see Exercise 4.9).

4.7.1 Equality

Equality is a binary relation and as such can be captured in first-order logic by a theory whose vocabulary $\mathcal{V}$ includes a specific binary predicate (=). The *theory of equality* $\mathcal{T}_E$ for $\mathcal{V}$ (also called *theory of equality and uninterpreted functions*, since functions are left completely unspecified) is defined by the following *axiom set*:[3]

- *reflexivity:* $\forall x \,.\, x = x$
- *symmetry:* $\forall x, y \,.\, x = y \rightarrow y = x$
- *transitivity:* $\forall x, y, z \,.\, x = y \wedge y = z \rightarrow x = z$
- *congruence for functions:* for every function $f \in \mathcal{V}$ with $\mathrm{ar}(f) = n$,

$$\forall \overline{x}, \overline{y} \,.\, x_1 = y_1 \wedge \cdots \wedge x_n = y_n \rightarrow f(x_1, \ldots, x_n) = f(y_1, \ldots, y_n)$$

- *congruence for predicates:* for every predicate $P \in \mathcal{V}$ with $\mathrm{ar}(P) = n$,

$$\forall \overline{x}, \overline{y} \,.\, x_1 = y_1 \wedge \cdots \wedge x_n = y_n \rightarrow P(x_1, \ldots, x_n) \rightarrow P(y_1, \ldots, y_n)$$

Any $\mathcal{V}$-structure that interprets the equality predicate as the equality relation in the interpretation domain clearly satisfies the axioms—these are called *normal models* (c.f. first-order logic with equality, Sect. 4.6.1). Conversely, given any $\mathcal{V}$-structure interpreting = by a relation $\sim$ and satisfying the axioms, we can always construct a normal interpretation over the equivalence classes of $\sim$ (the congruence axioms guarantee that the interpretations of functions and predicates remain well-defined). The following theorem summarises the relation between the theory of equality and normal models (first-order logic with equality).

Theorem 4.65 *A set Γ of $\mathcal{V}$-sentences is valid in all normal models if and only if $\mathcal{T}_E \models \Gamma$.*

---

[3]Note that in fact reflexivity is redundant as it follows from symmetry and transitivity.

Like unconstrained first-order logic validity, $\mathcal{T}_E$-validity is undecidable. However, the quantifier-free fragment is efficiently decidable and has found many applications in the verification of hardware and software systems.

A typical application of uninterpreted functions is to abstract complex formulas which would otherwise be difficult to reason about—if in a specific domain we treat a function symbol f in a formula ϕ as an uninterpreted function (i.e. ignoring its semantics except for congruence with respect to equality), we get an approximation formula ϕ^{UF} such that if ϕ^{UF} is valid then so is ϕ (but ϕ^{UF} may fail to be valid if validity of ϕ depends on the meaning of f).

4.7.2 Natural Numbers

Arithmetic is an important branch of mathematics, and hence a theory worth considering in first-order logic. Consider the vocabulary $\mathcal{V} = \{0, 1, +, \times, =, <\}$ with the usual arities. The *theory of natural numbers* $\mathcal{T}_\mathbb{N}$ is defined as the semantic theory $\text{Th}(\mathcal{M})$ where $\mathcal{M}$ interprets each symbol with its standard mathematical meaning in the interpretation domain $\mathbb{N}$. Being a semantic theory, $\mathcal{T}_\mathbb{N}$ is a complete theory. However it is *not axiomatisable*, not even by an infinite recursive set of axioms. This result was first proved by Kurt Gödel in his first *incompleteness theorem*, one of the most influential results of 20th century mathematics. The theorem states that no effectively generated (i.e. recursively enumerable) theory of arithmetic can be simultaneously consistent and complete.

The lack of an axiomatisation for arithmetic is a serious drawback for the programme of basing mathematical reasoning on first-order logic. In fact, without an axiomatisation it is not possible to construct a sound and complete proof system for arithmetic reasoning. The incompleteness result is indeed striking because, at the end of the 19th century, G. Peano had given a set of axioms that were shown to characterise natural numbers up to isomorphism. One of these axioms—the *axiom of induction*—involves quantification over arbitrary properties of natural numbers: "*for every unary predicate P, if $P(0)$ and $\forall n.\,P(n) \rightarrow P(n+1)$ then $\forall n.\,P(n)$*". This is of course a *second-order* axiom since it contains quantification over predicates (see Sect. 4.6.3). Nevertheless, it can be approximated by a first-order axiom scheme that replaces the arbitrary property P by a first-order formula ϕ with a free variable x:

$$\frac{\phi[0/x] \qquad \forall n.\,\phi[n/x] \rightarrow \phi[n+1/x]}{\forall n.\,\phi[n/x]}$$

It is however important to notice that in doing so we restrict reasoning to properties that are definable by first-order formulas, which can only capture a small fragment of all possible properties of natural number. Recall that the set of first-order formulas is countable while the set of arbitrary properties of natural numbers is $\mathcal{P}(\mathbb{N})$, which is uncountable.

The theory defined by the first-order formulation of the Peano axioms is named *Peano arithmetic* and is denoted by $\mathcal{T}_{PA}$. A possible axiomatisation is given in

$$\forall x . \neg(x + 1 = 0)$$

$$\forall x . x = 0 \lor \exists y . x = y + 1$$

$$\forall x, y . x + 1 = y + 1 \to x = y$$

$$\forall x, y . x \leq y \leftrightarrow \exists z . . x + z = y$$

$$\forall x . x + 0 = x$$

$$\forall x, y . x + (y + 1) = (x + y) + 1$$

$$\forall x . x \times 0 = 0$$

$$\forall x, y . x \times (y + 1) = (x \times y) + x$$

$$\phi[0/x] \land (\forall x . \phi \to \phi[x + 1/x]) \to \forall x . \phi, \quad \text{for every formula } \phi \text{ with } \mathsf{FV}(\phi) = \{x\}$$

Fig. 4.2 Peano axioms (in first-order logic with equality)

Fig. 4.2. An immediate consequence of the incompleteness theorem is that this theory is necessarily not complete. Moreover, it has also been shown to be undecidable, even for the quantifier-free fragment. These negative results are overcome when attention is restricted to the additive fragment (without multiplication), a theory known as *Presburger arithmetic*. This fragment is both complete and decidable (even if, for the quantified case, it has double-exponential time complexity).

4.7.3 Integers

What we have said for natural numbers apply equally for (possibly negative) integers. In fact, integer reasoning can be performed directly on $\mathcal{T}_{\mathbb{N}}$ with a suitable encoding. The trick is to replace each integer variable by the difference of two variables ranging over the natural numbers. Let us illustrate this with an example. Consider the formula

$$\forall x, d . \exists q, r . \neg(d = 0) \to x = q \times d + r \land 0 \leq r < d$$

where x, d, q, r range over the integers. Let us then replace x by $(x_p - x_n)$, and so on:

$$\forall x_p, x_n, d_p, d_n . \exists q_p, q_n, r_p, r_n . \neg((d_p - d_n) = 0) \to$$
$$(x_p - x_n) = (q_p - q_n) \times (d_p - d_n) + (r_p - r_n) \land$$
$$0 \leq (r_p - r_n) < (d_p - d_n)$$

After some algebraic manipulation we reach the following formula without subtraction

$$\forall x_p, x_n, d_p, d_n . \exists q_p, q_n, r_p, r_n . \neg(d_p = d_n) \to$$
$$x_p + q_p \times d_n + q_n \times d_p + r_n = x_n + q_p \times d_p + q_n \times d_n + r_p \land$$
$$r_n \leq r_p \land r_p + d_n < d_p + r_n$$

This formula is $\mathcal{T}_{\mathbb{N}}$-valid precisely when the original formula is valid in the integer interpretation. Since an encoding in the reverse direction can be obtained by simply adding a constraint $\neg(x < 0)$ for every variable ranging over the natural numbers, we conclude that the two domains have essentially the same logical strength.

Nevertheless it is useful to explicitly consider the *theory of integers* $T_{\mathbb{Z}}$, defined semantically over the vocabulary is $V = \{\ldots, -2, -1, 0, 1, 2, \ldots, +, \times, =, <\}$ with the standard mathematical interpretation in $\mathbb{Z}$. As with its natural numbers counterpart, it is complete but not axiomatisable or decidable.

For the theory of *linear arithmetic* T_{LA} we consider the vocabulary $V = \{\ldots, -2, -1, 0, 1, 2, \ldots -2\cdot, -1\cdot, 1\cdot, 2\cdot, \ldots, +, =, <\}$, where $n\cdot$ is a unary function that multiplies its argument by a constant (with this vocabulary, $x - y$ can be considered a shorthand for $x + -1 \cdot y$). This theory is both complete and decidable, and it is in fact one of the most widely used in the context of program verification.

4.7.4 Arrays

Arrays are pervasive data structures in programming. They are often modeled in logic as applicative data structures with operations $read(a, i)$ (that returns the element stored in the position with subscript i of array a) and $write(a, i, e)$ (that stores the value e in the position with subscript i of array a). The name *applicative arrays* is often used to emphasize the functional nature of these structures: note that the term $write(a, i, e)$ denotes a new array, distinct from a. These operations are constrained by the following axioms:

- *equality axioms of* T_{E};
- *array congruence*: $\forall a, i, j . i = j \rightarrow read(a, i) = read(a, j)$
- *read/write axiom 1*: $\forall a, i, j, v . i = j \rightarrow read(write(a, i, v), j) = v$
- *read/write axiom 2*: $\forall a, i, j, v . \neg(i = j) \rightarrow read(write(a, i, v), j) = read(a, j)$
- *array extensionality*: $\forall a, b . (\forall i . read(a, i) = read(b, i)) \rightarrow a = b$

The *theory of arrays* T_{A} is axiomatised by the axioms given above. It extends the theory T_{E} with axioms that capture the semantics of the *write* operation (read/write axioms). The last axiom asserts that two arrays with equal elements in all positions, are themselves equal arrays.

We remark that the axioms can be seen as an axiomatisation of arrays as mappings: $read(a, i)$ corresponds to the application of map a to the point i of its domain, and $write(a, i, e)$ to the patching $a[i \mapsto e]$. *read/write axiom 1* and *read/write axiom 2* indeed match the definition of patching (see Definition 4.9).

T_{A} shares with T_{E} the fact that it is decidable in the quantifier-free fragment but undecidable for unrestricted formulas. This is unfortunate since quantifiers are necessary to express useful array properties such as "all the elements in a given range are positive", or "the array is sorted". For this reason, alternative fragments of T_{A} are often preferred that subsume the quantifier-free fragment, while allowing syntactically restricted forms of index quantification.

4.7.5 Other Theories

Other theories that are relevant in the context of program verification are:

Theories of rational numbers: as is the case with integers, the full theory of rational numbers (with addition and multiplication) is undecidable, since the property of being a natural number can be encoded in it. But the theory of *linear arithmetic over rational numbers* T_{QLA} is decidable, and actually more efficiently than the corresponding theory of integers.

Theory of real numbers: when the interpretation domain of integer arithmetic is replaced by $\mathbb{R}$ we obtain the theories of reals $T_{\mathbb{R}}$. Surprisingly, this theory is decidable even in the presence of multiplication and quantifiers. However, the time complexity of the associated decision procedure may make its application prohibitive.

Theory of fixed-size bit vectors: fixed-sized bit vectors model bit-level operations of machine words, including 2^n-modular operations (where n is the word size), shift operations, etc. Since ultimately all high-level operations of a computing device will be implemented at the bit-level, theories of bit-vectors have attracted much attention recently. Decision procedures for the theory of fixed-sized bit vectors often rely on appropriate encodings in propositional logic.

Lists and other recursive data structures: another trend in program verification is to provide specific theories for high-level datatypes. An example of such a theory is that of *lists*, which axiomatises the behaviour of recursive lists. These theories are built around the theory of equality with uninterpreted functions, and are normally efficiently decidable for the quantifier-free case.

4.7.6 Combining Theories

A natural question that arises when dealing with theories is whether they can be combined in order to form new and richer theories.

Two vocabularies are said to be compatible if whenever a symbol is included in both vocabularies then its arity is the same in both of them. Given theories T_1 and T_2 defined over compatible vocabularies V_1 and V_2, its *union* $T_1 \cup T_2$ is defined as the theory $\mathsf{Th}(\mathsf{Mod}(T_1) \cup \mathsf{Mod}(T_2))$ with vocabulary $V_1 \cup V_2$. Clearly, $(T_1 \cup T_2)$-structures are simultaneously T_1-structures and T_2-structures (formally, the V_i-reduct of a $(T_1 \cup T_2)$-structure is a T_i-structure, for $i \in \{1, 2\}$).

As an example, consider the combination of the theory of integer linear arithmetic T_{LA} with uninterpreted functions. Can we rely on solvers for each of these individual theories? The aim of general theory combination methods is to investigate whether and how decision procedures for individual theories can be used to solve problems in the combined theory. Most methods available are based on the *Nelson-Oppen procedure*. This allows for decision procedures for theory satisfiability, in the conjunctive quantifier-free fragment of first-order logic with equality, to be combined under certain conditions, namely:

Disjoint vocabularies: no symbol is shared by the vocabularies of the theories (equality is treated as a logical symbol, since we are assuming first-order logic with equality).

Convex theories: each theory $\mathcal{T}$ being combined is such that for every finite conjunction of literals Γ and for all non-empty disjunctions of the form $\bigvee_{i \in I} x_i = y_i$, we have $\mathcal{T} \models \Gamma \rightarrow \bigvee_{i \in I} x_i = y_i$ iff $\mathcal{T} \models \Gamma \rightarrow x_i = y_i$ for some $i \in I$.

Stably-infinite theories: the theories $\mathcal{T}$ being combined are such that every $\mathcal{T}$-satisfiable quantifier-free formula has an infinite model.

Examples of convex theories include the theory of equality and uninterpreted functions $\mathcal{T}_E$, and the theory of rational linear arithmetic $\mathcal{T}_{\mathbb{Q}LA}$. On the other hand, the theory of integer linear arithmetic $\mathcal{T}_{LA}$ is not convex, since a formula such as $x_1 = 1 \wedge x_2 = 2 \wedge 1 \leq x_3 \wedge x_3 \leq 2$ clearly implies $x_1 = x_3 \vee x_2 = x_3$, but does not imply neither $x_1 = x_3$ or $x_2 = x_3$. Another example of a non-convex theory is the theory of arrays $\mathcal{T}_A$, since $read(write(a, i, x), j) = x$ implies $i = j \vee read(a, j) = x$, but does not imply either $i = j$ or $read(a, j) = x$. Non-stably-infinite theories restrict the size of their interpretations. A notable example is the theory of bit-vectors, whose models are clearly finite.

Several extensions of the Nelson-Oppen procedure relax these assumptions in different ways. As an example, relaxing the convexity requirement leads to a non-deterministic (backtracking) variant of the procedure.

4.8 To Learn More

Introductory books on logic include [9, 13, 21]. More targeted at computer science students are [3] and [14], which also treat other topics of interest not covered in this book, like binary decision diagrams and model checking. Books that cover automated theorem proving include [10–12, 20]. More oriented towards theory solving and verification, [5] gives an overview of algorithmic methods for the verification of a simple imperative language, and [15] offers a detailed account of decision procedures for theories relevant to program verification.

Concerning the definition of more expressive logics, it should be mentioned that second-order logic is not the end of the story. *Higher-order logic*, whose language admits higher-order predicates (resp. functions) that take predicates (resp. functions) as arguments, and of course quantification over these higher-order entities, would be the next step. This logic serves as the basis for the HOL family of proof assistants, of which Isabelle[4] [17] is a representative.

The Coq proof assistant[5] [4] is based on the *Calculus of Inductive Constructions*, which adds *inductive definitions* to a rich underlying type theory. See [1] for a general introduction to the relation between type systems and corresponding logics, following the Curry-Howard correspondence.

---

[4]http://isabelle.in.tum.de/.

[5]http://coq.inria.fr/.

Fig. 4.3 SMT-Solver basic
scheme

$$
\begin{aligned}
&\text{SMT-Solver } (\psi) \\
&A \leftarrow \text{prop}(\psi) \\
&\text{loop} \\
&\quad (r, \rho) \leftarrow SAT(A) \\
&\quad \text{if } r = \text{unsat then return unsat} \\
&\quad (r, \Upsilon) \leftarrow TSolver(\Phi(\rho)) \\
&\quad \text{if } r = \text{sat then return sat} \\
&\quad C \leftarrow \bigvee_{B \in \Upsilon} \neg\text{prop}(B) \\
&\quad A \leftarrow A \wedge C
\end{aligned}
$$

Following our remarks in Sect. 3.7, in the absence of decidability results the deductive method becomes essential for validity checking. Interactive *proof assistants* like those we have just mentioned are tools aimed precisely at assisting users in the construction of derivations in a given proof system.

The effort of formalising and developing proofs using a proof assistant is compensated by the ability to handle large-size problems. For such problems the proof construction process easily becomes extremely complicated; proof assistants offer "proof management" and bookkeeping facilities that may be of invaluable help in managing and reusing proofs. Furthermore, using such a tool forces the user to give the necessary attention to details that may apparently seem minor, and might be disregarded if proofs were constructed by hand.

The details of the different families of proof assistants and their internal functioning is outside the scope of this book (but see Sect. 2.3.2). [18] is a good reference on automated reasoning in general and on interactive proving in particular. The reader is also encouraged to read one of the tutorial introductions to either Isabelle or Coq, available from the respective websites. This should be sufficient to gain the minimal capabilities necessary for constructing simple derivations, as those proposed in the lists of exercises at the end of Chap. 3 and of the present chapter.

In Sect. 4.6.3 it was shown that interesting theories sometimes have efficiently decidable fragments, which indicates that automated semantics-based methods can be successfully used with these theories. Indeed a class of provers known as *SMT solvers* [15, 16] are very popular in the program verification community, either used as standalone tools or as proof engines, as part of verification frameworks. We will see in Sect. 10 that they play an important role in modern verification architectures for programs annotated with contracts.

Let us elaborate a little on the basic principles of SMT solvers. The SMT problem is a variation of the propositional SAT problem for first-order logic, with the interpretation of symbols constrained by (a combination of) specific theories. More precisely, SMT solvers address the issue of satisfiability of quantifier-free first-order CNF formulas, using as building blocks (i) a propositional SAT solver, and (ii) state-of-the-art theory solvers to address the satisfiability of sets of first-order literals, like $x + 4 * y \leq 5$ or $\neg(f(f(x-1)) = f(x+1))$.

More concretely, and following [16], let prop be a map from first-order formulas to propositional formulas that substitutes every atomic formula by a fresh propositional symbol P_i, and unprop the inverse mapping of prop. Consider a quantifier-free

CNF formula ϕ. Given a propositional valuation ρ for the formula $\mathsf{prop}(\phi)$, we let the set $\Phi(\rho)$ of first-order literals be defined as follows

$$\Phi(\rho) = \{\mathsf{unprop}(P_i) \mid \rho(P_i) = \mathbf{T}\} \cup \{\neg\mathsf{unprop}(P_i) \mid \rho(P_i) = \mathbf{F}\}$$

Intuitively, this contains the first-order literals that the valuation maps to true, and the negation of the literals that the valuation maps to false. This set will then be fed to the theory solver.

The basic SMT scheme for integration of propositional SAT solving with theory solvers is shown in Fig. 4.3. The algorithm assumes that the SAT solver invoked with a propositional CNF formula A returns whether A is satisfiable or not, and in the positive case also the valuation ρ that satisfies A. The theory solver *TSolver* invoked with a set of first-order literals Γ is expected to return whether Γ is satisfiable or not, and in the latter case also a justification, i.e. an unsatisfiable subset $\Upsilon \subseteq \Gamma$.

The main loop invokes the propositional SAT solver with a propositional formula A that is initialised with $\mathsf{prop}(\psi)$. If a valuation ρ satisfying A is found, the theory solver is invoked to check if $\Phi(\rho)$ is satisfiable. If not, it will add to A a clause which will have the effect of excluding ρ when the SAT solver is invoked again in the next iteration. The algorithm stops whenever the SAT solver returns "unsat", in which case ψ is unsatisfiable, or the theory solver returns "sat", in which case ψ is satisfiable. The literature on SMT solvers proposes many improvements on this basic scheme that fall outside the scope of the book. Let us only mention that the procedure *TSolver* must provide an adequate combination of decision procedures for the various decidable theories taken into account, as explained in Sect. 4.7.6.

SMT solvers have been applied with success in several domains in computer science and are the subject of very active research. One should refer the normative effort of the SMT-LIB initiative[6] associated with the competition SMT-COMP,[7] to provide a common input format and benchmarking framework for the evaluation and comparison of these systems, and to facilitate advances in the state of the art. Popular SMT solvers include Yices [8], CVC3 [2], Z3 [7], or Alt-Ergo [6]. A more complete list is available from the SMT-LIB website.

Finally, [19] is a recent survey on automated deduction for verification.

4.9 Exercises

4.1. Let $\mathcal{V}$ be the first-order vocabulary $\{0, +, =\}$ where 0 is a constant, $+$ a binary function and $=$ a binary predicate. Consider the following formulas (using infix notation for $+$ and $=$):
   1. $(0 + x) = x$
   2. $\forall x. (\exists y. (x + y) = z) \vee x = (y + z)$

---

[6]http://www.smtlib.org/.

[7]http://www.smtcomp.org/.

3. $\forall x, y . x = z \rightarrow y = z \rightarrow x = y$

For each formula, answer the following questions:

(a) Compute the sets of free and bound variables.

(b) Is the formula valid? Is it a contradiction?

(c) Is the formula satisfiable? In the affirmative case, provide a concrete $\mathcal{V}$-structure that validates it.

(d) Is the formula refutable? In the affirmative case, provide a concrete $\mathcal{V}$-structure that refutes it.

(e) Consider now that the equality symbol ($=$) is treated as a logical symbol (as in first-order logic with equality, see Sect. 4.6.1). Which of the previous answers are affected by this?

4.2. Prove that:

(a) If $x \notin \mathsf{Vars}(t)$, then $t[u/x] = t$.

(b) If $x \notin \mathsf{FV}(\phi)$, then $\phi[u/x] = \phi$.

4.3. Let ϕ be a first-order formula and t a term. Show that if $\mathsf{BV}(\phi) \cap \mathsf{Vars}(t) = \emptyset$, then is t *free for* x *in* ϕ, for every variable x.

4.4. Let ϕ be the formula

$$(\forall x . (\exists y . P(z, y, x)) \vee P(x, y, x)) \wedge (\forall z . P(z, z, x))$$

For each of the following substitutions:

$$\sigma_1 = [f(y)/x]$$
$$\sigma_2 = [g(z, w)/y]$$
$$\sigma_3 = [f(x)/z]$$
$$\sigma_4 = [w/y, x/z]$$

(a) Check if σ_i is *free for* ϕ.

(b) If not, perform the necessary renamings of bound variables to meet the criterion.

(c) Compute each substitution application $\phi \sigma_i$. Confirm that the bound occurrences of the result are exactly the same as in ϕ.

4.5. Prove the omitted cases from the proof of Lemma 4.22. Concretely,

(a) If for every variable $x \in \mathsf{Vars}(t)$, $\alpha(x) = \alpha'(x)$, then $[\![t]\!]_{\mathcal{M}}(\alpha) = [\![t]\!]_{\mathcal{M}}(\alpha')$.

(b) $[\![t[u/x]]\!]_{\mathcal{M}}(\alpha) = [\![t]\!]_{\mathcal{M}}(\alpha[x \mapsto [\![u]\!]_{\mathcal{M}}(\alpha)])$.

4.6. Consider a mapping $\lceil \cdot \rceil : \mathbf{Prop} \rightarrow \mathbf{Form}_{\mathcal{V}}$ and its corresponding extension to propositional formulas, as in Proposition 4.23. Let A be a propositional formula and $\lceil A \rceil$ be its first-order image (under $\lceil \cdot \rceil$).

(a) Show that if A is unsatisfiable, then $\lceil A \rceil$ is also unsatisfiable.

(b) Let ρ be a propositional valuation. The set of first-order formulas $\Phi(\rho)$ is defined as

$$\Phi(\rho) = \{\lceil P \rceil \mid P \in \mathsf{pSymbs}(A), \rho(P) = \mathbf{T}\}$$
$$\cup \{\neg \lceil P \rceil \mid P \in \mathsf{pSymbs}(A), \rho(P) = \mathbf{F}\}$$

where pSymbs(A) denotes the set of propositional symbols used in A. Show that if $[\![A]\!]_\rho = \mathbf{T}$ and the set $\Psi(\rho)$ is satisfiable, then $\lceil A \rceil$ is also satisfiable.

4.7. Prove that equivalence of first-order formulas $\equiv$ is indeed an equivalence relation (proof of Proposition 4.24). Concretely, show that for all formulas ϕ, ψ and θ,

$\phi \equiv \phi$.   (*reflexivity*)
If $\phi \equiv \psi$, then $\psi \equiv \phi$.   (*symmetry*)
If $\phi \equiv \psi$ and $\psi \equiv \theta$, then $\phi \equiv \theta$.   (*transitivity*)

4.8. Establish the following equivalences:
(a) $\forall x . \phi \wedge \psi \equiv (\forall x . \phi) \wedge (\forall x . \psi)$
(b) $\neg \exists x . \phi \equiv \forall x . \neg \phi$

4.9. Let ϕ be a first-order formula. Prove that:
(a) $\models \phi$ iff $\models \forall x . \phi$
(b) $\phi \not\equiv \forall x . \phi$ (and hence, $\phi \not\equiv \forall x . \phi$)
This exercise shows that, for formulas with free variables, logical equivalence distinguishes more formulas than validity (this is in fact the reason why some authors prefer to restrict the notions of logical equivalence and entailment to sentences).

4.10. Explain why the following equivalences do not hold. Relate the formulas using the entailment relation instead.
(a) $\forall x . \phi \vee \psi \not\equiv (\forall x . \phi) \vee (\forall x . \psi)$
(b) $\exists x . \phi \wedge \psi \not\equiv (\exists x . \phi) \wedge (\exists x . \psi)$

4.11. Assume that x does not occur free in ψ. Show that the following equivalences hold.
(a) $(\forall x . \phi) \rightarrow \psi \equiv \exists x . \phi \rightarrow \psi$
(b) $(\exists x . \phi) \rightarrow \psi \equiv \forall x . \phi \rightarrow \psi$
(c) $\psi \rightarrow (\forall x . \phi) \equiv \forall x . \psi \rightarrow \phi$
(d) $\psi \rightarrow (\exists x . \phi) \equiv \exists x . \psi \rightarrow \phi$
These equivalences are often used to convert formulas to prenex normal form, thus avoiding the intermediate NNF translation mentioned in the proof of Proposition 4.48.

4.12. Prove the cases omitted from the proof of Proposition 4.30. Specifically,
(a) Prove the following assertions, (1) and (2):

  – $\Gamma \models \neg \phi$ iff $\Gamma, \phi \models \bot$
  – $\Gamma \models \phi$ iff $\Gamma, \neg \phi \models \bot$

(b) Prove the following assertions:

  – $\Gamma \models \phi \vee \psi$ iff $\Gamma, \neg \phi, \neg \psi \models \bot$
  – $\Gamma, \phi \wedge \psi \models \theta$ iff $\Gamma, \phi, \psi \models \theta$

(c) Using the previously established assertions, prove (3) and (4):

  – $\Gamma \models \phi \wedge \psi$ iff $\Gamma \models \phi$ and $\Gamma \models \psi$
  – $\Gamma \models \phi \vee \psi$ iff $\Gamma \models \phi$ or $\Gamma \models \psi$

4.13. Compute Herbrand and Skolem normal forms for each of the following formulas:
  (a) $\exists x.\,(\forall y.\,\exists z.\,P(x,y,z)) \wedge \exists x.\,\forall y.\,\neg P(x,y,z)$
  (b) $\forall x.\,(\exists y.\,Q(x,y)) \vee \forall y.\,\exists z.\,R(x,y,z)$
  (c) $\forall x.\,(\exists y.\,P(x,y)) \rightarrow \exists y.\,Q(x,y)$
  (d) $\forall x.\,P(x) \wedge (\forall y.\,Q(y)) \rightarrow \exists y.\,R(x,f(x,y))$

4.14. Show that the following rules are admissible in $\mathcal{N}_{\mathsf{FOL}}$:
  (a) $$\dfrac{\Gamma,\forall x.\,\phi,\phi[t/x] \vdash \psi}{\Gamma,\forall x.\,\phi \vdash \psi}$$
  (b) $\dfrac{\Gamma,\phi[y/x] \vdash \psi}{\Gamma,\exists x.\,\phi \vdash \psi}$ if y does not occur free in Γ or ψ

4.15. Construct derivations in $\mathcal{N}_{\mathsf{FOL}}$ for the following sequents:
  (a) $\vdash (\exists x,y.\,R(x,y)) \rightarrow (\exists x,y.\,R(y,x))$
  (b) $\forall x.\,P(x) \wedge Q(x) \vdash (\forall x.\,P(x)) \wedge \forall x.\,Q(x)$
  (c) $\forall x.\,P(x), \exists x.\,Q(x) \vdash \exists x.\,\forall y.\,P(y) \wedge Q(x)$
  (d) $\forall x.\,R(x,x) \vdash \forall x.\,\exists y.\,R(y,x) \rightarrow R(x,y)$
  (e) $\exists x,y.\,P(x) \wedge P(y) \rightarrow R(x,y), \forall x.\,P(x) \vdash \exists x,y.\,R(y,x)$
  (f) $\vdash \neg((\exists x.\,P(x)) \rightarrow \forall x.\,P(x))$
  (g) $\vdash \exists x.\,(P(x) \rightarrow \forall y.\,P(y))$
  (h) $\forall x,y,z.\,R(x,y) \wedge R(y,z) \rightarrow R(x,z), \forall x,y.\,R(x,y) \rightarrow R(y,x) \vdash$ $\forall x.\,(\exists y.\,R(x,y)) \rightarrow R(x,x)$

4.16. Consider the extension of the $\mathcal{N}_{\mathsf{FOL}}$ system with the rules for equality given in Sect. 4.6.1. Construct derivations for the following sequents:
  (a) $\vdash \forall x.\,P(x) \rightarrow \exists y.\,x = y \wedge P(y)$
  (b) $\forall x.\,P(f(x)) \rightarrow Q(x), \forall x,y.\,Q(x) \wedge Q(y) \rightarrow x = y \vdash Q(a) \rightarrow P(f(b))$ $\rightarrow P(f(a))$

References

1. Barendregt, H.P.: Lambda calculi with types. In: Abramsky, S., Gabbay, D., Maibaum, T. (eds.) Handbook of Logic in Computer Science, vol. 2, pp. 117–310. Oxford University Press, New York (1992)
2. Barrett, C., Tinelli, C.: CVC3. In: Damm, W., Hermanns, H. (eds.) Proceedings of the 19th International Conference on Computer Aided Verification (CAV '07). Lecture Notes in Computer Science, vol. 4590, pp. 298–302. Springer, Berlin (2007)
3. Ben-Ari, M.: Mathematical Logic for Computer Science, 2nd edn. Springer, Berlin (2001)
4. Bertot, Y., Castran, P.: Interactive Theorem Proving and Program Development. Coq'Art: The Calculus of Inductive Constructions. Texts in Theoretical Computer Science. Springer, Berlin (2004)
5. Bradley, A.R., Manna, Z.: Calculus of Computation: Decision Procedures with Applications to Verification. Springer, Berlin (2007)
6. Conchon, S., Contejean, E., Kanig, J.: Ergo: A theorem prover for polymorphic first-order logic modulo theories (2006)
7. De Moura, L., Bjørner, N.: Z3: An efficient smt solver. In: Conference on Tools and Algorithms for the Construction and Analysis of Systems (TACAS), 2008
8. Dutertre, B., De Moura, L.: The Yices SMT solver. Technical report, SRI (2006)
9. Enderton, H.B.: A Mathematical Introduction to Logic, 2nd edn. Academic Press, New York (2001)

10. Gallier, J.: Logic for Computer Science: Foundations of Automatic Theorem Proving. Wiley, New York (1986). Out of print. Corrected online version available from the authors webpage http://www.cis.upenn.edu/~jean/gbooks/logic.html
11. Goubault-Larrecq, J., Mackie, I.: Proof Theory and Automated Deduction. Applied Logic Series, vol. 6. Kluwer Academic, Dordrecht (1997)
12. Harrison, J.: Handbook of Practical Logic and Automated Reasoning. Cambridge University Press, Cambridge (2009)
13. Hedman, S.: A First Course in Logic: An Introduction to Model Theory, Proof Theory, Computability, and Complexity. Oxford Texts in Logic, vol. 1. Oxford University Press, Oxford (2004)
14. Huth, M., Ryan, M.: Logic in Computer Science: Modelling and Reasoning About Systems, 2nd edn. Cambridge University Press, Cambridge (2004)
15. Kroening, D., Strichman, O.: Decision Procedures: An Algorithmic Point of View. Springer, Berlin (2008)
16. Mendonça de Moura, L., Bjørner, N.: Satisfiability modulo theories: An appetizer. In: Vinicius Medeiros Oliveira, M., Woodcock, J. (eds.) SBMF. Lecture Notes in Computer Science, vol. 5902, pp. 23–36. Springer, Berlin (2009)
17. Paulson, L.: Isabelle: A Generic Theorem Prover. Lecture Notes in Computer Science, vol. 828. Springer, Berlin (1994)
18. Robinson, J.A., Voronkov, A. (eds.): Handbook of Automated Reasoning (in 2 volumes). Elsevier and MIT Press (2001)
19. Shankar, N.: Automated deduction for verification. ACM Comput. Surv. 41(4), 1–56 (2009)
20. Smullyan, R.M.: First-Order Logic. Dover, New York (1995)
21. van Dalen, D.: Logic and Structure, 4th edn. Universitext. Springer, Berlin (2004)

Chapter 5
Hoare Logic

Hoare logic is the fundamental formalism introduced by C.A.R. Hoare in 1969 for reasoning about the correctness of imperative programs, building on first-order logic. We will now study Hoare logic for simple *While* programs.

The logic deals with the notion of correctness vis a vis a specification that consists of a *precondition* and a *postcondition*. The correctness of a program with respect to a given specification is asserted by constructing a derivation in the inference system of Hoare logic. While doing so, one must identify an *invariant* for every loop in the program.

There is a fundamental difference between the standard presentation of Hoare logic and what we do here: in our presentation loop invariants are not invented during the construction of proofs; we see them as parts of the programs, that must be given beforehand (but do not affect the operational semantics). So whereas in the standard presentation a program can be proved correct with respect to a specification if there exist adequate invariants for proving correctness, in our view a program can only be proved correct if it is *correctly annotated*.

The reason we take this view is that we are interested in automated verification, and invariants are notoriously difficult to infer automatically (this is a topic of current research). So in practice loop invariants are typically given by the programmer as an input to the program verification process. We remark that an immediate consequence of this is that our system cannot be complete in any sense: it may well be the case that a program is correct with respect to a specification, but its annotations are not, which means that a derivation for correctness with respect to that specification cannot be constructed.

This chapter also discusses the important problem of *adaptation* of specifications, since it has major implications on the design of practical verification systems based on Hoare logic.

J.B. Almeida et al., *Rigorous Software Development*,
Undergraduate Topics in Computer Science,
DOI 10.1007/978-0-85729-018-2_5, © Springer-Verlag London Limited 2011

5.1 Annotated *While* Programs

In this section we describe the general structure of the simple imperative languages
which are usually called *While* languages. Commands of these languages include a
do-nothing command, assignment, sequential composition, a while loop and (two-
branched) conditional execution.

Such languages are studied in a number of books on the semantics of programming languages (see Sect. 5.8). Two important aspects are characteristic of our presentation here:

1. We focus on the general structure of *While* languages, without fixing the underlying language of expressions. The syntax and semantics of these expressions
   are parameters of this general structure. In Sect. 5.1.2 we then go on to specify
   a concrete language based on integer-type expressions. The reason we follow
   a general approach is that this allows us to give more general results concerning the verification of programs, rather than results that apply only to a specific
   language.
2. Since we are interested in exploring methods for specifying programs and for
   proving the correctness of such programs with respect to their specifications, we
   consider that every while loop is annotated with a logical assertion, called a *loop
   invariant*. The role of invariants will be explained in Sect. 5.3. The language of
   assertions used to specify invariants, and also preconditions and postconditions
   for the programs, is given as an expansion of the language of boolean expressions, containing a few fixed connectives, first-order quantifiers, and also a set
   of user-provided functions and predicates. This user-provided vocabulary adds
   a second degree of parameterization to the language, required for verification
   purposes. This will be illustrated in Example 5.6.

   The language has a set of base types, **Type**, ranged over by τ, which include at least a Boolean type **bool** and is parameterized by a many-sorted vo-
cabulary $\mathcal{V}_{\mathsf{Prog}} = (\mathbf{Type}, \mathcal{F}^{\mathsf{Prog}}, \mathrm{rank})$ that is the union of the two vocabularies
$\mathcal{V}_{\mathsf{Exp}} = (\mathbf{Type}, \mathcal{F}^{\mathsf{Exp}}, \mathrm{rank})$, which describes the concrete syntax of program expressions, and $\mathcal{V}_{\mathsf{User}} = (\mathbf{Type}, \mathcal{F}^{\mathsf{User}}, \mathrm{rank})$, which describes the term and predicate language introduced by the programmer.

   $\mathcal{F}^{\mathsf{Prog}}$ is a **Type**-indexed set of sets $\mathcal{F}^{\mathsf{Prog}} = \{\mathcal{F}_\tau\}_{\tau \in \mathbf{Type}}$, with $\mathcal{F}_\tau$ the set of
function symbols with codomain τ, and rank is a **Type**-indexed set of mappings
$\mathrm{rank} = \{\mathrm{rank}_\tau : \mathcal{F}_\tau \to \mathbf{Type}^*\}_{\tau \in \mathbf{Type}}$, where $\mathrm{rank}_\tau : \mathcal{F}_\tau \to \mathbf{Type}^*$ assigns to each
function symbol the sorts of its arguments. We let $f^\tau, g^\tau, h^\tau, \ldots$ range over $\mathcal{F}_\tau$
with $\tau \neq \mathbf{bool}$ (the set of function symbols), and $P, Q, R, \ldots$ range over $\mathcal{F}_{\mathbf{bool}}$ (the
set of predicate symbols). Also, **bool** is not allowed as the type of an argument of a
predicate, since we are in a first-order framework.

   For each type $\tau \in \mathbf{Type}$, we assume a countable set $\mathbf{Var}_\tau$ of variables and we
let $x^\tau, y^\tau, z^\tau, \ldots$ range over $\mathbf{Var}_\tau$. Var is the **Type**-indexed set of sets $\mathbf{Var} = \{\mathbf{Var}_\tau\}_{\tau \in \mathbf{Type}}$. Variables will be used as program variables but also as logical variables, as will be seen later.

   The syntax of the language is based on the following phrase types: $\mathbf{Exp}_\tau$ is the
type of *expressions* of type τ; **Comm** is the type of *commands*; $\mathbf{Term}_\tau$ is the type

$$\mathbf{Exp}_\tau \quad \ni \quad e^\tau \quad ::= \quad x^\tau \mid f^\tau(e_1^{\tau_1}, \ldots, e_n^{\tau_n}) \text{ , for } f^\tau \in \mathcal{F}^{\mathsf{Exp}}, \text{ with } \tau \neq \mathbf{bool},$$
$$\text{and } \mathrm{rank}(f^\tau) = \langle \tau_1, \ldots, \tau_n \rangle$$

$$\mathbf{Exp_{bool}} \quad \ni \quad b \quad ::= \quad f^{\mathbf{bool}}(e_1^{\tau_1}, \ldots, e_n^{\tau_n}) \qquad \text{, for } f^{\mathbf{bool}} \in \mathcal{F}^{\mathsf{Exp}}$$
$$\text{and } \mathrm{rank}(f\mathbf{bool}) = \langle \tau_1, \ldots, \tau_n \rangle$$

$$\mathbf{Comm} \quad \ni \quad C \quad ::= \quad \mathbf{skip} \mid C_1 ; C_2 \mid x^\tau := e^\tau \mid \mathbf{if} \ b \ \mathbf{then} \ C_1 \ \mathbf{else} \ C_2 \mid$$
$$\mathbf{while} \ b \ \mathbf{do} \ \{\theta\} \ C$$

$$\mathbf{Term}_\tau \quad \ni \quad t^\tau \quad ::= \quad x^\tau \mid f^\tau(t_1^{\tau_1}, \ldots, t_n^{\tau_n}) \text{ , for } f^\tau \in \mathcal{F}^{\mathsf{Prog}}$$
$$\text{and } \mathrm{rank}(f^\tau) = \langle \tau_1, \ldots, \tau_n \rangle$$

$$\mathbf{Assert} \quad \ni \quad \phi, \psi, \theta \quad ::= \quad t^{\mathbf{bool}} \mid \top \mid \bot \mid \neg\phi \mid \phi \wedge \psi \mid \phi \vee \psi \mid \phi \rightarrow \psi \mid \phi \leftrightarrow \psi \mid$$
$$\forall x^\tau . \phi \mid \exists x^\tau . \phi$$

Fig. 5.1 Abstract syntax of the core language with assertions

of *terms* of type τ; and **Assert** is the type of *assertions*. The abstract syntax of these is defined in Fig. 5.1. Note that $\mathbf{Exp}_\tau \subseteq \mathbf{Term}_\tau$ for every type τ, since $\mathcal{V}_{\mathsf{Prog}}$ is an expansion of $\mathcal{V}_{\mathsf{Exp}}$.

We assume sequential composition to be left associative, and employ '{' and '}' in our concrete syntax to clarify precedence when necessary.[1] The type superscript τ will be dropped when clear from the context.

The syntactic class of assertions corresponds to formulas of first-order logic, and includes terms of type **bool**. Note that functions in $\mathcal{F}_{\mathbf{bool}}$ can in fact be seen as predicates; in particular, each $f^{\mathbf{bool}} \in \mathcal{F}^{\mathsf{User}}$ is a user-provided predicate.

Assertions will be used to describe the intended behaviour of a program using first-order formulas, which will be included in the program files as annotations. The current trend is to use the *same syntax* for these formulas as for the expressions of the language, which makes this annotation process much more accessible to programmers and software engineers. Note that indeed we have $\mathbf{Exp_{bool}} \subseteq \mathbf{Assert}$.

5.1.1 Program Semantics

The meaning of a grammatically correct program can be formalized in different ways. In the following we define semantic functions to interpret expressions, and an operational semantics (*natural* semantics) to describe the meaning of commands. We must also describe the meaning of the assertions embedded in the programs.

---

[1]In fact this is a *parsing* issue, related to the construction of an abstract syntax tree corresponding to a given program text. In our language the purpose of the above convention is to allow brackets to be omitted.

1. Since the language of expressions is a parameter of our general structure for *While* languages, the evaluation of programs is also parameterized by that language, since expression evaluation clearly affects program execution.
2. Program assertions are first-order formulas of a language obtained as an expansion of the boolean expressions of the programming language. The semantics of assertions thus depends on the semantics of the language of expressions used. On the other hand, since assertions may also contain occurrences of functions provided by the user, the semantics of such functions must also be given axiomatically by the user. This means that we will be reasoning in the context of a theory that is specified in part by the semantics of program expressions and in part by user-provided axioms.

Expressions of type τ are interpreted as values in the corresponding domain of interpretation D_τ. Let $D = \{D_\tau\}_{\tau \in \textbf{Type}}$ be a **Type**-indexed set of nonempty sets such that $D_{\textbf{bool}} = \{\textbf{F}, \textbf{T}\}$. The values of expressions depend on the values of variables that may occur in it. An essential notion is that of *state* which is a **Type**-indexed family of functions that map each variable of type τ into its current value. Note that the notion of state coincides with the notion of assignment given in Chap. 4 for many-sorted first-order logic.

We define the set of states (denoted Σ_D) as follows

$$\Sigma_D = \prod_{x^\tau \in \textbf{Var}} D_\tau$$

which denotes the set of functions s of domain **Var** such that $s(x^\tau) \in D_\tau$ if $x^\tau \in \textbf{Var}_\tau$.[2] Technically Σ_D is a generic function space.

We are now in the conditions of presenting the semantics of expressions, commands, and assertions. Given a many-sorted $\mathcal{V}_{\textsf{Prog}}$-structure, $\mathcal{M} = (D, I)$, the semantics of program expressions is given by a functional $[\![\cdot]\!]_\mathcal{M}$ that maps every $e^\tau \in \textbf{Exp}_\tau$ to a function $[\![e^\tau]\!]_\mathcal{M} : \Sigma_D \to D_\tau$ following Definition 4.60. We admit the following very important simplifying assumptions:

− Expressions are *free of side effects*, i.e. the evaluation of an expression is a purely functional process that does not alter the state in which the expression is evaluated.
− The meaning of an expression must be defined in every state, thus *expression evaluation does not go wrong*.

This second assumption implies that this is a very simplistic language—when machine errors occur, the language returns the error results to the user without producing any kind of error message. In Chap. 7 we will take a more realistic approach and admit the occurrence of errors.

For commands, we consider a standard operational, natural style semantics, based on an *evaluation relation*.

---

[2]This is distinguished from the set **Var** $\to D$ in that it excludes the functions that map elements of **Var**$_\tau$ to values in $D_{\tau'}$ with $\tau \neq \tau'$.

1. $(\textbf{skip}, s) \rightsquigarrow_{\mathcal{M}} s$.
2. $(x^{\tau} := e^{\tau}, s) \rightsquigarrow_{\mathcal{M}} s[x \mapsto [\![e]\!]_{\mathcal{M}}(s)]$.
3. If $(C_1, s) \rightsquigarrow_{\mathcal{M}} s'$ and $(C_2, s') \rightsquigarrow_{\mathcal{M}} s''$, then $(C_1 ; C_2, s) \rightsquigarrow_{\mathcal{M}} s''$.
4. If $[\![b]\!]_{\mathcal{M}}(s) = \textbf{T}$ and $(C_t, s) \rightsquigarrow_{\mathcal{M}} s'$, then $(\textbf{if } b \textbf{ then } C_t \textbf{ else } C_f, s) \rightsquigarrow_{\mathcal{M}} s'$.
5. If $[\![b]\!]_{\mathcal{M}}(s) = \textbf{F}$ and $(C_f, s) \rightsquigarrow_{\mathcal{M}} s'$, then $(\textbf{if } b \textbf{ then } C_t \textbf{ else } C_f, s) \rightsquigarrow_{\mathcal{M}} s'$.
6. If $[\![b]\!]_{\mathcal{M}}(s) = \textbf{T}$, $(C, s) \rightsquigarrow_{\mathcal{M}} s'$, and $(\textbf{while } b \textbf{ do } \{\theta\} C, s') \rightsquigarrow_{\mathcal{M}} s''$, then
   $(\textbf{while } b \textbf{ do } \{\theta\} C, s) \rightsquigarrow_{\mathcal{M}} s''$.
7. If $[\![b]\!]_{\mathcal{M}}(s) = \textbf{F}$, then $(\textbf{while } b \textbf{ do } C, s) \rightsquigarrow_{\mathcal{M}} s$.

Fig. 5.2 Natural semantics rules

Definition 5.1 (Evaluation relation) Let $\mathcal{M} = (D, I)$ be a $\mathcal{V}_{\mathsf{Prog}}$-structure. The relation $\rightsquigarrow_{\mathcal{M}} \subseteq \textbf{Comm} \times \Sigma_D \times \Sigma_D$ is defined inductively by the set of rules given in Fig. 5.2. $(C, s) \rightsquigarrow_{\mathcal{M}} s'$ denotes the fact that if C is executed in the initial state s, then its execution terminates, and the final state is s'.

We make the following remarks:

- If C does not terminate, then there exists no s' such that $(C, s) \rightsquigarrow_{\mathcal{M}} s'$.
- The loop invariant θ plays no role in the semantics of the loop $\textbf{while } b \textbf{ do } \{\theta\} C$.
- As usual, $s[x \mapsto r]$ denotes the state that maps x to r and any other variable y to $s(y)$.

The semantic interpretation of a command C can be seen as a partial function, since the binary relation on states induced by C satisfies the following property.

Proposition 5.2 (Determinacy) *If $(C, s) \rightsquigarrow_{\mathcal{M}} s'$ and $(C, s) \rightsquigarrow_{\mathcal{M}} s''$, then $s' = s''$.*

Proof By induction on the structure of C.   □

First-order logic is at the core of the approach to program verification that will be introduced in the present and subsequent sections. The reader may have noticed that the assertions embedded in while programs are first-order logic formulas over the vocabulary $\mathcal{V}_{\mathsf{Prog}}$. Similarly to expressions, terms and assertions are interpreted as functions from states to the corresponding domain of interpretation.

In logical terms, one needs to consider $\mathcal{V}_{\mathsf{Prog}}$-structures $\mathcal{M} = (D, I)$ which are expansions of the interpretation structure $\mathcal{M}_{\mathsf{Exp}}$ considered for program expressions. Specifically, we will be interested in structures $\mathcal{M} \in \mathsf{Mod}(\mathcal{T}_{\mathsf{Prog}})$, where $\mathcal{T}_{\mathsf{Prog}} = \mathsf{Th}(\mathcal{M}_{\mathsf{Exp}}) \cup \mathcal{T}_{\mathsf{User}}$ with $\mathcal{T}_{\mathsf{User}}$ being specified by a user-provided axiomatisation. For such an $\mathcal{M}$, the semantics of each term $t^{\tau} \in \textbf{Term}_{\tau}$ is given by the function $[\![t^{\tau}]\!]_{\mathcal{M}} : \Sigma_D \to D_{\tau}$ and the semantics of each assertion $\phi \in \textbf{Assert}$ is given by the function $[\![\phi]\!]_{\mathcal{M}} : \Sigma_D \to \{\textbf{F}, \textbf{T}\}$, following Definition 4.60.

We remark that the presence of a user-defined theory opens a door for possible inconsistencies: great care is required when writing axioms of this theory since the use of an inconsistent logical theory will of course result in a useless verification process.

$$\mathbf{Exp_{int}} \quad \ni \quad e \quad ::= \quad \dots \mid -1 \mid 0 \mid 1 \mid \dots \mid x \mid$$
$$-e \mid e_1 + e_2 \mid e_1 - e_2 \mid e_1 \times e_2 \mid e_1 \text{ div } e_2 \mid e_1 \text{ mod } e_2$$

$$\mathbf{Exp_{bool}} \quad \ni \quad b \quad ::= \quad \top \mid \bot \mid \neg b \mid b_1 \wedge b_2 \mid b_1 \vee b_2 \mid e_1 = e_2 \mid e_1 \neq e_2 \mid$$
$$e_1 < e_2 \mid e_1 \leq e_2 \mid e_1 > e_2 \mid e_1 \geq e_2$$

Fig. 5.3 Abstract syntax of program expressions of While^{int}

5.1.2 The While^{int} Programming Language

We now introduce a concrete example of a very simple language, called While^{int}, that will be used in the examples to come. The language has just two base types,

$$\mathbf{Type} \quad \ni \quad \tau \quad ::= \quad \mathbf{bool} \mid \mathbf{int}$$

and the abstract syntax of program expressions is given in Fig. 5.3. Let $\mathcal{V}_{\mathsf{Exp}}^{\mathsf{int}}$ be the vocabulary of expressions of this language. There is nothing remarkable to say about the program expressions except that for the sake of readability we adopt the usual mathematical syntax instead of using a programming language-like syntax for operators.

Expressions of type **int** are interpreted as values in $\mathbb{Z}$ and the interpretation structure $\mathcal{M}_{\mathsf{Exp}}^{\mathsf{int}}$ for $\mathcal{V}_{\mathsf{Exp}}^{\mathsf{int}}$ is the usual interpretation of integers. Operators of the language are interpreted as the corresponding operators in the semantic domain: $e_1 + e_2$ is interpreted as the integer addition of e_1 and e_2, $e_1 \leq e_2$ is interpreted as "e_1 is less than or equal to e_2", and so forth. Thus the meaning of an expression is calculated by evaluating the expression in a given state in the expected way.

Since we have fixed the interpretation structure $\mathcal{M}_{\mathsf{Exp}}^{\mathsf{int}}$ for the While^{int} language, we will drop subscripts in what follows. Moreover, we will simply write e (resp. b) to denote an integer (resp. Boolean) expression, and $(C, s) \rightsquigarrow s'$ to denote that if a While^{int}-program C is executed in the initial state s then its execution terminates in a final state s'.

Definition 5.3 (Semantics of expressions of While^{int}) The functional $[\![\cdot]\!]$ maps every $e \in \mathbf{Exp_{int}}$ to a function $[\![e]\!] : \Sigma \rightarrow \mathbb{Z}$ and every $b \in \mathbf{Exp_{bool}}$ to a function $[\![b]\!] : \Sigma \rightarrow \{\mathbf{F}, \mathbf{T}\}$

- $[\![e]\!] : \Sigma \to \mathbb{Z}$ is defined inductively by:

$$
\begin{aligned}
[\![n]\!](s) &= n \\
[\![x]\!](s) &= s(x) \\
[\![-e]\!](s) &= -[\![e]\!](s) \\
[\![e_1 + e_2]\!](s) &= [\![e_1]\!](s) + [\![e_2]\!](s) \\
[\![e_1 - e_2]\!](s) &= [\![e_1]\!](s) - [\![e_2]\!](s) \\
[\![e_1 \times e_2]\!](s) &= [\![e_1]\!](s) \times [\![e_2]\!](s) \\
[\![e_1 \operatorname{div} e_2]\!](s) &= \begin{cases} [\![e_1]\!](s) \div [\![e_2]\!](s) & \text{if } [\![e_2]\!](s) \neq 0 \\ 0 & \text{otherwise} \end{cases} \\
[\![e_1 \operatorname{mod} e_2]\!](s) &= \begin{cases} [\![e_1]\!](s) \operatorname{mod} [\![e_2]\!](s) & \text{if } [\![e_2]\!](s) \neq 0 \\ 0 & \text{otherwise} \end{cases}
\end{aligned}
$$

- $[\![b]\!] : \Sigma \to \{\mathbf{F}, \mathbf{T}\}$ is defined inductively by:

$$
\begin{aligned}
[\![\top]\!](s) &= \mathbf{T} \\
[\![\bot]\!](s) &= \mathbf{F} \\
[\![\neg e]\!](s) &= \begin{cases} \mathbf{T} & \text{if } [\![e]\!](s) = \mathbf{F} \\ \mathbf{F} & \text{if } [\![e]\!](s) = \mathbf{T} \end{cases} \\
[\![e_1 \wedge e_2]\!](s) &= \begin{cases} \mathbf{F} & \text{if } [\![e_1]\!](s) = \mathbf{F} \\ [\![e_2]\!](s) & \text{otherwise} \end{cases} \\
[\![e_1 \vee e_2]\!](s) &= \begin{cases} \mathbf{T} & \text{if } [\![e_1]\!](s) = \mathbf{T} \\ [\![e_2]\!](s) & \text{otherwise} \end{cases} \\
[\![e_1 \odot e_2]\!](s) &= [\![e_1]\!](s) \odot [\![e_2]\!](s), \quad \text{where } \odot \in \{=, \neq, <, \leq, >, \geq\}
\end{aligned}
$$

Following our simplifying assumption that expression evaluation does not go wrong, we consider here that division by zero produces a fixed integer error result (0).

Example 5.4 Let $s \in \Sigma$ be a state such that $s(x) = 5$ and $s(y) = 87$, then

- the integer expression $x + 3$ is interpreted as a function $[\![x + 3]\!] : \Sigma \to \mathbb{Z}$. It is interpreted in s as $[\![x + 3]\!](s) = [\![x]\!](s) + [\![3]\!](s) = 5 + 3 = 8$;
- the Boolean expression $(x < 10 \vee x > 20)$ is interpreted as a function $[\![x < 10 \vee x < 20]\!] : \Sigma \to \{\mathbf{F}, \mathbf{T}\}$, and thus $[\![x < 10 \vee x > 20]\!](s) = \mathbf{T}$ because $[\![x < 10]\!](s) = [\![x]\!](s) < [\![10]\!](s) = 5 < 10 = \mathbf{T}$.
- $[\![20 \operatorname{div} (x - 5)]\!](s) = 0$ since $[\![x - 5]\!](s) = 0$.
- $[\![y - (16 \times x) < x]\!](s) = 87 - 16 \times 5 < 5 = 7 < 5 = \mathbf{F}$.

Example 5.5 Let C be the command

$$
\begin{aligned}
&\textbf{if } x > 0 \textbf{ then } \{ \, x := 2 \times x \, ; \\
&\qquad\qquad\qquad \textbf{while } x < 10 \textbf{ do } \{x \leq 10\} \, x := 2 \times x \, \} \\
&\qquad\quad \textbf{else skip}
\end{aligned}
$$

and s be a state such that $s(x) = 3$. Then $(C, s) \rightsquigarrow s[x \mapsto 12]$.

In While^{int} we have fixed the vocabulary $\mathcal{V}_{\text{Exp}}$, but there is still scope to customize the user vocabulary $\mathcal{V}_{\text{User}}$. Reasoning about While^{int} programs will take place in the context of theories $\mathcal{T}_{\text{Prog}} = \text{Th}(\mathcal{M}_{\text{Exp}}^{\text{int}}) \cup \mathcal{T}_{\text{User}}$ for some user theory $\mathcal{T}_{\text{User}}$. The next example illustrates this.

Example 5.6 Suppose that we wish to encode in the logic a description of what the *factorial* of a number is. In first-order logic we do not have the means to encode inductive definitions, so all we can do is write axioms to be included in the theory $\mathcal{T}_{\text{User}}$. For example the following two axioms could be given for a predicate *isfact*$^{\text{bool}} \in \mathcal{F}^{\text{User}}$ with rank(*isfact*^{boll}) = $\langle \textbf{int}, \textbf{int} \rangle$

$$isfact(0, 1) \tag{5.1}$$

$$\forall n, r.\, n > 0 \rightarrow isfact(n - 1, r) \rightarrow isfact(n, n \times r) \tag{5.2}$$

Note that there is no guarantee from this axiomatisation that the validity of a formula *isfact*(n, r) implies that r is indeed the factorial of n. We would like to be able to work with the weakest predicate that satisfies these axioms.

It is possible to ensure that the factorial of a number is unique, by additionally declaring a factorial function *fact*$^{\text{int}} \in \mathcal{F}^{\text{User}}$ with rank(*fact*^{int}) = $\langle \textbf{int} \rangle$, and adding the axioms

$$\forall n.\, isfact(n, fact(n)) \tag{5.3}$$

$$\forall n, r.\, isfact(n, r) \rightarrow r = fact(n) \tag{5.4}$$

Alternatively, the factorial function could also be formalised directly without using the binary predicate *isfact*, by writing axioms with the equality predicate as follows

$$fact(0) = 1 \tag{5.5}$$

$$\forall n.\, n > 0 \rightarrow fact(n) = n \times fact(n - 1) \tag{5.6}$$

One last note: it may also be useful to have predicates that are simply defined (like macros); this may improve the conciseness and readability of assertions. For instance a predicate that tests if a number is even can be defined as

$$isEven(x) \stackrel{\text{def}}{=} x \bmod 2 = 0$$

Such definitions are also typically allowed by program verification tools.

A second example of a concrete programming language will be given in Sect. 5.5, where While^{int} will be extended with integer array variables.

5.2 Specifications and Hoare Triples

The correctness of a program is always defined relative to a given *specification* for that program. The basic ingredients to build specifications are *preconditions*—assertions that *are assumed* to hold when execution of the program is started—and *postconditions*—assertions that *must* hold when execution stops. A specification can thus be written as a pair of assertions (ϕ, ψ), where ϕ is the precondition and ψ is the postcondition.

Correctness properties come in two different strengths:

– A *total correctness* property for a program C relative to specification (ϕ, ψ) has the following meaning: if ϕ holds in a given state and C is executed in that state, then execution of C *will stop*, and moreover ψ will hold in the final state of execution.

– A *partial correctness* property for a program C relative to specification (ϕ, ψ) has the meaning that if ϕ holds in a given state and C is executed in that state, then either execution of C does not stop, or *if it does*, ψ will hold in the final state.

Let us start with an example of reasoning about the behaviour of a program using assertions.

Example 5.7 Consider the following While^{int}-program that sorts two numbers stored in variables x and y: the input values will be stored in z and w on exit, in such a way that $z \leq w$ holds. Let **sort2** be

$$\textbf{if } x \leq y \textbf{ then } \{ z := x \,;\, w := y \} \textbf{ else } \{ w := x \,;\, z := y \}$$

The program is so simple that we can affirm with a high degree of confidence that it satisfies the informal specification stated above. But let us try to present an argument for this.

First of all, we need to write an adequate formal specification. We have not mentioned any conditions under which the program will be executed, and indeed it looks like it will always sort two numbers. This means that the precondition in our specification is simply $\top$, the weakest of all possible assertions. The postcondition, on the other hand, can be written as follows

$$z \leq w \wedge ((z = x \wedge w = y) \vee (w = x \wedge z = y))$$

So our goal is now to reason about the above program in order to prove that it is correct with respect to the specification

$$(\top, z \leq w \wedge ((z = x \wedge w = y) \vee (w = x \wedge z = y)))$$

Such reasoning always follows the structure of the program. The precondition $\top$ is not informative, but once we enter the conditional command, we have more information available: the first (resp. second) branch corresponds to the situation in which the Boolean condition is true (resp. false). This allows us to strengthen the initial precondition. Our correctness proof can be decomposed in the proofs of the two following correctness properties, which must *both* be pursued, since they correspond to two possible executions of the program:

1. the program $z := x \,;\, w := y$ is correct with respect to the specification $(x \leq y, z \leq w \wedge ((z = x \wedge w = y) \vee (w = x \wedge z = y)))$

2. the program $w := x$; $z := y$ is correct with respect to the specification $(x > y, z \leq w \wedge ((z = x \wedge w = y) \vee (w = x \wedge z = y)))$

Let us continue our reasoning for the first of these proofs. When the command $z := x$ is executed in a state in which the precondition $x \leq y$ holds, then clearly $z \leq y \wedge z = x$ will hold after this assignment, so we reach the next assignment command with a stronger precondition. It is now sufficient to prove that

− the program $w := y$ is correct with respect to the specification $(z \leq y \wedge z = x, z \leq w \wedge ((z = x \wedge w = y) \vee (w = x \wedge z = y)))$.

After execution of the assignment the assertion $z \leq w \wedge z = x \wedge w = y$ will hold (note that we choose to keep $z = x$ in this condition, since it is part of the postcondition of the program).

Program execution is now completed, and we have arrived at a postcondition for our program. It is still necessary to prove that the postcondition $z \leq w \wedge z = x \wedge w = y$ is stronger than the one that is present in the specification, $z \leq w \wedge ((z = x \wedge w = y) \vee (w = x \wedge z = y))$. It is clearly true that the former implies the latter.

The proof corresponding to the other branch of the conditional command is very similar; together these proofs give evidence that the program **sort2** is correct with respect to its specification. This is true irrespectively of whether this is taken to be a partial or a total correctness specification, since this program will always terminate.

It is immediately clear from this small example that this kind of reasoning needs to be performed in a more formal setting in order to be mechanised. In Sect. 5.4 we will introduce an inference system that will allow us to reason rigorously about the correctness of programs; we remark that although in the above example we have reasoned by propagating preconditions forward until no commands were left to be executed (we have informally performed a quite intuitive *symbolic execution* of the program), the standard way to reason about correctness is to start from postconditions and to propagate them backwards until the first command is reached. This will become clear when the inference system is studied.

To be able to reason formally about correctness, we add to the abstract syntax of Fig. 5.1 two new classes of formulas called *Hoare triples* as follows

$$\textbf{Spec} \ni S ::= \{\phi\} C \{\psi\} \mid [\phi] C [\psi]$$

These formulas correspond to *partial* and *total* correctness respectively: the Hoare triple $\{\phi\} C \{\psi\}$ is valid when the program C is partially correct with respect to the specification (ϕ, ψ), and $[\phi] C [\psi]$ is valid when C is totally correct with respect to the specification (ϕ, ψ). This is stated formally by defining the semantic interpretation of Hoare triples.

Definition 5.8 (Semantics of Hoare triples) Let $\mathcal{M} = (D, I)$ be a $\mathcal{V}_{\mathsf{Prog}}$-structure.

- The semantics of a partial correctness Hoare triple $\{\phi\}\,C\,\{\psi\}$ is given by a function $[\![\{\phi\}\,C\,\{\psi\}]\!]_{\mathcal{M}} : \Sigma_D \to \{\mathbf{F}, \mathbf{T}\}$ defined as follows

$$[\![\{\phi\}\,C\,\{\psi\}]\!]_{\mathcal{M}}(s) = \mathbf{T} \quad \text{iff} \quad \text{if } [\![\phi]\!]_{\mathcal{M}}(s) = \mathbf{T} \text{ and } (C,s) \rightsquigarrow_{\mathcal{M}} s',$$
$$\text{then } [\![\psi]\!]_{\mathcal{M}}(s') = \mathbf{T}$$

- The semantics of a total correctness Hoare triple $[\phi]\,C\,[\psi]$ is given by a function $[\![[\phi]\,C\,[\psi]]\!]_{\mathcal{M}} : \Sigma_D \to \{\mathbf{F}, \mathbf{T}\}$ defined as follows

$$[\![[\phi]\,C\,[\psi]]\!]_{\mathcal{M}}(s) = \mathbf{T} \quad \text{iff} \quad \text{if } [\![\phi]\!]_{\mathcal{M}}(s) = \mathbf{T}, \text{ then there exists } s' \text{ such that}$$
$$(C,s) \rightsquigarrow_{\mathcal{M}} s' \text{ and } [\![\psi]\!]_{\mathcal{M}}(s') = \mathbf{T}$$

Notice that $[\![\{\phi\}\,C\,\{\psi\}]\!]_{\mathcal{M}}(s) = \mathbf{T}$ whenever $[\![\phi]\!]_{\mathcal{M}}(s) = \mathbf{F}$ or whenever there is no $s' \in \Sigma_D$ such that $(C,s) \rightsquigarrow_{\mathcal{M}} s'$. That is why $\{\phi\}\,C\,\{\psi\}$ expresses partial correctness. Termination is required for $[\![[\phi]\,C\,[\psi]]\!]_{\mathcal{M}}(s)$ to be true.

Analogously to what was done for first-order logic, let us now introduce the concepts of validity and logical consequence for Hoare triples in the context of a program vocabulary $\mathcal{V}_{\mathsf{Prog}}$.

Definition 5.9

- A Hoare triple $\{\phi\}\,C\,\{\psi\}$ is said to be *valid in a* $\mathcal{V}_{\mathsf{Prog}}$-*structure* $\mathcal{M} = (D, I)$, denoted $\mathcal{M} \models \{\phi\}\,C\,\{\psi\}$, if $[\![\{\phi\}\,C\,\{\psi\}]\!]_{\mathcal{M}}(s) = \mathbf{T}$ for all states $s \in \Sigma_D$.
- A Hoare triple $\{\phi\}\,C\,\{\psi\}$ is said to be *logically valid*, denoted $\models \{\phi\}\,C\,\{\psi\}$, if $\mathcal{M} \models \{\phi\}\,C\,\{\psi\}$ for all $\mathcal{V}_{\mathsf{Prog}}$-structures $\mathcal{M}$.
- A Hoare triple $\{\phi\}\,C\,\{\psi\}$ is said to be a *logical consequence* of a set of assertions Γ, denoted $\Gamma \models \{\phi\}\,C\,\{\psi\}$, if $\mathcal{M} \models \{\phi\}\,C\,\{\psi\}$ whenever $\mathcal{M} \models \Gamma$.

Similar concepts can be introduced for total correctness triples in the obvious way. Observe that the semantics of a Hoare triple do not depend on the annotations contained in the program.

It is important to distinguish different categories of variables that may occur in a Hoare triple.

- First of all, the presence of quantifiers in assertions imposes the usual notions of free and bound variables. Given a triple $\{\phi\}\,C\,\{\psi\}$, the variables that are bound by some quantifier in ϕ or ψ, or in annotations contained in C will be called *logical variables*, and can of course be freely renamed as far as variable capture is avoided.
- Variables that are used in the program C are called *program variables*. By this we mean that they occur in actual commands of C, not just in loop annotations, although in fact program variables can also occur (*free*) in annotations, as well as in the precondition ϕ and postcondition ψ.
- Variables that occur free in ϕ or ψ (and possibly in annotations inside C) but do not occur as program variables in C are called *auxiliary variables*.

Auxiliary variables play an important role in specifications. To understand why they are useful, consider the following alternative version of the While^{int}-program given at the beginning of this section. Let **sort2$'$** be

$$\textbf{if } x \leq y \textbf{ then skip else } \{z := y; \; y := x; \; x := z\}$$

Clearly this is still a program that sorts the pair of numbers stored in x and y; the difference with respect to the previous version is that the output of the program is stored in the very same variables that contained the input, so in the postcondition one has to be able to refer to the value of a variable in both the initial state and the final state. The way to achieve this is to use auxiliary variables to record (or *freeze*) the initial values of the input variables, and then use them in the postcondition. The specification of **sort2$'$** can be written as

$$(x = x_0 \wedge y = y_0, x \leq y \wedge ((x = x_0 \wedge y = y_0) \vee (y = x_0 \wedge x = y_0)))$$

This only works because auxiliary variables are *not allowed to occur in the program*—if this was not the case, they could not be guaranteed to record the initial values of x and y to be used in the postcondition.

Note that if a variable occurs in a triple both as program or auxiliary variable and as logical variable, the latter variable can always be renamed to avoid confusion.

5.3   Loop Invariants

You have probably met the notion of *invariant* before. This key concept is prevalent in computer science: an invariant of a program C is any assertion θ such that the triple $\{\theta\} C \{\theta\}$ is valid, i.e. it is preserved by executions of the program.

For instance the assertion $x > 0$ is an invariant of the program **sort2**: the Hoare triple $\{x > 0\} \textbf{ sort2} \{x > 0\}$ since the program does not modify the value of x. The same is true of the assertion $x \leq 0$. Note however that neither of these assertions is an invariant of program **sort2$'$**.

A *loop invariant* is any property that is preserved by executions of the loop's *body*. Since these executions may only take place when the loop condition is true, an invariant of the loop **while** b **do** C is any assertion θ such that $\{\theta \wedge b\} C \{\theta\}$ is valid, in which case of course it also holds that $\{\theta\}$ **while** b **do** $C \{\theta\}$ is valid: if the truth of θ is preserved by individual executions of the loop body, then it is also preserved by any finite sequence of such executions. Note that an assertion may be an invariant of the program **while** b **do** C without actually being a loop invariant (see Exercise 5.4).

Observe that the validity of $[\theta \wedge b] C [\theta]$ *does not* imply the validity of $[\theta]$ **while** b **do** $C [\theta]$. Proving such a total correctness specification implies proving the *termination* of the loop, which requires the use of specific techniques—loop invariants are inappropriate. The required notion here is a quantitative one: a *loop variant* is any numerical expression whose value is initially positive and strictly decreases from iteration to iteration. The existence of a valid variant for a given loop implies the termination of all possible executions of the loop. We will return to this issue in Sect. 5.6.

Used informally, loop invariants are an essential tool in algorithm design. Ideally, the identification of appropriate invariants should be done as the programs are written; invariants should play a dual role as guiding principles for the design of algorithms and as code documentation. From our point of view in this book, invariants must be provided as inputs to the program verification process (this is why we work with a language of annotated programs). Common verification tools will not only check the validity of loop invariants but also use them to help assert the correctness of programs.

It is time to illustrate informally the use of loop invariants and variants in proving the correctness of a program.

Example 5.10 Consider the following While^{int}-program for calculating the factorial of n. Let **fact** be

$$f := 1 \,;\, i := 1 \,;$$
$$\textbf{while } i \leq n \textbf{ do} \{f = fact(i-1) \wedge i \leq n+1\}\{$$
$$\quad f := f \times i \,;$$
$$\quad i := i+1$$
$$\}$$

The value of i is incremented at the end of each iteration, and the value of variable f at the beginning of an iteration is always the factorial of the previous value of i, so that when the loop finishes f contains the factorial of n.

Loop invariants were explicitly included in the program syntax introduced in Sect. 5.1. They play no role in the execution semantics of programs, and are used simply as program *annotations* to be used in verification. In this example the loop has been annotated with an invariant that captures the above description rigorously. The invariant states that when execution of an arbitrary iteration of the loop begins, the value of variable f is equal to the factorial of $i-1$, and moreover the value of i cannot be greater than $n+1$ (but it can be equal to $n+1$, since the invariant must also be true on exit, when the loop condition fails).

What would be a suitable specification of a program that calculates the factorial of n? The formalisation of factorial was already given as Example 5.6, so we may now simply use either the predicate *isfact* or the function *fact* to write a specification. Clearly the postcondition is simply $f = fact(n)$, and the precondition that we require just states that n must be non-negative. So we want to prove the validity of the following Hoare triple (we start by considering a partial correctness specification):

$$\{n \geq 0\} \, \textbf{fact} \, \{f = fact(n)\}$$

Proving the correctness of this program is immediate after the invariant has been identified. Note that when the loop terminates the Boolean condition no longer holds, thus $i > n$. But the condition $i \leq n+1$ which is part of the invariant still holds, and thus $i = n + 1$ must hold. Substituting in the other invariant conjunct, we have $f = fact(n + 1 - 1) = fact(n)$, which is the postcondition sought.

We have proved that if the invariant is valid then the program is correct. But it still remains to prove that it *is* valid. This is an inductive proof on the length of the sequence of executions of the loop body, that requires proving two things.

1. Initial conditions: the invariant is valid at the beginning of the first iteration.
2. Preservation: if the invariant is assumed to be valid at the beginning of an arbitrary iteration, then it will be valid at the beginning of the next iteration.

In our example, the initial conditions are $f = 1$ and $i = 1$, and indeed $fact(1 - 1) = 1$. Moreover the precondition ensures $n \geq 0$, thus $1 \leq n + 1$ so the invariant is initially valid. Moreover it is preserved by executions of the loop body: if we assume $f = fact(i - 1)$ at the beginning of some iteration, and let f', i' denote the values of variables f and i at the beginning of the next iteration, then it holds that $i' = i + 1$ and $f' = f \times i = fact(i - 1) \times i = fact(i' - 2) \times (i' - 1)$. And by definition of factorial, $f' = fact(i' - 1)$, so the first part of the invariant has been preserved. For the second part, notice that an iteration will only be executed if the loop condition $i \leq n$ holds, in which case $i' = i + 1 \leq n + 1$.

This concludes the proof of partial correctness: if the loop (and the program) terminates, then the value of $fact(n)$ is calculated; consequently the Hoare triple $\{n \geq 0\}$ **fact** $\{f = fact(n)\}$ is indeed valid. A proof of total correctness is also easy to obtain: it suffices to additionally prove that the loop, and thus the program, always terminates. We note that since i is initialized with 1 and incremented in each iteration, it will eventually be greater than the value of n, which is not modified. We can give as evidence of termination the loop variant $n - i$, which strictly decreases in each iteration, since $n - i - (n - i') = n - i - (n - (i + 1)) = 1 > 0$.

You may be wondering why invariants are necessary at all. Why not simply apply to loops the kind of reasoning based on propagating preconditions that was illustrated in Example 5.7? It is well-known that the behaviour of a loop can be understood by unfolding it, using the following equivalence (that can be proved using the operational semantics of the language)

$$\textbf{while } b \textbf{ do } C \equiv \textbf{if } b \textbf{ then } \{C\,;\ \textbf{while } b \textbf{ do } C\} \textbf{ else skip}$$

So for instance, to calculate the factorial of 10 the loop in our example program would behave equivalently to 11 nested conditional commands, so it would apparently suffice to propagate conditions along this sequence of commands to check the correctness of the program without having to resort to loop invariants. At the beginning of the first iteration the condition $f = 1 \wedge i = 1$ would be valid; at the beginning of the second iteration these conditions would have been propagated as $f = 1 \wedge i = 2$, and so on.

The number of iterations may be huge, but it could be argued that this is only a small problem since we are considering mechanised, computer-assisted verification. The problem lies elsewhere: our goal is to reason about programs *statically*, considering all possible executions, and not dynamically, for a single execution trace. The kind of reasoning just outlined is not feasible since it is not in general possible to determine statically the number of iterations of a given loop that will be executed—in the factorial example, suppose that **fact** is a subroutine invoked by a main program; then one would have to be able to determine every value of n with which the subroutine is called, without running the program.

Ideally one would start a verification with a specification that indeed captures the informal specification one has in mind (let us call this a *validated* specification), and a (possibly incorrect) program that has been annotated with adequate invariants. By *adequate* we mean that the assertions given as invariants should not only indeed be invariants, but they should also be *sufficiently strong* to allow for the program to be verified. For instance in the factorial example, if one took the invariant to be simply $i \leq n + 1$, this would certainly be insufficiently strong and correctness would not be established.

Remark 5.11 There is a problem with the specification of factorial used in the previous example: although our program does compute factorial, the specification $(n \geq 0, f = fact(n))$ admits many other solutions that do not calculate the factorial of n. Any program can trivially meet the specification by simply modifying the value of the input variable n. For instance the triple $\{n \geq 0\} n := 0; \ f := 1 \{f = fact(n)\}$ is a valid Hoare triple, whatever the initial value of n may be.

The problem can be avoided by using an *auxiliary variable* to record the initial value of n, as mentioned in Sect. 5.2. The specification

$$(n \geq 0 \wedge n = n_0, f = fact(n_0))$$

solves the problem as long as n_0 is indeed auxiliary, i.e. it is not used as a program variable. A proof of correctness with respect to this specification will be given in Sect. 6.1.

Another possibility would be the following specification, which explicitly forces the preservation of the value of n.

$$(n \geq 0 \wedge n = n_0, f = fact(n) \wedge n = n_0)$$

5.4 Hoare Calculus

Program verification can be succinctly described as the activity that attempts to establish the correctness (or otherwise) of programs with respect to specifications, which in the setting introduced in this chapter amounts to studying the validity of Hoare triples. We now describe a theoretical tool that is widely used for this purpose.

The Hoare calculus is an inference system for reasoning about Hoare triples. Its rules are given in Fig. 5.4. We will in the rest of the book refer to it as system $\mathcal{H}$. This system describes in a formal way the use of preconditions, postconditions, and invariants in order to verify programs, and is usually described as an alternative semantics of the underlying programming language. This style of semantics is known as *axiomatic*.

$$(skip) \quad \overline{\{\phi\} \, \textbf{skip} \, \{\phi\}} \qquad (assign) \quad \overline{\{\psi[e/x]\} \, x := e \, \{\psi\}}$$

$$(seq) \quad \frac{\{\phi\} \, C_1 \, \{\theta\} \qquad \{\theta\} \, C_2 \, \{\psi\}}{\{\phi\} \, C_1 \, ; \, C_2 \, \{\psi\}}$$

$$(while) \quad \frac{\{\theta \wedge b\} \, C \, \{\theta\}}{\{\theta\} \, \textbf{while} \, b \, \textbf{do} \, \{\theta\} \, C \, \{\theta \wedge \neg b\}}$$

$$(if) \quad \frac{\{\phi \wedge b\} \, C_t \, \{\psi\} \qquad \{\phi \wedge \neg b\} \, C_f \, \{\psi\}}{\{\phi\} \, \textbf{if} \, b \, \textbf{then} \, C_t \, \textbf{else} \, C_f \, \{\psi\}}$$

$$(conseq) \quad \frac{\{\phi\} \, C \, \{\psi\}}{\{\phi'\} \, C \, \{\psi'\}} \quad \text{if } \phi' \to \phi \text{ and } \psi \to \psi'$$

Fig. 5.4 Inference system of Hoare logic: system $\mathcal{H}$

Before proceeding, note that the (*conseq*) rule has a side condition stating that two first-order formulas must hold. This means that this rule relates the Hoare triples to the first-order logic formulas over the vocabulary under consideration.

Definition 5.12 Let $\Gamma \subseteq$ **Assert** and $\{\phi\} \, C \, \{\psi\} \in$ **Spec**. We say that $\{\phi\} \, C \, \{\psi\}$ is *derivable in $\mathcal{H}$ assuming Γ*, denoted $\Gamma \vdash_{\mathcal{H}} \{\phi\} \, C \, \{\psi\}$, if there exists a proof of $\{\phi\} \, C \, \{\psi\}$ in $\mathcal{H}$ which uses as assumptions (for the side conditions) assertions from Γ.

Let us now examine in turn each rule of the system. The first five rules concern each of the program constructs; we will call them *program rules*.

— The axiom for **skip** states that this command preserves the truth of assertions. Assertions that are true before execution will be true after execution.
— Assignment commands are also handled by an axiom, which is indeed one of the crucial elements of this system. This axiom states that a postcondition ψ can be ensured for an assignment $x := e$ by taking the assertion that results from substituting e for x in ψ as precondition. A trivial example in our While^{int} language is the triple $\{x + 1 > 10\} \, x := x + 1 \, \{x > 10\}$.

   This axiom has a very important consequence: this is a system that propagates postconditions backwards, i.e. it calculates preconditions from postconditions. We will return to this point later.
— The rule for a sequence of commands introduces an intermediate assertion. A Hoare triple $\{\phi\} \, C_1 \, ; \, C_2 \, \{\psi\}$ for the composed command can be inferred if there exists an assertion that can be used as postcondition for C_1 (with precondition ϕ) and also as precondition for C_2 (with postcondition ψ).

- The inference rule for while-loops formalises the discussion in Sect. 5.3: it states that in order to prove that the assertion θ is a loop invariant, one needs to prove that it is preserved by the *body* of the loop when the loop condition also holds as a precondition. Additionally, the rule states the obvious fact that the negation of the loop condition is a postcondition of the loop.
- The command **if** b **then** C_t **else** C_f with precondition ϕ has ψ as postcondition if it can be inferred that C_t and C_f have the same postcondition with the preconditions $\phi \wedge b$ and $\phi \wedge \neg b$ respectively. This is in accordance with the fact that C_t is carried out exactly when b holds and C_f when $\neg b$ holds.

Unlike program rules, the (*conseq*) rule can be applied to infer triples containing programs of any shape. In particular it allows us to infer a Hoare triple from another triple containing the same program, in which either the precondition has been weakened or the postcondition has been strengthened. An immediate consequence of this is that derivations for a given goal are not unique. The utility of the (*conseq*) rule will become clear by looking at an example.

We remark that given a vocabulary $\mathcal{V}_{\mathsf{Prog}}$, the Hoare calculus is not designed for a specific $\mathcal{V}_{\mathsf{Prog}}$-structure, but includes only rules which are valid in all models. However, pragmatically, we are interested in Hoare triples which are valid in a class of models induced by a specific theory, as discussed in Sect. 5.1.1. This means that the first-order formulas that appear in the side condition of the (*conseq*) rule should hold in those models.

The following example illustrates these points. Throughout this chapter a numbered, indented style is used for displaying proof trees. For each goal, the name of the rule applied at that point is shown between brackets, and its sub-goals are shown below with higher indentation.

Example 5.13 Consider the $\mathsf{While}^{\mathsf{int}}$-code fragment

$$\textbf{if } x < 0 \textbf{ then } x := x + 100 \textbf{ else skip};$$
$$y := 2 \times x$$

We want to prove that if the initial value of x stands between -100 and 100, then the final value of y will not be negative, and will not be greater than 300. The derivation in Fig. 5.5 has as conclusion a Hoare triple that expresses this specification formally.

The above construction, while syntactically well-formed, is not guaranteed to be an actual proof tree until the side conditions (corresponding to three instances of the (*conseq*) rule) have been checked. It turns out that these are straightforward in the model $\mathcal{M}_{\mathsf{Exp}}^{\mathsf{int}}$ (or any expansion of it, in the presence of a user-provided theory).

1.1. $x \geq -100 \wedge x \leq 100 \wedge x < 0 \rightarrow x + 100 \geq 0 \wedge x + 100 \leq 150$ and $x \geq 0 \wedge x \leq 150 \rightarrow x \geq 0 \wedge x \leq 150$

1.2. $x \geq -100 \wedge x \leq 100 \wedge \neg(x < 0) \rightarrow x \geq 0 \wedge x \leq 150$ and $x \geq 0 \wedge x \leq 150 \rightarrow x \geq 0 \wedge x \leq 150$

2.1. $x \geq 0 \wedge x \leq 150 \rightarrow 2 \times x \geq 0 \wedge 2 \times x \leq 300$ and $y \geq 0 \wedge y \leq 300 \rightarrow y \geq 0 \wedge y \leq 300$

(in general we will use the numbering of the nodes of a proof tree to associate side conditions with rule instances).

$\{x \geq -100 \wedge x \leq 100\}$
if $x < 0$ **then** $x := x + 100$ **else skip**; $y := 2 \times x$
$\{y \geq 0 \wedge y \leq 300\}$

(*seq*)

1. $\{x \geq -100 \wedge x \leq 100\}$ **if** $x < 0$ **then** $x := x + 100$ **else skip** $\{x \geq 0 \wedge x \leq 150\}$ (*if*)
   1.1. $\{x \geq -100 \wedge x \leq 100 \wedge x < 0\} x := x + 100 \{x \geq 0 \wedge x \leq 150\}$ (*conseq*)
      1.1.1. $\{x + 100 \geq 0 \wedge x + 100 \leq 150\} x := x + 100 \{x \geq 0 \wedge x \leq 150\}$ (*assign*)
   1.2. $\{x \geq -100 \wedge x \leq 100 \wedge \neg(x < 0)\}$ **skip** $\{x \geq 0 \wedge x \leq 150\}$ (*conseq*)
      1.2.1. $\{x \geq 0 \wedge x \leq 150\}$ **skip** $\{x \geq 0 \wedge x \leq 150\}$ (*skip*)
2. $\{x \geq 0 \wedge x \leq 150\} y := 2 \times x \{y \geq 0 \wedge y \leq 300\}$ (*conseq*)
   2.1. $\{2 \times x \geq 0 \wedge 2 \times x \leq 300\} y := 2 \times x \{y \geq 0 \wedge y \leq 300\}$ (*assign*)

Fig. 5.5 Example of Hoare logic derivation

$\{x \geq -100 \wedge x \leq 100\}$
if $x < 0$ **then** $x := x + 100$ **else skip**; $y := 2 \times x$
$\{y \geq 0 \wedge y \leq 300\}$

(*conseq*)

1. $\{x \geq -100 \wedge x \leq 100\}$
   if $x < 0$ **then** $x = x + 100$ **else skip**; $y := 2 \times x$
   $\{y \geq 0 \wedge y \leq 200\}$
   (*seq*)
   1.1. $\{x \geq -100 \wedge x \leq 100\}$ **if** $x < 0$ **then** $x := x + 100$ **else skip** $\{x \geq 0 \wedge x \leq 100\}$ (*if*)
      1.1.1. $\{x \geq -100 \wedge x \leq 100 \wedge x < 0\} x := x + 100 \{x \geq 0 \wedge x \leq 100\}$ (*conseq*)
         1.1.1.1. $\{x + 100 \geq 0 \wedge x + 100 \leq 100\} x := x + 100 \{x \geq 0 \wedge x \leq 100\}$ (*assign*)
      1.1.2. $\{x \geq -100 \wedge x \leq 100 \wedge \neg(x < 0)\}$ **skip** $\{x \geq 0 \wedge x \leq 100\}$ (*conseq*)
         1.1.2.1. $\{x \geq 0 \wedge x \leq 100\}$ **skip** $\{x \geq 0 \wedge x \leq 100\}$ (*skip*)
   1.2. $\{x \geq 0 \wedge x \leq 100\} y := 2 \times x \{y \geq 0 \wedge y \leq 200\}$ (*conseq*)
      1.2.1. $\{2 \times x \geq 0 \wedge 2 \times x \leq 200\} y := 2 \times x \{y \geq 0 \wedge y \leq 200\}$ (*assign*)

Fig. 5.6 A second derivation for the same goal

Let us now look at a second derivation for the same goal, shown in Fig. 5.6. It is easy to see that the final value of y can never be greater than 200; the first step in this derivation is an application of the (*conseq*) rule that strengthens the postcondition to this tighter bound. In this second derivation the instances of the (*conseq*) rule introduce the following side conditions (one additional condition exists for each rule, which we omit since it is trivial). Again, all of these are valid in the models under consideration.

    0. $y \geq 0 \wedge y \leq 200 \rightarrow y \geq 0 \wedge y \leq 300$
1.1.1. $x \geq -100 \wedge x \leq 100 \wedge x \leq 0 \rightarrow x + 100 \geq 0 \wedge x + 100 \leq 100$
1.1.2. $x \geq -100 \wedge x \leq 100 \wedge \neg(x < 0) \rightarrow x \geq 0 \wedge x \leq 100$
  1.2. $2 \times x \geq 0 \wedge 2 \times x \leq 200 \rightarrow x \geq 0 \wedge x \leq 100$

To exhibit an example of a construction that is syntactically well-formed but generates side conditions that are not valid in the model under consideration (and is thus *not* a derivation of our goal), it suffices for instance to strengthen the postcondition *too much*. The reader can try to reproduce the construction above, but strengthening the postcondition in the first step to, say, $y \geq 0 \wedge y \leq 100$. This will create the following side conditions, of which the second and fourth are not valid in the models we are considering.

$\quad$ 0. $\;y \geq 0 \wedge y \leq 100 \rightarrow y \geq 0 \wedge y \leq 300$
1.1.1. $\;x \geq -100 \wedge x \leq 100 \wedge x < 0 \rightarrow x + 100 \geq 0 \wedge x + 100 \leq 50$
1.1.2. $\;x \geq -100 \wedge x \leq 100 \wedge \neg(x < 0) \rightarrow x \geq 0 \wedge x \leq 50$
$\quad$ 1.2. $\;x \geq 0 \wedge x \leq 50 \rightarrow 2 \times x \geq 0 \wedge 2 \times x \leq 100$

This example shows that when an attempt to construct a proof tree fails because some of the side conditions are not valid, this does not mean that such a tree does not exist. Suppose one tries to construct a derivation for an invalid Hoare triple, such as

$\{x \geq -100 \wedge x \leq 100\}$
if $x < 0$ **then** $x := x + 100$ **else skip** ; $y := 2 \times x$
$\{y \geq 0 \wedge y \leq 100\}$

This will generate side conditions that will not hold for *whichever* derivation one tries to construct, since this is clearly a program that is not correct with respect to the specification.

It is extremely important that the inference system cannot be used to derive invalid Hoare triples—it would be useless if it could. Formally, this last point corresponds to a fundamental property, which justifies the use of the Hoare calculus for proving the correctness of programs.

Proposition 5.14 (Soundness of $\mathcal{H}$) *Let* $\{\phi\} C \{\psi\} \in \mathbf{Spec}$ *and* $\Gamma \subseteq \mathbf{Assert}$.

$$\text{If} \quad \Gamma \vdash_{\mathcal{H}} \{\phi\} C \{\psi\}, \quad \text{then} \quad \Gamma \models \{\phi\} C \{\psi\}.$$

Proof By induction on the derivation of $\Gamma \vdash_{\mathcal{H}} \{\phi\} C \{\psi\}$. For the while case we also proceed by induction on the definition of the evaluation relation. $\qquad \square$

The purpose of the Hoare calculus is to derive Hoare triples that are valid in the interpretation structure implemented by the programming language and induced by the user-provided axiomatization. In particular, from the soundness result for system $\mathcal{H}$ it follows that

$$\text{if} \quad \mathcal{T}_{\mathsf{Prog}} \vdash_{\mathcal{H}} \{\phi\} C \{\psi\}, \quad \text{then} \quad \mathcal{T}_{\mathsf{Prog}} \models \{\phi\} C \{\psi\}$$

where $\mathcal{T}_{\mathsf{Prog}} = \mathsf{Th}(\mathcal{M}_{\mathsf{Exp}}) \cup \mathcal{T}_{\mathsf{User}}$ is the first-order logic theory described in Sect. 5.1.1.

The soundness result relates derivability in the calculus, which depends on the invariants annotated in the programs, with the meaning of a triple, which does not. Note that the reverse implication does not hold: it may be that $\Gamma \models \{\phi\} C \{\psi\}$ but the annotations in C do not allow for a derivation of $\{\phi\} C \{\psi\}$ to be constructed. The absence of this *completeness* property is a consequence of the fact that invariants are part of the programs. In fact, even the original system of Hoare logic (for programs without annotations) is only complete in a relative sense (see Sect. 5.8 for references).

It is important to understand that the annotations in a program cannot be dissociated from its specification; suppose that one has written two programs C_1 and C_2 that have exactly the same code, i.e. they differ only in the invariants added as annotations. It may well be that $\Gamma \vdash_{\mathcal{H}} \{\phi_1\} C_1 \{\psi_1\}$ and $\Gamma \vdash_{\mathcal{H}} \{\phi_2\} C_2 \{\psi_2\}$ for some $\phi_1, \psi_1, \phi_2, \psi_2$, but $\Gamma \nvdash_{\mathcal{H}} \{\phi_1\} C_2 \{\psi_1\}$ and $\Gamma \nvdash_{\mathcal{H}} \{\phi_2\} C_1 \{\psi_2\}$.

On the other hand more than one suitable annotated program may exist for the same base code, i.e. $\{\phi_1\} C_1 \{\psi_1\}$ and $\{\phi_1\} C_2 \{\psi_1\}$ may both be derivable. In this situation one should bear in mind that stronger invariants will naturally be more difficult to prove (i.e. the conditions required for establishing that the given assertions are indeed invariants will be harder to discharge). It may well be sufficient to use weaker notions, that describe only partially how a loop transforms the program state.

The bottom line is that given an initial program without annotations, producing an annotated program from it is crucial. Let C_0 be the code of program C stripped of invariants (or equivalently with all annotations set to $\top$). The fact that $\Gamma \nvdash_{\mathcal{H}} \{\phi\} C \{\psi\}$ does not necessarily mean that $\Gamma \nvDash \{\phi\} C_0 \{\psi\}$; it may also indicate that the annotations in C are not adequate. Of course if $\Gamma \nvDash \{\phi\} C_0 \{\psi\}$ there is no choice of annotations that will lead to $\Gamma \vdash_{\mathcal{H}} \{\phi\} C \{\psi\}$. Program verification systems should help the user understand if the problem is in the initial code or in the annotations (or even in the specifications, when these have not been sufficiently validated).

But how exactly should we proceed in order to prove the correctness of a program? We simply attempt to construct a proof tree with $\{\phi\} C \{\psi\}$ as conclusion. If it can be shown that all the side conditions of this tree are valid, then we will have succeeded in proving correctness. Otherwise we should try to construct another tree with a valid set of side conditions, or else convince ourselves that no such tree exists. The next chapter is devoted to mechanising this process.

5.5 The While^{array} Programming Language

In this section we extend the While^{int}-language with integer-subscript arrays. Array variables are clearly of a different kind with respect to integer variables: they can be seen as having as value functions that map subscripts (also called indexes) into integers. We will assume unrealistically that arrays are unbounded, i.e. every integer can be used as an index (in Chap. 7 we will address bounded arrays).

The new language While^{array} has three base types

$$\textbf{Type} \; \ni \; \tau \; ::= \; \textbf{bool} \,|\, \textbf{int} \,|\, \textbf{array}$$

$$\textbf{Exp}_{\textbf{int}} \quad \ni \quad e \quad ::= \quad \ldots \mid -1 \mid 0 \mid 1 \mid \ldots \mid x \mid$$
$$-e \mid e_1 + e_2 \mid e_1 - e_2 \mid e_1 \times e_2 \mid e_1 \text{ div } e_2 \mid e_1 \text{ mod } e_2 \mid$$
$$a[e]$$

$$\textbf{Exp}_{\textbf{array}} \quad \ni \quad a \quad ::= \quad u \mid a[e \triangleright e']$$

$$\textbf{Exp}_{\textbf{bool}} \quad \ni \quad b \quad ::= \quad \top \mid \bot \mid \neg b \mid b_1 \wedge b_2 \mid b_1 \vee b_2 \mid e_1 = e_2 \mid e_1 \neq e_2 \mid$$
$$e_1 < e_2 \mid e_1 \leq e_2 \mid e_1 > e_2 \mid e_1 \geq e_2$$

Fig. 5.7 Abstract syntax of program expressions of While^{array}

and the abstract syntax of program expressions is given in Fig. 5.7. Instead of annotating expressions with their types, we use the letters a to represent array expressions, e for integer expressions, and b for Boolean expressions. Distinguished letters $(u, v, \ldots)$ will also be used for array-type variables.

In While^{array} we consider only integer-valued arrays. Therefore we have a new form of integer expression $a[e]$, which represents the integer value associated in a to the subscript given by the value of e (or in other words, the contents of array a in position e). In addition to array variables, expressions of array type can also be array updates. The expression $a[e \triangleright e']$ represents an array that maps the subscript e to e' and any other subscript to the same value as the array a does. Concerning Boolean expressions note that the equality predicate is only defined for integer (not array) expressions.

Let us turn to the semantics of the language with arrays. The view of array values as functions leads us to interpret array expressions as functions in $\mathbb{Z} \to \mathbb{Z}$. So we have $D_{\textbf{array}} = \mathbb{Z} \to \mathbb{Z}$, $D_{\textbf{int}} = \mathbb{Z}$ and $D_{\textbf{bool}} = \{\textbf{F}, \textbf{T}\}$. Now that the interpretation structure for While^{array} has been fixed, this index can be dropped.

Definition 5.15 (Semantics of expressions of While^{array}) The semantics of While^{array} expressions is given by extending the semantics of While^{int} expressions (Definition 5.3) as follows

- $\llbracket \cdot \rrbracket$ maps every array $a \in \textbf{Exp}_{\textbf{array}}$ to a function $\llbracket a \rrbracket : \Sigma \to (\mathbb{Z} \to \mathbb{Z})$ defined inductively by

$$\llbracket u \rrbracket(s) \quad = \quad s(u)$$

$$\llbracket a[e \triangleright e'] \rrbracket(s) \quad = \quad \llbracket a \rrbracket(s)[\llbracket e \rrbracket(s) \mapsto \llbracket e' \rrbracket(s)]$$

- the definition of $\llbracket e \rrbracket : \Sigma \to \mathbb{Z}$ has the following additional case for integer expressions of the form $a[e]$:

$$\llbracket a[e] \rrbracket(s) = \llbracket a \rrbracket(s)(\llbracket e \rrbracket(s))$$

The reader will have noticed that we did not introduced a specific assignment command for arrays; this could easily be added as syntactic sugar, as follows

$$u[e] := e' \quad \text{is an abbreviation of} \quad u := u[e \rhd e'] \tag{5.7}$$

Thus assignment to array positions in this language is performed as an assignment to the corresponding array variable, in which the assigned value is an update of the current array.

A consequence of this is that no specific rule of the Hoare calculus is required: once arrays have been introduced in our programming language, we can simply use system $\mathcal{H}$ to derive Hoare triples involving array programs and profit from the generality of the soundness property (Proposition 5.14).

5.5.1 A Rule of Hoare Logic for Array Assignment

From the point of view of axiomatic reasoning, an alternative to expanding the syntactic sugar and then reasoning with the rules of system $\mathcal{H}$ is to extend that system with a specific rule for array assignment. This is just a special case of the assignment rule, obtained by expanding the syntactic sugar following (5.7):

$$(array\ assign) \quad \frac{}{\{\psi[u[e \rhd e']/u]\}\, u[e] := e'\, \{\psi\}}$$

This rule allows us to reason about array assignment in a direct fashion. We remark that the new rule is different from what would be obtained by trying to adapt the variable assignment axiom of system $\mathcal{H}$, replacing variables by array positions:

$$\frac{}{\{\psi[e'/u[e]]\}\, u[e] := e'\, \{\psi\}}$$

The latter axiom, in which $\psi[e'/u[e]]$ denotes ψ with e' substituted for $u[e]$, is *wrong*. To see this consider for instance the triple $\{u[j] > 100\}\, u[i] := 10\, \{u[j] > 100\}$. Since $u[i]$ does not occur in the postcondition, this triple would be derived by the latter axiom, and it is clearly not valid: if the command is executed in a state in which $i = j$ holds, then the precondition will not be preserved by execution. This phenomenon is called *subscript aliasing*. Depending on whether the integer expressions i and j have equal values, the expressions $u[i]$ and $u[j]$ may refer to the same positions in the array or not.

We remark that the correct axiom (*array assign*) would derive the following valid triple, given the same command and postcondition:

$$\{u[i \rhd 10][j] > 100\}\, u[i] := 10\, \{u[j] > 100\}$$

This is a valid triple, since the interpretation of $u[i \rhd 10]$ correctly handles aliasing.

Aliasing in general is a phenomenon that occurs in programming whenever the same object can be accessed through more than one name (or some other reference). In Chap. 8 we will look at another form of aliasing that occurs in the presence of procedures.

5.6 Loop Termination and Total Correctness

As discussed in Sect. 5.3, while the validity of the Hoare triple $\{\theta \wedge b\} C \{\theta\}$ implies the validity of $\{\theta\}$ **while** b **do** $\{\theta\} C \{\theta\}$, the validity of $[\theta \wedge b] C [\theta]$ *does not* imply the validity of $[\theta]$ **while** b **do** $\{\theta\} C [\theta]$. Proving such a total correctness specification implies proving the *termination* of the loop, for which loop invariants are inappropriate.

The required notion here is a *loop variant*: any program expression (or more generally some function on the state) whose value strictly decreases with each iteration, with respect to some well-founded relation. The existence of a valid variant for a given loop implies the termination of all possible executions of the loop. The natural choice in our language (but also in realistic specification languages) is to use *non-negative integer* expressions with strictly decreasing values.

In a total correctness setting the Hoare logic rule for while loops becomes

$$\frac{[\theta \wedge b \wedge V = v_0] C [\theta \wedge V < v_0]}{[\theta]\ \textbf{while}\ V\ \textbf{do}\ \{\theta, V\}\, b\, [\theta \wedge \neg b]} \quad \text{if } \theta \wedge b \rightarrow V >= 0$$

where V is the loop variant (note that similarly to the invariant θ it is also *annotated* in the while loop) and v_0 is a fresh auxiliary variable used to remember its initial value.

If the partial correctness of a command has been proven, its total correctness can be established by simply finding variants for every loop. Program verification tools usually consider loop variants as being in an additional optional layer on top of invariants. Variants are introduced as distinguished annotations, and tools generate the proof obligations required to prove that each such annotated expression is indeed a variant, which amounts to it being strictly decreasing. If no variants are provided however, no proof obligations are generated for termination, which means that the tools default to partial correctness mode. We will illustrate this is Chap. 10.

5.7 Adaptation

The importance of the use of auxiliary variables in specifications was demonstrated in Remark 5.11. There is however a difficulty with the use of these variables, that will now be described.

A crucial aspect of reasoning about programs is the ability to do so *modularly*, and to reuse correctness results. For instance, once we have constructed a derivation of the triple

$$\{n \geq 0\} \, \textbf{fact} \, \{f = fact(n)\}$$

We expect to be able to use this result to prove the correctness of the same program with respect to weaker specifications. For instance the triple

$$\{n = 10\} \, \textbf{fact} \, \{f = fact(n)\}$$

should be derivable from the above without reconstructing the entire derivation following the structure of the program. It is easy to see that this can indeed be done, using the (*conseq*) rule of Hoare logic.

Similarly, once we have constructed a derivation of the triple

$$\{n \geq 0 \wedge n = n_0\} \, \textbf{fact} \, \{f = fact(n) \wedge n = n_0\} \tag{5.8}$$

we expect of course to be able to prove in one step the correctness of

$$\{n = 10\} \, \textbf{fact} \, \{f = fact(10)\} \tag{5.9}$$

The latter triple cannot however be derived from the former using the (*conseq*) rule, since none of the side conditions hold.

The problem of matching the proved specification of a program with a weaker specification, required for reusing the program at a given point, is in general called the *adaptation* problem. The problem is particularly important when routines are added to the language, and will be dealt with in detail in Chap. 8. For now we briefly formalise the problem and mention a possible fix based on an alternative version of the consequence rule.

Definition 5.16 (Satisfiable specification) Let $\mathcal{M}$ be a $\mathcal{V}_{\text{Prog}}$-structure and ϕ, $\psi \in \textbf{Assert}$. A specification (ϕ, ψ) is $\mathcal{M}$-*satisfiable* if there exists a program $C \in \textbf{Comm}$ such that $\mathcal{M} \models \{\phi\} \, C \, \{\psi\}$.

Definition 5.17 (Adaptation completeness) Let $\mathcal{M}$ be a $\mathcal{V}_{\text{Prog}}$-structure and $\phi, \psi, \phi', \psi' \in \textbf{Assert}$ such that (ϕ, ψ) is $\mathcal{M}$-satisfiable, and for all $C \in \textbf{Comm}$, $\mathcal{M} \models \{\phi'\} \, C \, \{\psi'\}$ whenever $\mathcal{M} \models \{\phi\} \, C \, \{\psi\}$. An inference system for Hoare triples is said to be *adaptation complete* iff, for any program $C \in \textbf{Comm}$ the following rule is derivable

$$\frac{\{\phi\} \, C \, \{\psi\}}{\{\phi'\} \, C \, \{\psi'\}}$$

The previous example show that Hoare logic is not adaptation complete, and this clearly has to do with the presence of auxiliary variables. Informally, auxiliary variables can be seen as universally quantified over Hoare triples, establishing a connection between preconditions and postconditions. The side conditions in the (*conseq*) rule do not take this into consideration—in fact auxiliary variables are not given any special status in Hoare logic, they are treated in the same way as program variables.

The problem can be fixed by considering a stronger consequence rule, as proposed by Kleymann in a setting that formalises the difference between program variables and auxiliary variables.

Let us observe the consequence rule of the Hoare calculus:

$$\frac{\{\phi\}\,C\,\{\psi\}}{\{\phi'\}\,C\,\{\psi'\}} \quad \text{if } \phi' \to \phi \text{ and } \psi \to \psi'$$

Note that the first side condition is interpreted in the pre-state, whereas the second is interpreted in the post-state. We should seek an alternative side condition that combines both conditions, allowing for communication between both through auxiliary variables. We write this informally as

$$\frac{\{\phi\}\,C\,\{\psi\}}{\{\phi'\}\,C\,\{\psi'\}} \quad \text{if } (\phi'_{pre} \to \phi_{pre}) \wedge (\psi_{post} \to \psi'_{post})$$

But bear in mind that auxiliary variables occurring in ψ_{post} should always be interpreted in the pre-state, and moreover they should be existentially quantified; in the factorial example $n = 10 \to n \geq 0 \wedge n = n_0$ does not hold, but $n = 10 \to \exists n_0.\, n \geq 0 \wedge n = n_0$ does.

Kleymann [9] shows that the adequate side condition is weaker than this, and looks more like

$$(\phi'_{pre} \to \phi_{pre}) \wedge (\phi'_{pre} \to \psi_{post} \to \psi'_{post})$$

or equivalently

$$\phi'_{pre} \to \phi_{pre} \wedge (\psi_{post} \to \psi'_{post})$$

And if we let $\overline{y}$ represent the auxiliary variables in $\{\phi\}\,C\,\{\psi\}$ quantification can be introduced as follows.

$$\phi'_{pre} \to \exists \overline{y_f}.\, \phi_{pre}[\overline{y_f}/\overline{y}] \wedge (\psi_{post}[\overline{y_f}/\overline{y}] \to \psi'_{post})$$

Note the substitution of $\overline{y}$ by fresh variables $\overline{y_f}$ to distinguish between occurrences of auxiliary variables in $\{\phi\}\,C\,\{\psi\}$ and $\{\phi'\}\,C\,\{\psi'\}$.

All that is missing now is a way to handle the interpretation of program variables in the pre-state and post-state. This is achieved by substituting program variables in the post-state by freshly introduced variables and quantifying universally over these. The resulting rule is then written as

(*conseq-auxvars-total*)

$$\frac{[\phi]\,C\,[\psi]}{[\phi']\,C\,[\psi']} \quad \text{if } \phi' \to \forall \overline{x_f}.\, \exists \overline{y_f}.\, \phi[\overline{y_f}/\overline{y}] \wedge (\psi[\overline{y_f}/\overline{y}, \overline{x_f}/\overline{x}] \to \psi'[\overline{x_f}/\overline{x}])$$

where $\overline{y}$ are the auxiliary variables of $[\phi]\,C\,[\psi]$
$\overline{x}$ are the program variables of C
$\overline{x_f}$ and $\overline{y_f}$ are fresh variables

Note that the rule is written for total correctness triples; in fact Kleymann also shows that there is a weaker side condition that results in a correct consequence rule for partial correctness formulas. Schematically this can be written as

$$\phi'_{pre} \to (\phi_{pre} \to \psi_{post}) \to \psi'_{post}$$

This condition makes sense if you look at the triple $\{\phi\}\,C\,\{\psi\}$ as an implication $\phi_{pre} \to \psi_{post}$. And auxiliary variables should now be quantified universally:

$$\phi'_{pre} \to \forall \overline{y}.\ (\phi_{pre} \to \psi_{post}) \to \psi'_{post}$$

The resulting version of the consequence rule is

(*conseq-auxvars*)

$$\frac{\{\phi\}\,C\,\{\psi\}}{\{\phi'\}\,C\,\{\psi'\}} \quad \text{if } \phi' \to \forall \overline{x_f}.\ (\forall \overline{y_f}.\ \phi[\overline{y_f}/\overline{y}] \to \psi[\overline{y_f}/\overline{y}, \overline{x_f}/\overline{x}]) \to \psi'[\overline{x_f}/\overline{x}]$$

where $\overline{y}$ are the auxiliary variables of $\{\phi\}\,C\,\{\psi\}$
$\overline{x}$ are the program variables of C
$\overline{x_f}$ and $\overline{y_f}$ are fresh variables

For the previous example (5.9) of reusing the factorial program based on the Hoare triple (5.8), one would get the following side condition:

$$n = 10 \to \forall n_f, f_f.\ (\forall n_{0f}.\ n \ge 0 \wedge n = n_{0f} \to f_f = fact(n_f) \wedge n_f = n_{0f})$$
$$\to f_f = fact(10)$$

In practice arbitrary modular reasoning is not required; in Chap. 8 we will introduce in the programming language procedures annotated with contracts, which are simply specifications meant for external use. Modularity will be required at this level only, so we will be interested in the adaptation of procedure contracts, not of arbitrary Hoare triples. This is in accordance with modern specification languages, which are in general contract-oriented.

5.8 To Learn More

Hoare logic was introduced in a classic paper by Hoare [5]. It is the subject of a number of very good survey papers. A historical description of the development of Hoare Logic, its impact, and derived formalisms is given in [8]. The papers [1, 3] are more formal surveys, which also consider many programming language features that were not covered in this chapter. Both cover the issue of completeness of the original Hoare calculus (for program without annotations), and explain the relative sense in which it can be considered complete (recall that the calculus studied in this chapter cannot be considered complete in any sense, due to the presence of annotations in the programs).

Good books on the semantics of programming languages include [4, 11, 12, 14]. [10] is an excellent book that covers the foundations of program verification, including a discussion of program semantics and Hoare logic.

An excellent, modern introduction to specification and verification using Hoare Logic is Tennent's book [13], where a wealth of examples can be found. Identifying adequate invariants is an extremely important activity in software development (in particular in *formal* software development). Backhouse [2] covers this aspect in great detail.

The topic of adaptation of specifications has been considered since the very early days of Hoare logic [6], because of the obstacle that it poses to constructing proofs of correctness of recursive procedures. Kleymann [9] shows that the obstacle is not a specific problem of recursive procedures, but comes from the fact that Hoare logic is not adaptation complete. His paper offers an excellent overview of the subject and shows how the author's consequence rules subsume previous work on specific adaptation rules for procedure calls.

Finally, a very recent article by Hoare [7] looks at Hoare logic and axiomatic reasoning in retrospective, and considers its impact over the last 40 years.

5.9 Exercises

5.1. Write partial correctness specifications (i.e. preconditions and postconditions) for:
   (a) A program that computes the maximum of x and y and stores the result in z.
   (b) A program that calculates the greatest common divisor of integers x and y and stores the result in variable z.
   (c) A program that calculates the integer division of x by y, and stores the quotient in z and the remainder in w.

5.2. Discuss the meaning of the following Hoare triples. Are any of them necessarily valid?
   (a) $\{\top\} C \{\psi\}$
   (b) $\{\bot\} C \{\psi\}$
   (b) $\{\psi\} C \{\top\}$
   (c) $\{\psi\} C \{\bot\}$

5.3. Repeat the previous exercise considering the total correctness version of each Hoare triple.

5.4. Write a while-loop of the form **while** b **do** $\{\theta\} C$ and an assertion ϕ such that $\{\phi\}$ **while** b **do** $\{\theta\} C \{\phi\}$ is valid but $\{\phi \wedge b\} C \{\phi\}$ is not.

5.5. Consider again the specification of factorial in Example 5.10. Write a formal derivation that replicates in Hoare logic the reasoning used in the example to prove the total correctness of the program.

5.6. Derive the following alternative versions of the loop and conditional rules of
Hoare Logic:

$$\frac{\{\theta \wedge b\}\, C\, \{\theta\}}{\{\theta\}\, \textbf{while}\, b\, \textbf{do}\, \{\theta\}\, C\, \{\psi\}} \quad \text{if } \theta \wedge \neg b \to \psi$$

$$\frac{\{\phi_t\}\, C_t\, \{\psi\} \qquad \{\phi_f\}\, C_f\, \{\psi\}}{\{\phi\}\, \textbf{if}\, b\, \textbf{then}\, C_t\, \textbf{else}\, C_f\, \{\psi\}} \quad \text{if } \phi \to (b \to \phi_t) \wedge (\neg b \to \phi_f)$$

5.7. Write programs to match the specifications of Exercise 5.1, and annotate them
with loop invariants if appropriate.

5.8. Construct Hoare logic derivations to show that each program in the previous
exercise satisfies the specification written in Exercise 5.1.

5.9. Repeat the previous two exercises considering the total correctness version
of each specification in Exercise 5.1. You will now need to introduce a loop
variant for each loop in the code.

5.10. In this exercise you are asked to add to our programming language a concept
that has been left out—the notion of *locality* of a variable. The command

$$\textbf{local}\ x := e\ \textbf{in}\ C$$

has the intended meaning that the scope of variable x (initialised to e) is local
to the command C, i.e. if there also exists a global variable x, its value does
not affect execution of C in any way, and moreover the final value of x during
execution of C is not visible after execution of $\textbf{local}\ x := e\ \textbf{in}\ C$ terminates.
The natural semantics rule for this command is

– If $(C, s[x \mapsto [\![e]\!]_\mathcal{M}]) \leadsto_\mathcal{M} s'$, then $(\textbf{local}\ x := e\ \textbf{in}\ C, s) \leadsto_\mathcal{M} s'[x \mapsto s(x)]$.

The following rule of Hoare logic

$$\frac{\{\phi \wedge y = e\}\, C[y/x]\, \{\psi\}}{\{\phi\}\, \textbf{local}\ x := e\ \textbf{in}\ C\, \{\psi\}} \quad \text{if } y \text{ does not occur free in } \phi, C, \text{ or } \psi$$

captures the semantics of locality, but introduces (i) a new form of side condition (involving variable occurrence in programs and assertions); and (ii)
variable substitution at the command level. See [1] for a detailed discussion
of locality rules.

(a) Define the variable substitution operation on commands (and do not for-
get that $\textbf{local}\ x := e\ \textbf{in}\ C$ is now a command).

(b) Recall the factorial program of Example 5.10. Rewrite it with variable i
converted to a local variable and repeat Exercise 5.5 for this alternative
version.

5.11. Calculate the side condition that would be required to derive the total correctness version of the triple (5.9) from the total correctness version of (5.8). Use
rule (*conseq-auxvars-total*).

References

1. Apt, K.R.: Ten years of Hoare's logic: A survey—part I. ACM Trans. Program. Lang. Syst. **3**(4), 431–483 (1981)
2. Backhouse, R.: Program Construction—Calculating Implementations from Specifications. Wiley, New York (2003)
3. Cousot, P.: Methods and logics for proving programs. In: Handbook of Theoretical Computer Science, Volume B: Formal Models and Semantics (B), pp. 841–994. Elsevier/MIT Press, Cambridge (1990)
4. Hennessy, M.: The Semantics of Programming Languages. Wiley, New York (1990)
5. Hoare, C.A.R.: An axiomatic basis for computer programming. Commun. ACM **12**, 576–580 (1969)
6. Hoare, C.A.R.: Procedures and parameters: an axiomatic approach. In: Proceedings of Symposium on Semantics of Algorithmic Languages. Lecture Notes in Mathematics, vol. 188. Springer, Berlin (1971)
7. Hoare, C.A.R.: Viewpoint retrospective: An axiomatic basis for computer programming. Commun. ACM **52**(10), 30–32 (2009)
8. Jones, C.B.: The early search for tractable ways of reasoning about programs. IEEE Ann. Hist. Comput. **25**(2), 26–49 (2003)
9. Kleymann, T.: Hoare logic and auxiliary variables. Form. Asp. Comput. **11**(5), 541–566 (1999)
10. Loeckx, J., Sieber, K.: The Foundations of Program Verification, 2nd edn. Wiley, New York (1987)
11. Nielson, H.R., Nielson, F.: Semantics with Applications: An Appetizer. Undergraduate Topics in Computer Science. Springer, Berlin (2007)
12. Reynolds, J.C.: Theories of Programming Languages. Cambridge University Press, Cambridge (1998)
13. Tennent, R.D.: Specifying Software—A Hands-on Introduction. Cambridge University Press, Cambridge (2002)
14. Winskel, G.: The Formal Semantics of Programming Languages: An Introduction. Foundations of Computing. MIT Press, Cambridge (1993)

Chapter 6
Generating Verification Conditions

In this chapter we consider the problem of mechanising the construction of derivations in Hoare logic having a given Hoare triple as conclusion. We are thus concerned with the backwards application of rules of the logic, which will eventually produce a derivation, i.e. a tree in which all leaves correspond to instances of axioms, and all side conditions hold.

Another way to look at this, which should have become clear from the previous chapter, is that we apply rules backwards *assuming* that the side conditions hold, and we later check whether they do indeed. If all side conditions hold, then we have a proof of the goal Hoare triple; if at least one side condition does not hold, then the constructed tree is not a derivation.

This is an undesirable situation: if we construct a tree and later find out that it is not a well-formed derivation because at least one side condition does not hold, there is no way to know if an alternative derivation exists or not. The goal of this chapter is to show that there exists a strategy for conducting the proofs such that, if some of the side conditions required do not hold, then no derivation exists for the goal at hand. This strategy results in the definition of what is usually known as a *verification conditions generator*.

6.1 Mechanising Hoare Logic

Most of the rules of the Hoare calculus have the desirable *subformula property*: all the assertions that occur in the premises of a rule also occur in its conclusion. A system in which all rules satisfy this condition is particularly amenable to being mechanised, since there is no need to 'guess' assertions that appear in the new goals generated when a rule is applied. The exceptions are the (*seq*) rule, which requires an intermediate condition θ to be guessed, and the (*conseq*) rule, where the precondition ϕ and the postcondition ψ must be guessed.

A second desirable property is the absence of ambiguity in the choice of rule: given a goal $\{\phi\} C \{\psi\}$ it would be desirable that a single rule could be applied to produce a derivation with this goal as conclusion. This is true for the subset of

J.B. Almeida et al., *Rigorous Software Development*,
Undergraduate Topics in Computer Science,
DOI 10.1007/978-0-85729-018-2_6, © Springer-Verlag London Limited 2011

$(skip)$
$$\frac{}{\{\phi\}\,\textbf{skip}\,\{\psi\}}\quad\text{if }\phi\to\psi$$

$(assign)$
$$\frac{}{\{\phi\}\,x := e\,\{\psi\}}\quad\text{if }\phi\to\psi[e/x]$$

(seq)
$$\frac{\{\phi\}\,C_1\,\{\theta\}\qquad\{\theta\}\,C_2\,\{\psi\}}{\{\phi\}\,C_1\,;\,C_2\,\{\psi\}}$$

$(while)$
$$\frac{\{\theta\wedge b\}\,C\,\{\theta\}}{\{\phi\}\,\textbf{while } b\,\textbf{do}\,\{\theta\}\,C\,\{\psi\}}\quad\text{if }\phi\to\theta\text{ and }\theta\wedge\neg b\to\psi$$

(if)
$$\frac{\{\phi\wedge b\}\,C_t\,\{\psi\}\qquad\{\phi\wedge\neg b\}\,C_f\,\{\psi\}}{\{\phi\}\,\textbf{if } b\,\textbf{then } C_t\,\textbf{else } C_f\,\{\psi\}}$$

Fig. 6.1 Inference system of Hoare logic without consequence rule: system $\mathcal{H}_g$

program rules of Hoare Logic, but the (*conseq*) rule introduces ambiguity, since it can be applied with any arbitrary goal.

The second property is easy to obtain. We give in Fig. 6.1 an alternative inference system that we will call system $\mathcal{H}_g$. This system contains program rules only—the (*conseq*) rule has been eliminated, and so has ambiguity. The side conditions, which are introduced by the (*conseq*) rule in the original system of Hoare logic, are in this new system present in the **skip**, assignment, and loop rules. In the original system, the (*conseq*) rule might have to be applied (as was shown in Example 5.13) in order to modify the precondition or postcondition before a program rule could be applied. This is no longer the case.

Take for instance the assignment axiom: whereas in the original system it could only be applied if the precondition resulted from performing a substitution in the postcondition, in the modified system it can be applied with stronger preconditions, and the relation between precondition and postcondition is introduced as a side condition. Thus it is not necessary to use the consequence rule in order to weaken the precondition.

It is easy to see that the (*conseq*) rule is admissible in $\mathcal{H}_g$, and that the systems $\mathcal{H}$ and $\mathcal{H}_g$ are equivalent.

Definition 6.1 Let $\Gamma \subseteq \textbf{Assert}$, and $\{\phi\}\,C\,\{\psi\} \in \textbf{Spec}$. We say that $\{\phi\}\,C\,\{\psi\}$ is *derivable in* $\mathcal{H}_g$ *assuming* Γ, denoted $\Gamma \vdash_{\mathcal{H}_g} \{\phi\}\,C\,\{\psi\}$, if there exists a proof of $\{\phi\}\,C\,\{\psi\}$ in $\mathcal{H}_g$ which uses as assumptions (for the side conditions) assertions from Γ.

$\{n \geq 0\}\, \mathbf{fact}\, \{f = fact(n)\}$

1. $\{n \geq 0\}\, f := 1\, ;\, i := 1\, \{n \geq 0 \wedge f = 1 \wedge i = 1\}$
   1.1 $\{n \geq 0\}\, f := 1\, \{n \geq 0 \wedge f = 1\}$
   1.2 $\{n \geq 0 \wedge f = 1\}\, i := 1\, \{n \geq 0 \wedge f = 1 \wedge i = 1\}$
2. $\{n \geq 0 \wedge f = 1 \wedge i = 1\}$
   $\mathbf{while}\ i \leq n\ \mathbf{do}\ \{f = fact(i - 1) \wedge i \leq n + 1\}\, C_w$
   $\{f = fact(n)\}$
   2.1. $\{f = fact(i - 1) \wedge i \leq n + 1 \wedge i \leq n\}\, C_w\, \{f = fact(i - 1) \wedge i \leq n + 1\}$
   2.1.1. $\{f = fact(i - 1) \wedge i \leq n + 1 \wedge i \leq n\}\, f := f \times i\, \{f = fact(i - 1) \times i \wedge i \leq n\}$
   2.1.2. $\{f = fact(i - 1) \times i \wedge i \leq n\}\, i := i + 1\, \{f = fact(i - 1) \wedge i \leq n + 1\}$

where C_w represents the command $f := f \times i\, ;\, i := i + 1$.

Fig. 6.2 Derivation for correctness of the factorial program

Lemma 6.2 *Let* $\Gamma \subseteq$ **Assert**, $\phi', \psi' \in$ **Assert**, *and* $\{\phi\}\, C\, \{\psi\} \in$ **Spec**. *If* $\Gamma \vdash_{\mathcal{H}_g} \{\phi\}\, C\, \{\psi\}$ *and both* $\Gamma \models \phi' \to \phi$ *and* $\Gamma \models \psi \to \psi'$ *hold, then* $\Gamma \vdash_{\mathcal{H}_g} \{\phi'\}\, C\, \{\psi'\}$.

Proof By induction on the derivation of $\Gamma \vdash_{\mathcal{H}_g} \{\phi\}\, C\, \{\psi\}$. □

Proposition 6.3 *Let* $\{\phi\}\, C\, \{\psi\} \in$ **Spec** *and* $\Gamma \subseteq$ **Assert**.

$$\Gamma \vdash_{\mathcal{H}_g} \{\phi\}\, C\, \{\psi\} \quad \textit{iff} \quad \Gamma \vdash_{\mathcal{H}} \{\phi\}\, C\, \{\psi\}$$

Proof $\Rightarrow$: By induction on the derivation of $\Gamma \vdash_{\mathcal{H}_g} \{\phi\}\, C\, \{\psi\}$.
   $\Leftarrow$: By induction on the derivation of $\Gamma \vdash_{\mathcal{H}} \{\phi\}\, C\, \{\psi\}$, using Lemma 6.2. □

Example 6.4 Figure 6.2 contains a derivation that formalises the proof about the factorial program of Example 5.10. Note that, since the choice of rule is not ambiguous in this system, we no longer need to label nodes with the name of the inference rule applied: it is immediate from the current goal.

Recall that for this example we resort to the possibility of expanding the vocabulary with $fact^{\mathbf{int}} \in \mathcal{F}^{\mathsf{User}}$ and $isfact^{\mathbf{bool}} \in \mathcal{F}^{\mathsf{User}}$, with $\mathsf{rank}(fact^{\mathbf{int}}) = \langle \mathbf{int} \rangle$ and $\mathsf{rank}(isfact^{\mathbf{bool}}) = \langle \mathbf{int}, \mathbf{int} \rangle$, whose meaning is given by the axioms (5.5) and (5.6), repeated below for convenience:

$$fact(0) = 1$$

$$\forall n.\ n > 0 \to fact(n) = n \times fact(n - 1)$$

The following side conditions are required for each node of the tree:

1.1 $n \geq 0 \to (n \geq 0 \wedge f = 1)[1/f]$
1.2 $n \geq 0 \wedge f = 1 \to (n \geq 0 \wedge f = 1 \wedge i = 1)[1/i]$
2. $n \geq 0 \wedge f = 1 \wedge i = 1 \to f = fact(i - 1) \wedge i \leq n + 1$ and $f = fact(i - 1) \wedge i \leq n + 1 \wedge \neg(i \leq n) \to f = fact(n)$
2.1.1. $f = fact(i - 1) \wedge i \leq n + 1 \wedge i \leq n \to (f = fact(i - 1) \times i \wedge i \leq n)[f \times i/f]$

2.1.2. $f = fact(i - 1) \times i \wedge i \leq n \rightarrow (f = fact(i - 1) \wedge i \leq n + 1)[i + 1/i]$

The validity of these conditions is fairly obvious in the current theory. In particular 2. and 2.1.2. are consequences of the axioms about factorial.

From the point of view of mechanisation, this system improves over the original system of Hoare logic: by eliminating the *(conseq)* rule we have eliminated ambiguity, and also one of the rules lacking the sub-formula property. However, the presence of the *(seq)* rule means that certain intermediate assertions still have to be guessed, and it is still possible to construct different trees for a given Hoare triple. Such trees are all equally shaped, but may contain different intermediate assertions. Observe that if some side condition turns out to be invalid for some rule application, this does not imply that no derivation exists for our goal, since we may have chosen inadequate intermediate assertions.

In what follows we will study a method for systematically determining the intermediate assertions required by the *(seq)* rule. The resulting tree will be such that *if it is not a proof tree (because some side conditions do not hold), then no derivation exists for the goal.* Thus this method can be used for mechanically verifying the correctness of programs.

Remark 6.5 As explained in Sect. 5.7, it is important to be able to reason modularly about programs and to reuse correctness results. Turning back to our running factorial example, once we have constructed a derivation of the triple

$$\{n \geq 0\} \mathbf{fact} \{f = fact(n)\}$$

we expect to be able to use this result to prove the correctness of the same program with respect to weaker specifications, like

$$\{n = 10\} \mathbf{fact} \{f = fact(n)\}$$

This should be derivable in one step without reconstructing the entire derivation following the structure of the program. In the explicit presence of the *(conseq)* rule (or its modified version discussed in Sect. 5.7, if auxiliary variables are used) this can be done in a single proof step, but in system $\mathcal{H}_g$ it is no longer possible to do so, since the *(conseq)* rule has been removed.

This ability will be restored in Chap. 8 when we extend the programming language with procedures annotated with *contracts*. The need for adaptation (between the specification of a procedure, annotated as a contract, and an actual specification required for a procedure invocation) will be concentrated in the procedure call rule.

6.2 The Weakest Precondition Strategy

The system $\mathcal{H}_g$ of Fig. 6.1 contains an implicit strategy for constructing proofs in a deterministic way. When the sequence rule is applied with a current goal, two subgoals are created, the first of which with an unknown postcondition, and the second with an unknown precondition. The first key point is to focus on the latter, in order to obtain the unknown condition that can then be propagated to the former sub-goal. The second key point is that three of the rules in our system have side conditions that involve the goal's precondition; when one of these is applied with an unknown precondition, we will simply choose the weakest of all preconditions that satisfy the side conditions.

To illustrate this point, consider the following triple with a sequence of assignment commands

$$\{\phi\} \, x := e_1 \,; \, y := e_2 \,; \, z := e_3 \, \{\psi\}$$

Applying the (*seq*) rule produces two sub-goals. In what follows, we will be depicting *partially constructed trees* (proof tree candidates), in which the currently open goals will be identified by placing the corresponding number inside a square. Note that if a goal has some open sub-goals, then the goal itself is still open. Applying the sequence rule to the current goal yields

$\{\phi\} \, x := e_1 \,; \, y := e_2 \,; \, z := e_3 \, \{\psi\}$

$\boxed{1.}$ $\{\phi\} \, x := e_1 \,; \, y := e_2 \, \{\theta\}$
$\boxed{2.}$ $\{\theta\} \, z := e_3 \, \{\psi\}$

Now the second sub-goal is an assignment, which means that the corresponding axiom can be applied by simply taking the precondition to be the one that trivially satisfies the side condition, i.e. $\theta = \psi[e_3/z]$. Now of course this can be substituted globally in the current proof construction, and we get

$\{\phi\} \, x := e_1 \,; \, y := e_2 \,; \, z := e_3 \, \{\psi\}$

$\boxed{1.}$ $\{\phi\} \, x := e_1 \,; \, y := e_2 \, \{\psi[e_3/z]\}$
2. $\{\psi[e_3/z]\} \, z := e_3 \, \{\psi\}$

The same process can be applied to the first goal:

$\{\phi\} \, x := e_1 \,; \, y := e_2 \,; \, z := e_3 \, \{\psi\}$

$\boxed{1.}$ $\{\phi\} \, x := e_1 \,; \, y := e_2 \, \{\psi[e_3/z]\}$
     $\boxed{1.1.}$ $\{\phi\} \, x := e_1 \, \{\psi[e_3/z][e_2/y]\}$
      1.2. $\{\psi[e_3/z][e_2/y]\} \, y := e_2 \, \{\psi[e_3/z]\}$
2. $\{\psi[e_3/z]\} \, z := e_3 \, \{\psi\}$

The proof is closed by applying the assignment axiom to the first command:

$\{\phi\} \, x := e_1 \,; \, y := e_2 \,; \, z := e_3 \, \{\psi\}$

1. $\{\phi\} \, x := e_1 \,; \, y := e_2 \, \{\psi[e_3/z]\}$
     1.1. $\{\phi\} \, x := e_1 \, \{\psi[e_3/z][e_2/y]\}$,
     1.2. $\{\psi[e_3/z][e_2/y]\} \, y := e_2 \, \{\psi[e_3/z]\}$
2. $\{\psi[e_3/z]\} \, z := e_3 \, \{\psi\}$

In this last step we were not free to choose the precondition for the assignment since this is now the first command in the sequence. Thus the side condition $\phi \rightarrow \psi[e_3/z][e_2/y][e_1/x]$ is introduced.

In general the strategy proceeds as follows. Consider that a derivation is being constructed for the Hoare triple $\{\phi\} C \{\psi\}$, where ϕ may be either known or unknown (in this case the triple will be written as $\{?\} C \{\psi\}$).

1. If ϕ is known, the derivation is constructed by simply applying the unique applicable rule from Fig. 6.1. If the sequence rule is applied with goal $\{\phi\} C_1 ; C_2 \{\psi\}$, then the *second* sub-derivation will be first constructed with a goal of the form $\{?\} C_2 \{\psi\}$. Eventually when the construction of this sub-derivation is concluded, ? will be instantiated with some assertion θ and the first sub-derivation may then be constructed for the goal $\{\phi\} C_1 \{\theta\}$.
2. If ϕ is unknown, the construction proceeds as before, except that, in the rules for **skip**, assignment, and loops, with a side condition $\phi \rightarrow \theta$ for some θ, we take the precondition ϕ to be *equal to* θ (the condition is thus trivially satisfied).

Note that this strategy ensures that a *weakest precondition* will be determined for goals that have unknown preconditions: a derivation for the triple $\{\phi\} C_1 ; C_2 \{\psi\}$ is constructed by taking as postcondition for C_1 the weakest precondition θ that will ensure the validity of ψ after execution of C_2. By this it is meant that $\{\theta\} C_2 \{\psi\}$ is derivable, and for any other θ' such that $\{\theta'\} C_2 \{\psi\}$ is derivable it holds that $\theta' \rightarrow \theta$.

Example 6.6 Let us use the weakest precondition strategy to verify the factorial program. The result will be a derivation with the same structure as the one in Example 6.4, but with different intermediate assertions.

The first step in the proof construction consists in applying an instance of the sequence rule as follows

$$\{n \geq 0\} \textbf{fact} \{f = fact(n)\}$$

$\boxed{1.}$ $\{n \geq 0\} f := 1 ; i := 1 \{?_1\}$
$\boxed{2.}$ $\{?_1\} \textbf{while } i \leq n \textbf{ do} \{f = fact(i-1) \wedge i \leq n+1\} C_w \{f = fact(n)\}$

where as before C_w represents $f := f \times i ; i := i + 1$. We further expand the second branch applying the while rule and taking $?_1$ to be equal to the invariant $f = fact(i-1) \wedge i \leq n+1$, following the strategy outlined above. This is then substituted globally.

$$\{n \geq 0\} \textbf{fact} \{f = fact(n)\}$$

$\boxed{1.}$ $\{n \geq 0\} f := 1 ; i := 1 \{f = fact(i-1) \wedge i \leq n+1\}$
$\boxed{2.}$ $\{f = fact(i-1) \wedge i \leq n+1\} \textbf{while } i \leq n \textbf{ do} \{f = fact(i-1) \wedge i \leq n+1\} C_w \{f = fact(n)\}$
  $\boxed{2.1.}$ $\{f = fact(i-1) \wedge i \leq n+1 \wedge i \leq n\} C_w \{f = fact(i-1) \wedge i \leq n+1\}$

The application of the loop rule introduces two side conditions, the first of which is trivially valid due to our choice of $?_1$. The second, which is also valid in the models of interest, is

$$f = fact(i-1) \wedge i \leq n+1 \wedge \neg(i \leq n) \rightarrow f = fact(n)$$

The construction now continues by applying the sequence rule inside the loop body.

$\{n \geq 0\}$ **fact** $\{f = fact(n)\}$

$\boxed{1.}$ $\{n \geq 0\}\, f := 1\,;\, i := 1\,\{f = fact(i-1) \land i \leq n+1\}$

$\boxed{2.}$ $\{f = fact(i-1) \land i \leq n+1\}$ **while** $i \leq n$ **do** $\{f = fact(i-1) \land i \leq n+1\}\, C_w\, \{f = fact(n)\}$

    $\boxed{2.1.}$ $\{f = fact(i-1) \land i \leq n+1 \land i \leq n\}\, C_w\, \{f = fact(i-1) \land i \leq n+1\}$

        $\boxed{2.1.1.}$ $\{f = fact(i-1) \land i \leq n+1 \land i \leq n\}\, f := f \times i\, \{?_2\}$

        $\boxed{2.1.2.}$ $\{?_2\}\, i := i+1\, \{f = fact(i-1) \land i \leq n+1\}$

In the next step we apply the assignment axiom by taking $?_2$ to be the weakest precondition for $f = fact(i-1) \land i \leq n+1$ to hold after execution of $i := i+1$, i.e. $?_2 = (f = fact(i-1) \land i \leq n+1)[i+1/i] = (f = fact((i+1)-1) \land i+1 \leq n+1)$. This may now be substituted in goal 2.1.1.

$\{n \geq 0\}$ **fact** $\{f = fact(n)\}$

$\boxed{1.}$ $\{n \geq 0\}\, f := 1\,;\, i := 1\,\{f = fact(i-1) \land i \leq n+1\}$

$\boxed{2.}$ $\{f = fact(i-1) \land i \leq n+1\}$ **while** $i \leq n$ **do** $\{f = fact(i-1) \land i \leq n+1\}\, C_w\, \{f = fact(n)\}$

    $\boxed{2.1.}$ $\{f = fact(i-1) \land i \leq n+1 \land i \leq n\}\, C_w\, \{f = fact(i-1) \land i \leq n+1\}$

        $\boxed{2.1.1.}$ $\{f = fact(i-1) \land i \leq n+1 \land i \leq n\}\, f := f \times i\, \{f = fact((i+1)-1) \land i+1 \leq n+1\}$

        2.1.2. $\{f = fact((i+1)-1) \land i+1 \leq n+1\}\, i := i+1\, \{f = fact(i-1) \land i \leq n+1\}$

Goal 2.1.1. will be closed by again applying the assignment axiom. Note that, since the goal is a Hoare triple in which the precondition is present, application of the axiom introduces the following side condition:

$$f = fact(i-1) \land i \leq n+1 \land i \leq n \to (f = fact((i+1)-1) \land i+1 \leq n+1)[f \times i/f]$$

which is

$$f = fact(i-1) \land i \leq n+1 \land i \leq n \to f \times i = fact((i+1)-1) \land i+1 \leq n+1$$

This is a valid condition in the models of interest since $fact(i-1) \times i = fact(i)$, as a consequence of the axiom (5.6).

It remains to close goal 1. Again the sequence rule is applied:

$\{n \geq 0\}$ **fact** $\{f = fact(n)\}$

$\boxed{1.}$ $\{n \geq 0\}\, f := 1\,;\, i := 1\,\{f = fact(i-1) \land i \leq n+1\}$

    $\boxed{1.1}$ $\{n \geq 0\}\, f := 1\, \{?_3\}$

    $\boxed{1.2}$ $\{?_3\}\, i := 1\, \{f = fact(i-1) \land i \leq n+1\}$

2. $\{f = fact(i-1) \land i \leq n+1\}$ **while** $i \leq n$ **do** $\{f = fact(i-1) \land i \leq n+1\}\, C_w\, \{f = fact(n)\}$

    2.1. $\{f = fact(i-1) \land i \leq n+1 \land i \leq n\}\, C_w\, \{f = fact(i-1) \land i \leq n+1\}$

        2.1.1. $\{f = fact(i-1) \land i \leq n+1 \land i \leq n\}\, f := f \times i\, \{f = fact((i+1)-1) \land i+1 \leq n+1\}$

        2.1.2. $\{f = fact((i+1)-1) \land i+1 \leq n+1\}\, i := i+1\, \{f = fact(i-1) \land i \leq n+1\}$

$\{n \geq 0\}$ **fact** $\{f = fact(n)\}$

1. $\{n \geq 0\}$ $f := 1$; $i := 1$ $\{f = fact(i - 1) \wedge i \leq n + 1\}$
    1.1 $\{n \geq 0\}$ $f := 1$ $\{f = fact(0) \wedge 1 \leq n + 1\}$
    1.2 $\{f = fact(0) \wedge 1 \leq n + 1\}$ $i := 1$ $\{f = fact(i - 1) \wedge i \leq n + 1\}$
2. $\{f = fact(i - 1) \wedge i \leq n + 1\}$ **while** $i \leq n$ **do** $\{f = fact(i - 1) \wedge i \leq n + 1\}$ C_w $\{f = fact(n)\}$
    2.1. $\{f = fact(i - 1) \wedge i \leq n + 1 \wedge i \leq n\}$ C_w $\{f = fact(i - 1) \wedge i \leq n + 1\}$
        2.1.1. $\{f = fact(i-1) \wedge i \leq n+1 \wedge i \leq n\}$ $f := f \times i$ $\{f = fact((i+1)-1) \wedge i+1 \leq n + 1\}$
        2.1.2. $\{f = fact((i+1)-1) \wedge i+1 \leq n+1\}$ $i := i+1$ $\{f = fact(i-1) \wedge i \leq n+1\}$

Fig. 6.3 Derivation for factorial using the weakest precondition strategy

This is now quite straightforward: $?_3$ is taken to be $(f = fact(i - 1) \wedge i \leq n + 1)[1/i]$; applying the assignment axiom to goal 1.1. introduces the side condition

$$n \geq 0 \rightarrow (f = fact(1 - 1) \wedge 1 \leq n + 1)[1/f]$$

which is

$$n \geq 0 \rightarrow 1 = fact(1 - 1) \wedge 1 \leq n + 1$$

This is valid in the models under consideration according to axiom (5.5).

The proof tree thus constructed is shown in Fig. 6.3. We remark that since the program does not assign to variable n, it is straightforward to adapt this proof tree to have as conclusion the triple $\{n \geq 0 \wedge n = n_0\}$ **fact** $\{f = fact(n) \wedge n = n_0\}$ that forces the preservation of the value of n (see Remark 5.11). It suffices to strengthen the loop invariant to $f = fact(i - 1) \wedge i \leq n + 1 \wedge n = n_0$.

The weakest precondition strategy used with system $\mathcal{H}_g$ results in a mechanical method for proving programs correct. If some side condition in the resulting tree can be shown not to be valid (for instance because an automatic prover succeeded in finding a counterexample), then it is guaranteed that no proof tree can be constructed for the current goal. If all conditions are valid, then the method has produced a proof tree and the goal is a valid Hoare triple. This property will be formalised in Sect. 6.4.

We remark that although we have chosen to express the weakest precondition strategy in the context of system $\mathcal{H}_g$, there is no reason why it could not be expressed in the original system $\mathcal{H}$ of Hoare logic (see Exercise 6.1).

Finally, observe that the weakest precondition of a command C with respect to a given postcondition ψ can be calculated by a very simple function that takes C and ψ as arguments. In fact we will see that the set of side conditions of a derivation can also be generated by an algorithm, thus the explicit construction of proof trees can be avoided altogether.

6.3 An Architecture for Program Verification

At this point we may outline a method for program verification as follows. Given a Hoare triple $\{\phi\} C \{\psi\}$ and a theory T,

1. We apply the principles studied in the previous two sections to mechanically produce a derivation with $\{\phi\} C \{\psi\}$ as conclusion, assuming that all the side conditions created in this process hold.
2. Each first-order formula generated as a side condition in step 1 must now be checked. To that effect, it is exported to some proof tool. In this context such a formula is called a *Verification Condition* (VC).
3. If all verification conditions are shown to be T-valid by a proof tool, then $T \vdash_{\mathcal{H}_g} \{\phi\} C \{\psi\}$. If at least one condition is shown not to be T-valid, then this is evidence that $T \not\vdash_{\mathcal{H}_g} \{\phi\} C \{\psi\}$.

Concerning step 1, as we have already hinted at the end of Sect. 6.2, since the construction of proof trees is mechanical and involves no choices, we can do without constructing them at all. A *Verification Conditions Generator* (VCGen for short) is an algorithm that takes a Hoare Triple and produces a set of first-order verification conditions such that the triple is derivable in Hoare logic if and only if all the conditions are valid.

Let us now turn to step 2. If (as it most likely will be the case!) machine assistance is required to establish the validity or otherwise of the verification conditions, then an automatic theorem prover or interactive proof assistant can be used. We refer to all such tools generically as "proof tools" or simply "provers". Verification conditions are sometimes said to be *discharged* by a prover when successfully proved.

We remark that the verification process (and in particular the degree of automation) can only be as good as the proof tools employed. T may be undecidable, in which case automatic provers may not be able to tell if a formula is valid or not. It can also be decidable but not efficiently, so it may not be realistic to obtain results in reasonable time. Interactive proof may well be the last resort in these situations. Note that, following the discussion in Sect. 4.7.3, the theory of interest for our example language While[int] is indeed undecidable (since it includes at least $T_{\mathbb{Z}}$).

The architecture for program verification outlined in this section is illustrated in Fig. 6.4. This is an interactive process: the verification of an incorrect program will fail, but will hopefully shed some light on what is wrong with the program, which will in turn help correct the errors. In particular, provers may be able to issue *counterexamples* that illustrate situations in which an assertion is false.

The advantage of this approach, which is followed by most modern program verification tools, has to do with flexibility. Any proof tool can be used to try to discharge the verification conditions. Later in this book we will use a *multi-prover* VCGen tool, which can export proof obligations to several different provers. Indeed, this is becoming an essential feature of modern program verification platforms: since automatic proof is in general hard, and different tools have different capabilities, using more than one tool increases the chances of successfully performing fully automatic verifications.

Fig. 6.4 General architecture
of a program verification
system

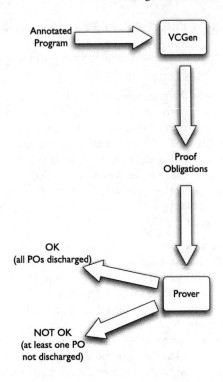

6.4 A VCGen Algorithm

Let us now study a VCGen algorithm based on the weakest precondition strategy.
In a nutshell, given a Hoare triple $\{\phi\}\,C\,\{\psi\}$, we calculate the weakest precondition $\mathsf{wp}\,(C, \psi)$ which is required for ψ to hold after terminating executions of C.
Then one verification condition that must be satisfied in order for the program to
be correct is $\phi \to \mathsf{wp}\,(C, \psi)$. For programs not containing loops this is indeed the
only verification condition to be taken into account. Loops introduce additional side
conditions that must be recursively collected, as explained below.

6.4.1 Calculating the Weakest Precondition

The weakest precondition of a program is calculated by the function wp defined in
Fig. 6.5, top. Notice that the clause for a sequence command $C_1 ; C_2$ states that the
weakest precondition of C_2 is first calculated, and used as postcondition for C_1,
which is precisely what the weakest precondition strategy for the construction of
derivations prescribes. The weakest precondition of a conditional command, on the
other hand, composes as a conjunction the weakest preconditions of both branches,
guarded respectively by the success and the failure of the Boolean condition.

$$\text{wp}(\textbf{skip}, \psi) = \psi$$

$$\text{wp}(x := e, \psi) = \psi[e/x]$$

$$\text{wp}(C_1; C_2, \psi) = \text{wp}(C_1, \text{wp}(C_2, \psi))$$

$$\text{wp}(\textbf{if } b \textbf{ then } C_t \textbf{ else } C_f, \psi) = (b \rightarrow \text{wp}(C_t, \psi)) \wedge (\neg b \rightarrow \text{wp}(C_f, \psi))$$

$$\text{wp}(\textbf{while } b \textbf{ do }\{\theta\} C, \psi) = \theta$$

$$\text{VC}(\textbf{skip}, \psi) = \emptyset$$

$$\text{VC}(x := e, \psi) = \emptyset$$

$$\text{VC}(C_1; C_2, \psi) = \text{VC}(C_1, \text{wp}(C_2, \psi)) \cup \text{VC}(C_2, \psi)$$

$$\text{VC}(\textbf{if } b \textbf{ then } C_t \textbf{ else } C_f, \psi) = \text{VC}(C_t, \psi) \cup \text{VC}(C_f, \psi)$$

$$\text{VC}(\textbf{while } b \textbf{ do }\{\theta\} C, \psi) = \{(\theta \wedge b) \rightarrow \text{wp}(C, \theta), (\theta \wedge \neg b) \rightarrow \psi\} \cup \text{VC}(C, \theta)$$

$$\text{VCG}(\{\phi\} C \{\psi\}) = \{\phi \rightarrow \text{wp}(C, \psi)\} \cup \text{VC}(C, \psi)$$

Fig. 6.5 VCGen algorithm

For a loop **while** b **do** $\{\theta\} C$, we take the weakest precondition to be simply its invariant θ, since it is easy to see from the rules in system $\mathcal{H}_g$ that it is not possible to derive triples of the form $\{\phi\}$ **while** b **do** $\{\theta\} C \{\psi\}$ such that ϕ is weaker than θ.

Proposition 6.7 *Given a command* $C \in$ **Comm** *and assertion* $\psi \in$ **Assert**, *if* $\Gamma \vdash_{\mathcal{H}_g}$ $\{\phi\} C \{\psi\}$ *for some precondition* $\phi \in$ **Assert**, *then*

1. $\Gamma \vdash_{\mathcal{H}_g} \{\text{wp}(C, \psi)\} C \{\psi\}$
2. $\Gamma \models \phi \rightarrow \text{wp}(C, \psi)$

Proof By induction on the structure of C.   □

Note that $\text{wp}(C, \psi)$ is weaker than any ϕ such that $\Gamma \vdash_{\mathcal{H}_g} \{\phi\} C \{\psi\}$, which does not mean that it is weaker than any ϕ such that $\Gamma \models \{\phi\} C \{\psi\}$. In fact, the original definition of weakest precondition, for programs without annotations, is based on an operational understanding. If one calculates the weakest precondition of a loop by considering its expansion as a conditional command one is led to a recursive equation; the weakest precondition is in fact the fixpoint of this equation (see Exercise 6.2), which may of course be weaker than the annotated invariant. In Sect. 5.3 we discussed the inadequacy of reasoning about loops based on the operational understanding, instead of using invariants. This is in fact the reason why we work with annotated programs: in this way we have a syntactic characterisation of weakest preconditions which is workable for verification purposes.

6.4.2 Calculating Verification Conditions

Our next task is to define a recursive function VC that produces a set of verification conditions from a program and a postcondition. This is given in the second part of Fig. 6.5. The function simply follows the structure of the rules of system $\mathcal{H}_g$ to collect the union of all sets of verification conditions. Observe that in the case of the sequence rule, the wp function is invoked with the second command, to calculate the postcondition for the first command in the sequence.

Recall that according to the weakest precondition strategy for the construction of proof trees, whenever a precondition ϕ occurs in a side condition of the form $\phi \rightarrow \psi$ for some ψ, the strategy prescribes that ϕ is taken to be equal to ψ (the weakest precondition), and so this particular verification condition is trivially satisfied. In fact it turns out that only the loop rule actually introduces verification conditions that need to be checked.

The clause for while loops introduces two verification conditions. One of these corresponds to the second side condition of the while rule (the first side condition is trivially satisfied by the use of weakest preconditions). To understand the other verification condition, note that the loop rule in system $\mathcal{H}_g$ has as premise a Hoare triple with a different precondition from that in the conclusion triple, which forces the inclusion of a verification condition stating that the intended precondition $\theta \wedge b$ must be stronger than the calculated weakest precondition of the loop body, with respect to the invariant θ. Thus this verification condition corresponds to the *preservation of the loop invariant*. It may help to observe that this clause is just an expansion of

$$\mathsf{VC}(\textbf{while } b \textbf{ do } \{\theta\}\, C, \psi) = \{(\theta \wedge \neg b) \rightarrow \psi\} \cup \mathsf{VCG}(\{\theta \wedge b\}\, C\, \{\theta\})$$

6.4.3 Putting It All Together

Weakest preconditions and verification conditions are calculated from a Hoare triple without taking the precondition into consideration. An additional verification condition must be added to the set of generated conditions to ensure that this is stronger than the calculated weakest precondition. The function VCG in Fig. 6.5 calculates the set of verification conditions using the previous two functions and adds this extra condition.

The following result implies that the VCGen can indeed be used to establish the validity of a Hoare triple: if all the verification conditions are valid then so is the triple, since there exists a derivation for it, and moreover if the triple is derivable in Hoare logic then all verification conditions are valid.

Proposition 6.8 (Adequacy of VCGen) *Let* $\{\phi\}\, C\, \{\psi\} \in$ **Spec** *and* $\Gamma \subseteq$ **Assert**.

$$\Gamma \models \mathsf{VCG}(\{\phi\}\, C\, \{\psi\}) \quad \textit{iff} \quad \Gamma \vdash_{\mathcal{H}_g} \{\phi\}\, C\, \{\psi\}$$

$\mathsf{VC}(\textbf{fact},\ f = fact(n))$

$=\quad \mathsf{VC}(f := 1\,;\ i := 1, \mathsf{wp}\,(\textbf{while}\ i \leq n\ \textbf{do}\ \{\theta\}\,C_w,\ f = fact(n)))$
$\qquad \cup\ \mathsf{VC}(\textbf{while}\ i \leq n\ \textbf{do}\ \{\theta\}\,C_w,\ f = fact(n))$

$=\quad \mathsf{VC}(f := 1\,;\ i := 1, \theta)\ \cup\ \{\theta \wedge i \leq n \rightarrow \mathsf{wp}\,(C_w, \theta)\}\ \cup\ \{\theta \wedge i > n \rightarrow f = fact(n)\}$
$\qquad \cup\ \mathsf{VC}(C_w, \theta)$

$=\quad \mathsf{VC}(f := 1, \mathsf{wp}\,(i := 1, \theta))\ \cup\ \mathsf{VC}(i := 1, \theta)$
$\qquad \cup\ \{f = fact(i - 1) \wedge i \leq n + 1 \wedge i \leq n \rightarrow \mathsf{wp}\,(f := f \times i, \mathsf{wp}\,(i := i + 1, \theta))\}$
$\qquad \cup\ \{f = fact(i - 1) \wedge i \leq n + 1 \wedge i > n \rightarrow f = fact(n)\}$
$\qquad \cup\ \mathsf{VC}(f := f \times i, \mathsf{wp}\,(i := i + 1, \theta))\ \cup\ \mathsf{VC}(i := i + 1, \theta)$

$=\quad \emptyset \cup \emptyset \cup \{f = fact(i - 1) \wedge i \leq n + 1 \wedge i \leq n$
$\qquad\qquad\qquad\qquad\qquad \rightarrow \mathsf{wp}\,(f := f \times i,\ f = fact(i + 1 - 1) \wedge i + 1 \leq n + 1)\}$
$\qquad \cup\ \{f = fact(i - 1) \wedge i \leq n + 1 \wedge i > n \rightarrow f = fact(n)\}\ \cup\ \emptyset \cup \emptyset$

$=\quad \{f = fact(i - 1) \wedge i \leq n + 1 \wedge i \leq n \rightarrow f \times i = fact(i + 1 - 1) \wedge i + 1 \leq n + 1,$
$\qquad f = fact(i - 1) \wedge i \leq n + 1 \wedge i > n \rightarrow f = fact(n)\}$

$\mathsf{VCG}(\{n \geq 0\}\,\textbf{fact}\,\{f = fact(n)\})$

$=\quad \{n \geq 0 \rightarrow \mathsf{wp}\,(\textbf{fact},\ f = fact(n))\}\ \cup\ \mathsf{VC}(\textbf{fact},\ f = fact(n))$

$=\quad \{n \geq 0 \rightarrow \mathsf{wp}\,(f := 1\,;\ i := 1, \mathsf{wp}\,(\textbf{while}\ i \leq n\ \textbf{do}\ \{\theta\}\,C_w,\ f = fact(n)))\}$
$\qquad \cup\ \{f = fact(i - 1) \wedge i \leq n + 1 \wedge i \leq n \rightarrow f \times i = fact(i + 1 - 1) \wedge i + 1 \leq n + 1,$
$\qquad f = fact(i - 1) \wedge i \leq n + 1 \wedge i > n \rightarrow f = fact(n)\}$

$=\quad \{n \geq 0 \rightarrow \mathsf{wp}\,(f := 1\,;\ i := 1, \theta),$
$\qquad f = fact(i - 1) \wedge i \leq n + 1 \wedge i \leq n \rightarrow f \times i = fact(i + 1 - 1) \wedge i + 1 \leq n + 1,$
$\qquad f = fact(i - 1) \wedge i \leq n + 1 \wedge i > n \rightarrow f = fact(n)\}$

$=\quad \{n \geq 0 \rightarrow 1 = fact(1 - 1) \wedge 1 \leq n + 1,$
$\qquad f = fact(i - 1) \wedge i \leq n + 1 \wedge i \leq n \rightarrow f \times i = fact(i + 1 - 1) \wedge i + 1 \leq n + 1,$
$\qquad f = fact(i - 1) \wedge i \leq n + 1 \wedge i > n \rightarrow f = fact(n)\}$

Fig. 6.6 Applying the VCGen algorithm

Proof $\Rightarrow$: By induction on the structure of C.
$\Leftarrow$: By induction on the derivation of $\Gamma \vdash_{\mathcal{H}_g} \{\phi\}\,C\,\{\psi\}$. $\qquad\qquad\qquad \square$

In particular, given $\mathcal{T}_{\mathsf{Prog}} = \mathsf{Th}(\mathcal{M}_{\mathsf{Exp}}) \cup \mathcal{T}_{\mathsf{User}}$ for a given programming language interpretation of expressions and set of user axioms, we have

$$\mathcal{T}_{\mathsf{Prog}} \models \mathsf{VCG}(\{\phi\}\,C\,\{\psi\}) \quad \text{iff} \quad \mathcal{T}_{\mathsf{Prog}} \vdash_{\mathcal{H}_g} \{\phi\}\,C\,\{\psi\}$$

To illustrate the functioning of the VCGen, we now return to our running example and obtain the verification conditions of factorial without constructing a proof tree.

Example 6.9 We apply the VCGen algorithm to the factorial program and specification. For simplicity we leave the condition $n = n_0$ out of the precondition and postcondition. Its inclusion would not interfere with the validity of the verification conditions generated (see Exercise 6.4).

We start by calculating in Fig. 6.6, top, $VC(\textbf{fact}, f = fact(n))$. Recall that θ is $f = fact(i - 1) \wedge i \leq n + 1$ and C_w is $f := f \times i\,; i := i + 1$. The same figure, bottom, then shows the calculation of $VCG(\{n \geq 0\}\,\textbf{fact}\,\{f = fact(n)\})$.

The resulting conditions are all valid formulas in our model; in fact they are the side conditions of the proof trees constructed in previous sections for the same Hoare triple. Note that we have chosen to leave them unsimplified to stress that the VCGen does not apply any kind of simplifications. The end result is thus the following set of proof obligations.

1. $n \geq 0 \rightarrow 1 = fact(1 - 1) \wedge 1 \leq n + 1$
2. $f = fact(i - 1) \wedge i \leq n + 1 \wedge i \leq n \rightarrow f \times i = fact(i + 1 - 1) \wedge i + 1 \leq n + 1$
3. $f = fact(i - 1) \wedge i \leq n + 1 \wedge i > n \rightarrow f = fact(n)$

6.5 Verification Conditions for While^{array} Programs

Let us now give an example of calculating verification conditions for a program that uses arrays.

Example 6.10 Consider the program **maxarray** that determines the position of the largest element in an array between indexes 0 and $size - 1$, where $size \geq 1$. Let **maxarray** be

$$
\begin{aligned}
&max := 0\,; \\
&i := 1\,; \\
&\textbf{while } i < size \textbf{ do } \{1 \leq i \leq size \wedge 0 \leq max < i \wedge \\
&\qquad\qquad\qquad\qquad \forall a.\, 0 \leq a < i \rightarrow u[a] \leq u[max]\} \\
&\{ \\
&\quad \textbf{if } u[i] > u[max] \textbf{ then } max := i \textbf{ else skip}\,; \\
&\quad i := i + 1 \\
&\}
\end{aligned}
$$

Our goal is to show that this program indeed meets its specification, i.e. at the end of execution the variable *max* contains an index of the array within the desired range and moreover the element in that position is larger than or equal to the contents of any other position in the range. This can be written as the following Hoare triple.

$$\text{wp}\,(C, \theta) = \big(u[i] > u[max] \rightarrow (1 \leq i + 1 \leq size \wedge 0 \leq i < i + 1 \wedge$$
$$\forall a.\, 0 \leq a < i + 1 \rightarrow u[a] \leq u[i])\big) \wedge$$
$$\big(\neg(u[i] > u[max]) \rightarrow (1 \leq i + 1 \leq size \wedge 0 \leq max < i + 1 \wedge$$
$$\forall a.\, 0 \leq a < i + 1 \rightarrow u[a] \leq u[max])\big)$$

$$\text{wp}\,(\textbf{maxarray}, \psi) = 1 \leq 1 \leq size \wedge 0 \leq 0 < 1 \wedge \forall a.\, 0 \leq a < 1 \rightarrow u[a] \leq u[0]$$

$\text{VC}(\textbf{maxarray}, \psi) = \{$
$\quad (1 \leq i < size \wedge 0 \leq max < i \wedge \forall a.\, 0 \leq a < i \rightarrow u[a] \leq u[max]) \rightarrow \text{wp}\,(C, \theta),$
$\quad (1 \leq i = size \wedge 0 \leq max < i \wedge \forall a.\, 0 \leq a < i \rightarrow u[a] \leq u[max]) \rightarrow \psi\}$

$\text{VCG}(\{size \geq 1\}\,\textbf{maxarray}\,\{\psi\}) = \{size \geq 1 \rightarrow \text{wp}\,(\textbf{maxarray}, \psi)\} \cup$
$\qquad\qquad\qquad\qquad\qquad\qquad\qquad\quad \text{VC}(\textbf{maxarray}, \psi)$

Fig. 6.7 Calculating the verification conditions for **maxarray**

$\{size \geq 1\}$
maxarray
$\{0 \leq max < size \wedge \forall a.\, 0 \leq a < size \rightarrow u[a] \leq u[max]\}$
    Assume

θ	is	$1 \leq i \leq size \wedge 0 \leq max < i \wedge \forall a.\, 0 \leq a < i \rightarrow u[a] \leq u[max]$
C	is	**if** $u[i] > u[max]$ **then** $max := i$ **else skip** ;
		$i := i + 1$
ψ	is	$0 \leq max < size \wedge \forall a.\, 0 \leq a < size \rightarrow u[a] \leq u[max]$

The verification conditions can be calculated by applying the VCGen as shown in Fig. 6.7. The reader can check that the resulting conditions are all valid first-order formulas. Finally, we note that it is straightforward to prove that the loop terminates ($size - i$ is a loop variant), and thus the program **maxarray** is in fact totally correct.

This program contained no occurrences of array assignments. Recall from Chap. 5 that the instruction $u[e] := e'$ is just syntactic sugar for the assignment $u := u[e \triangleright e']$, so the VCGen works without modifications in the presence of array assignments. Similarly to what we did for the inference system, it is easy to extend the VCGen with clauses for array assignment that are just special cases obtained by expanding the syntactic sugar:

$$\text{wp}\,(u[e] := e', \psi) = \psi[u[e \triangleright e']/u]$$
$$\text{VC}(u[e] := e', \psi) = \emptyset$$

so for instance

$$\text{wp}\,(u[i] := 10, u[j] > 100) = u[i \triangleright 10][j] > 100$$

Note that it is also easy to convert the (*array assign*) rule of Sect. 5.5.1 to be used with system $\mathcal{H}_g$ instead of system $\mathcal{H}$. It suffices to expand the syntactic sugar in the (*assign*) rule of system $\mathcal{H}_g$:

(*array assign*, $\mathcal{H}_g$)

$$\frac{}{\{\phi\}\, u[e] := e'\, \{\psi\}} \qquad \text{if } \phi \to \psi[u[e \triangleright e']/u]$$

Clearly the clause given above for $\mathsf{wp}\,(u[e] := e', \psi)$ is in accordance with this rule, as would be expected.

Let us illustrate the use of the VCGen through a few simple examples:

$$\mathsf{VCG}(\{u[j] > 100\}\, u[i] := 10\, \{u[j] > 100\})$$
$$= \{u[j] > 100 \to (u[j] > 100)[u[i \triangleright 10]/u]\}$$
$$= \{u[j] > 100 \to (u[i \triangleright 10][j] > 100)\} \qquad (6.1)$$

$$\mathsf{VCG}(\{i \neq j \wedge u[j] > 100\}\, u[i] := 10\, \{u[j] > 100\})$$
$$= \{i \neq j \wedge u[j] > 100 \to (u[j] > 100)[u[i \triangleright 10]/u]\}$$
$$= \{i \neq j \wedge u[j] > 100 \to (u[i \triangleright 10][j] > 100)\} \qquad (6.2)$$

$$\mathsf{VCG}(\{i = 70\}\, u[i] := 10\, \{u[i] = 10\})$$
$$= \{i = 70 \to (u[i] = 10)[u[i \triangleright 10]/u]\}$$
$$= \{i = 70 \to (u[i \triangleright 10][i] = 10)\} \qquad (6.3)$$

Note that when reasoning about programs of While^{array} the theory $\mathcal{T}_{\mathsf{Prog}}$ includes the interpretation of arrays of Definition 5.15. Equalities involving array lookup are usually decided using the *theory of arrays* $\mathcal{T}_{\mathsf{A}}$, see Sect. 4.7.4. We assume that the proof tool being used supports this theory (this is the case for instance if an SMT solver is used). Note that in our syntax we write $u[e]$ instead of $read(u, e)$, and $u[e \triangleright e']$ instead of $write(u, e, e')$.

The verification condition (6.1) cannot be proved using the axioms of Sect. 4.7.4 (since i and j may have the same value); (6.2) is discharged using *read/write axiom 2*, and (6.3) can be proved using *read/write axiom 1*. Let us now look at a more interesting example.

Example 6.11 The following program calculates the factorial of all the numbers stored in the positions with indexes 0 to $size - 1$ of a given array *in*, and stores the results in the corresponding positions of array *out*. Let **factab** be

$$k := 0;$$
$$\textbf{while } k < size \textbf{ do } \{\theta_2^0 \wedge \theta_2\} \{$$
$$\quad f := 1; \ i := 1; \ n := in[k];$$
$$\quad \textbf{while } i \leq n \textbf{ do } \{\theta_1^0 \wedge \theta_1\} \{$$
$$\quad\quad f := f \times i;$$
$$\quad\quad i := i + 1$$
$$\quad \}$$
$$\quad out[k] := f;$$
$$\quad k := k + 1$$
$$\}$$

where

θ_2^0 is $size \geq 0 \wedge \forall a. \ 0 \leq a < size \rightarrow in[a] \geq 0$

θ_2 is $0 \leq k \leq size \wedge \forall a. \ 0 \leq a < k \rightarrow out[a] = fact(in[a])$

θ_1^0 is $\theta_2^0 \wedge n = in[k] \wedge 0 \leq k < size \wedge \ \forall a. \ 0 \leq a < k \rightarrow out[a] = fact(in[a])$

θ_1 is $1 \leq i \leq n + 1 \wedge f = fact(i - 1)$

It is well worth understanding in detail the invariants involved in this example, which are not related to arrays, but involve a general issue which we now discuss.

The invariant of the outermost loop has two components: θ_2 is typical of an array-traversal loop: it states the bounds of k and contains also a universally quantified assertion that concerns the array positions that have already been visited by the loop (in the current case, it states that factorial has been calculated for the positions with indexes 0 to $k - 1$). θ_2^0 on the other hand is just the precondition of the program, which will be preserved throughout execution. Note that this component of the invariant is trivially preserved by loop iterations, since it contains no occurrences of the variables assigned by the loop body (for this reason such invariants are sometimes called *continuous invariants*).

The invariant of the innermost loop also has two components: θ_1 concerns the factorial calculation itself and is straightforward to understand (see Example 5.10). θ_1^0 on the other hand does not concern what the innermost loop does: it is a continuous invariant containing information that can be inferred from the invariant of the outermost loop and also from the assignment instruction that precedes the loop. In particular, it contains θ_2^0.

The need for continuous invariants comes from the verification condition that relates the loop invariant (together with the negated loop condition) and the calculated weakest precondition ψ of the subsequent command. The weakest precondition of the loop 'forgets' ψ (the postcondition with respect to which it was calculated), and the continuous invariant plays the role of transporting information between the initial and final states of the loop execution. This will hopefully become clear by looking at the verification conditions calculated below. We should also mention that tools for realistic languages (like the VCGen studied in Chap. 10) are capable of keeping this transported information in the context automatically; there is no need to explicitly include continuous invariants.

$$\mathsf{wp}\,(C_1, \theta_1^0 \wedge \theta_1) \;=\; \theta_1^0 \wedge 1 \le i+1 \le n+1 \wedge f \times i = fact(i+1-1)$$

$$\begin{aligned}
\mathsf{wp}\,(C_2, \theta_2^0 \wedge \theta_2) \;=\;\; & \theta_2^0 \wedge in[k] = in[k] \wedge 0 \le k < size\; \wedge \\
& \forall a.\, 0 \le a < k \to out[a] = fact(in[a])\; \wedge \\
& 1 \le 1 \le in[k]+1 \wedge 1 = fact(1-1)
\end{aligned}$$

$$\mathsf{wp}\,(\mathbf{factab}, \psi) \;=\; 0 \le 0 \le size \wedge \forall a.\, 0 \le a < 0 \to out[a] = fact(in[a])$$

$$\begin{aligned}
\mathsf{VC}(C_2, \theta_2^0 \wedge \theta_2) \;=\;\; & \mathsf{VC}(\mathbf{while}\ i \le n\ \mathbf{do}\ \{\theta_1^0 \wedge \theta_1\}\,C_1, \theta_2^0 \wedge 0 \le k+1 \le size\; \wedge \\
& \qquad\qquad \forall a.\, 0 \le a < k+1 \to out[k \rhd f][a] = fact(in[a])) \\
\;=\;\; & \{\, \theta_1^0 \wedge 1 \le i \le n \wedge f = fact(i-1) \to \mathsf{wp}\,(C_1, \theta_1^0 \wedge \theta_1), \\
& \quad \theta_1^0 \wedge 1 \le i = n+1 \wedge f = fact(i-1) \\
& \quad \to (\theta_2^0 \wedge 0 \le k+1 \le size\; \wedge \\
& \qquad \forall a.\, 0 \le a < k+1 \to out[k \rhd f][a] = fact(in[a]))\,\}
\end{aligned}$$

$$\begin{aligned}
\mathsf{VC}(\mathbf{while}\ & k < size\ \mathbf{do}\ \{\theta_2^0 \wedge \theta_2\}\,C_2, \psi) = \\
& \{\, \theta_2^0 \wedge 0 \le k < size \wedge (\forall a.\, 0 \le a < k \to out[a] = fact(in[a])) \to \mathsf{wp}\,(C_2, \theta_2^0 \wedge \theta_2), \\
& \;\; \theta_2^0 \wedge 0 \le k = size \wedge (\forall a.\, 0 \le a < k \to out[a] = fact(in[a])) \to \psi \,\} \cup \\
& \mathsf{VC}(C_2, \theta_2^0 \wedge \theta_2)
\end{aligned}$$

$$\begin{aligned}
\mathsf{VCG}(\{\phi\}\,\mathbf{factab}\,\{\psi\}) \;=\;\; & \{\phi \to \mathsf{wp}\,(\mathbf{factab}, \psi)\} \cup \\
& \mathsf{VC}(\mathbf{while}\ k < size\ \mathbf{do}\ \{\theta_2^0 \wedge \theta_2\}\,C_2, \psi)
\end{aligned}$$

Fig. 6.8 Calculating the VCs for the **factab** example

We now want to show the validity of the triple $\{\phi\}\,\mathbf{factab}\,\{\psi\}$, where

$$\begin{aligned}
\phi \quad &\text{is} \quad size \ge 0 \wedge \forall a.\, 0 \le a < size \to in[a] \ge 0 \\
\psi \quad &\text{is} \quad \forall a.\, 0 \le a < size \to out[a] = fact(in[a])
\end{aligned}$$

The verification conditions are calculated in Fig. 6.8, where we use the abbreviations

$$\begin{aligned}
C_1 \quad &\text{is} \quad f := f \times i;\, i := i+1 \\
C_2 \quad &\text{is} \quad f := 1;\, i := 1;\, n := in[k]; \\
& \qquad \mathbf{while}\ i \le n\ \mathbf{do}\ \{\theta_1^0 \wedge \theta_1\}\,C_1; \\
& \qquad out[k] := f;\, k := k+1
\end{aligned}$$

The reader will have noticed that as the examples grow slightly more complex, it becomes more difficult not only to generate the verification conditions by hand, but also to convince ourselves that all of them are valid first-order formulas. Clearly the existence of tools to support us in theses tasks is of great interest; one such tool will be studied in Chap. 10.

Equality of Arrays   Although we did not include array equality in the language While[array], there would be advantages in doing so; in particular this would allow us to use auxiliary variables for the purposes explained in Sect. 5.2, to relate the contents of an array in the post-state and in the pre-state.

The reader may recall from Sect. 4.7.4 that in addition to the *read/write* axioms, the array theory T_A includes the theory of equality T_E, an *array congruence* axiom, and an *array extensionality axiom*. The extensionality axiom, which in our syntax would be written as

$$\forall u, v. \ (\forall i. \ u[i] = v[i]) \rightarrow u = v$$

states that two arrays are equal if their contents are the same.

Note that the equality symbol is overloaded, since it was already used in While^{int} with rank $\langle \mathbf{int}, \mathbf{int} \rangle$ for integer equality, and we are now using it with rank $\langle \mathbf{array}, \mathbf{array} \rangle$.

6.6 To Learn More

The idea of calculating the weakest precondition which ensures that a given postcondition holds after termination of a program was advanced by Dijkstra in his *Weakest Liberal Precondition* (WLP) calculus. His approach was based on a *guarded commands* language and the notion of *predicate transformer* associated to a program. Given a program C, the predicate transformer *wlp.C* is a function on assertions, that maps a postcondition ψ into the weakest precondition sufficient for ψ to be true after execution of C, if C terminates.

Dijkstra's work has practically started a whole branch of formal methods, the *correct-by-construction* approach to software development already mentioned in Chap. 2. The WLP calculus is extensively employed in this approach, but it is also used for program verification: in addition to the influence on VCGens like the one studied in this chapter, a variant of this calculus is also used *directly as a VCGen* in a family of tools that includes ESC/Java [9] and Boogie [2]. Typically, in this approach the guarded commands language does not include any form of iteration. Programs of the source language to be verified are translated as guarded commands programs, and loops are encoded as guarded commands that are operationally different but have the same weakest preconditions as the original code.

The classic reference for Dijkstra's WLP calculus is [4]. Many other textbooks [1, 6, 8] cover the *correct-by-construction* approach to program development, either using guarded commands and predicate transformers or Hoare triples and inference rules.

For richer programming languages, it becomes more difficult to prove the soundness of a VCGen. An interesting topic that is outside the scope of this book is the use of proof tools to formalise and prove the soundness of the verification infrastructure itself. A formalisation and proof of soundness using a tool like HOL or Coq becomes invaluable. Homeier and Martin [7] have used HOL to prove the soundness of a VCGen for a While language with side effects. Bertot [3] has produced a Coq development that is a general formalisation of the semantics of a While language; this includes an embedding of the inference system of Hoare logic, in which the rules are encoded as an inductive predicate definition. The system is proved correct with respect to the operational semantics, also formalised in the same development.

Gordon uses a VCGen [5] that is slightly different from the one studied in this chapter in that it does not require weakest preconditions to be calculated by an auxiliary function, since sequence commands are partially annotated in a way that makes them easy to guess. This requires users to annotate programs not only with loop invariants but also with assertions that are valid at intermediate points in sequence commands.

Gordon's VCGen is proposed in the context of a mechanisation of Hoare logic using the HOL proof assistant. An interesting aspect of this mechanisation is that the VCGen is implemented as a *tactic* to construct derivations of Hoare logic: a function that takes the current proof construction and extends it step-by-step until a HOL proof is obtained.

6.7 Exercises

6.1. Formulate informally the weakest precondition strategy for the construction of derivations in the original inference system $\mathcal{H}$ of Hoare logic (Fig. 5.4). Note that you will have to prescribe how the consequence rule should be applied.

6.2. Consider calculating the weakest precondition of a *while* loop for which no invariant is provided. Show that using the following equivalence (provable using the operational semantics)

$$\textbf{while}\, b \,\textbf{do}\, C \equiv \textbf{if}\, b \,\textbf{then}\, (C \,;\, \textbf{while}\, b \,\textbf{do}\, C) \,\textbf{else}\, \textbf{skip}$$

together with the definition of weakest precondition of the conditional command, leads to a recursive equation.

6.3. Define an alternative VCGen algorithm for the inference system $\mathcal{H}_g$ of Fig. 6.1 that does not resort to the auxiliary function VC, i.e. the function VCG that takes as input a Hoare triple should itself be recursive and use only wp as auxiliary. Apply this VCGen to a suitable example and compare its output verification conditions with those that would be generated by the VCGen of Fig. 6.5.

6.4. Recalculate the verification conditions of the factorial program (Example 6.9), including the condition $n = n_0$ in the precondition and postcondition. You will need to include the same condition in the loop invariant.

6.5. Recall Exercise 5.10, which extended the programming language with local scope variables. Extend the VCGen studied in the present chapter to cope with this feature.

6.6. In this chapter we have considered the generation of verification conditions for partial correctness. Taking as point of departure the rule given in Sect. 5.6, adapt the inference system $\mathcal{H}_g$ of Fig. 6.1 to derive total correctness Hoare triples.

6.7. Define a VCGen for total correctness, taking into consideration that loops are annotated with variants as well as invariants.

6.8. Prove that the axioms of the theory of arrays of Sect. 4.7.4 are valid in the model of Definition 5.15.

6.9. We have assumed that the proof tool used for checking verification conditions related to array assignment contains a theory of applicative arrays. However, if this is not the case, then the VCGen can always expand the conditions according to the theory of arrays, for instance

$$u[i \rhd 10][j] > 100 \quad \text{stands for} \quad (i = j \to 10 > 100) \land (i \neq j \to u[j] > 100)$$

Write an algorithm that transforms an assertion containing occurrences of the array update operator into an equivalent assertion where no array updates occur (so no theory of arrays is required for reasoning).

References

1. Backhouse, R.: Program Construction—Calculating Implementations from Specifications. Wiley, New York (2003)
2. Barnett, M., Chang, B.-Y.E., DeLine, R., Jacobs, B., Leino, K.R.M.: Boogie: A modular reusable verifier for object-oriented programs. In: de Boer, F.S., Bonsangue, M.M., Graf, S., de Roever, W.P. (eds.) FMCO. Lecture Notes in Computer Science, vol. 4111, pp. 364–387. Springer, Berlin (2005)
3. Bertot, Y.: Theorem proving support in programming language semantics. CoRR, abs/0707.0926 (2007)
4. Dijkstra, E.W.: A Discipline of Programming. Prentice-Hall International, Englewood Cliffs (1976)
5. Gordon, M.J.C.: Mechanizing programming logics in higher order logic. In: Birtwistle, G., Subrahmanyam, P.A. (eds.) Current Trends in Hardware Verification and Automated Theorem Proving, pp. 387–439. Springer, New York (1989)
6. Gries, D.: The Science of Programming. Springer, Secaucus (1987)
7. Homeier, P.V., Martin, D.F.: A mechanically verified verification condition generator. Comput. J. **38**(2), 131–141 (1995)
8. Kaldewaij, A.: Programming: The Derivation of Algorithms. Prentice-Hall International, Upper Saddle River (1990)
9. Leino, K.R.M., Saxe, J.B., Stata, R.: Checking Java programs via guarded commands. In: Proceedings of the Workshop on Object-Oriented Technology, London, UK, 1999, pp. 110–111. Springer, Berlin (1999)

Chapter 7
Safety Properties

This chapter considers a method for dealing with the occurrence of *errors* during the execution of programs. Recall that to this point we have been considering that evaluation of an expression could never go wrong, and neither could the execution of a command. In particular in While^{int} division by zero did not go wrong, and the range of arrays in While^{array} was $\mathbb{Z}$. Finite ranges could not be supported, since there was no way to signal an error when some command tried to access a position of an array outside its range.

This is clearly unsatisfying—in real-world languages runtime errors do occur, and one of the main uses of verification methods is precisely to ensure the *safety* of programs, i.e. the absence of such error situations. We will now present a general framework for reasoning with errors and safety.

Following our previous approach, the framework will be independent from the language of program expressions. We will illustrate its application with the languages While^{int} and While$^{\text{array[N]}}$, a language with bounded arrays.

7.1 Error Semantics and Safe Programs

Until this point we have assumed a simple semantics in which all expressions evaluate to some value, and all commands can be executed without ever "going wrong". Of course in realistic programming languages this is not the case. We will now address this extra degree of realism by

- incorporating in the language semantics a special **error** value in the interpretation domains of expressions; expressions of type τ will be interpreted as values in the corresponding domain of interpretation D_τ extended with the distinguished value **error** that will be used to interpret undefined expressions (like those involving division by zero in While^{int} for instance).
- modifying the evaluation relation to admit evaluation of commands to a special **error** state.

J.B. Almeida et al., *Rigorous Software Development*,
Undergraduate Topics in Computer Science,
DOI 10.1007/978-0-85729-018-2_7, © Springer-Verlag London Limited 2011

1. $(\mathbf{skip}, s) \rightsquigarrow_{\mathcal{M}_e} s$.
2. If $[\![e]\!]_{\mathcal{M}_e}(s) = \mathbf{error}$, then $(x := e, s) \rightsquigarrow_{\mathcal{M}_e} \mathbf{error}$.
3. If $[\![e]\!]_{\mathcal{M}_e}(s) \neq \mathbf{error}$, then $(x := e, s) \rightsquigarrow_{\mathcal{M}_e} s[x \mapsto [\![e]\!]_{\mathcal{M}_e}(s)]$.
4. If $(C_1, s) \rightsquigarrow_{\mathcal{M}_e} \mathbf{error}$, then $(C_1 ; C_2, s) \rightsquigarrow_{\mathcal{M}_e} \mathbf{error}$.
5. If $(C_1, s) \rightsquigarrow_{\mathcal{M}_e} s'$, $s' \neq \mathbf{error}$, and $(C_2, s') \rightsquigarrow_{\mathcal{M}_e} s''$, then $(C_1 ; C_2, s) \rightsquigarrow_{\mathcal{M}_e} s''$.
6. If $[\![b]\!]_{\mathcal{M}_e}(s) = \mathbf{error}$, then $(\mathbf{if}\ b\ \mathbf{then}\ C_t\ \mathbf{else}\ C_f, s) \rightsquigarrow_{\mathcal{M}_e} \mathbf{error}$.
7. If $[\![b]\!]_{\mathcal{M}_e}(s) = \mathbf{T}$ and $(C_t, s) \rightsquigarrow_{\mathcal{M}_e} s'$, then $(\mathbf{if}\ b\ \mathbf{then}\ C_t\ \mathbf{else}\ C_f, s) \rightsquigarrow_{\mathcal{M}_e} s'$.
8. If $[\![b]\!]_{\mathcal{M}_e}(s) = \mathbf{F}$ and $(C_f, s) \rightsquigarrow_{\mathcal{M}_e} s'$, then $(\mathbf{if}\ b\ \mathbf{then}\ C_t\ \mathbf{else}\ C_f, s) \rightsquigarrow_{\mathcal{M}_e} s'$.
9. If $[\![b]\!]_{\mathcal{M}_e}(s) = \mathbf{error}$, then $(\mathbf{while}\ b\ \mathbf{do}\ \{\theta\}\ C, s) \rightsquigarrow_{\mathcal{M}_e} \mathbf{error}$.
10. If $[\![b]\!]_{\mathcal{M}_e}(s) = \mathbf{T}$ and $(C, s) \rightsquigarrow_{\mathcal{M}_e} \mathbf{error}$, then $(\mathbf{while}\ b\ \mathbf{do}\ \{\theta\}\ C, s) \rightsquigarrow_{\mathcal{M}_e} \mathbf{error}$.
11. If $[\![b]\!]_{\mathcal{M}_e}(s) = \mathbf{T}$, $(C, s) \rightsquigarrow_{\mathcal{M}_e} s'$, $s' \neq \mathbf{error}$, and $(\mathbf{while}\ b\ \mathbf{do}\ \{\theta\}\ C, s') \rightsquigarrow_{\mathcal{M}_e} s''$, then $(\mathbf{while}\ b\ \mathbf{do}\ \{\theta\}\ C, s) \rightsquigarrow_{\mathcal{M}_e} s''$.
12. If $[\![b]\!]_{\mathcal{M}_e}(s) = \mathbf{F}$, then $(\mathbf{while}\ b\ \mathbf{do}\ \{\theta\}\ C, s) \rightsquigarrow_{\mathcal{M}_e} s$.

Fig. 7.1 Natural semantics rules with error state

Definition 7.1 (Error semantics of program expressions) Let $\mathcal{M}_e = (D_e, I_e)$ be a $\mathcal{V}_{\mathsf{Exp}}$-structure, where $D_e = \{(D_\tau \cup \{\mathbf{error}\})\}_{\tau \in \mathbf{Type}}$. The semantics of program expressions is given by a functional $[\![\cdot]\!]_{\mathcal{M}_e}$ that maps every $e^\tau \in \mathbf{Exp}_\tau$ to a function $[\![e^\tau]\!]_{\mathcal{M}_e} : \Sigma_D \to (D_\tau \cup \{\mathbf{error}\})$ inductively defined by

$$
\begin{aligned}
[\![x^\tau]\!]_{\mathcal{M}_e}(s) &= s(x^\tau) \\
[\![f^\tau(e^{\tau_1}, \ldots, e^{\tau_n})]\!]_{\mathcal{M}_e}(s) &= I_e(f^\tau)([\![e^{\tau_1}]\!]_{\mathcal{M}_e}(s), \ldots, [\![e^{\tau_n}]\!]_{\mathcal{M}_e}(s))
\end{aligned}
$$

The natural semantics is a simple extension of the previous version. A command will evaluate to an error state, also denoted **error**, if the evaluation of some of the program expressions involved evaluate to **error**.

Definition 7.2 (Evaluation relation with error state) Let $\mathcal{M}_e = (D_e, I_e)$ be a $\mathcal{V}_{\mathsf{Exp}}$-structure. The relation $\rightsquigarrow_{\mathcal{M}_e} \subseteq \mathbf{Comm} \times \Sigma_D \times (\Sigma_D \cup \{\mathbf{error}\})$ is defined inductively by the set of rules given in Fig. 7.1. $(C, s) \rightsquigarrow_{\mathcal{M}_e} s'$ denotes the fact that if C is executed in the initial state s, then its execution will terminate, and the final state will be s', which may now be the error state.

The binary relation on states induced by C satisfies the following property, which means that the semantic interpretation of a command C can be seen as a partial function

Proposition 7.3 (Determinacy) If $(C, s) \rightsquigarrow_{\mathcal{M}_e} s'$ and $(C, s) \rightsquigarrow_{\mathcal{M}_e} s''$, then $s' = s''$.

Proof By induction on the structure of C. $\square$

Recall that in Chap. 5 the interpretation structures for terms and assertions were obtained as expansions of the interpretation structure $\mathcal{M}_{\mathsf{Exp}}$ implemented by the programming language. In the present context this will no longer be the case, since terms and assertions will still be interpreted as total functions that always produce some fixed value in Σ_D. Instead we impose a relation between $\mathcal{V}_{\mathsf{Exp}}$-structures and $\mathcal{V}_{\mathsf{Prog}}$-structures as follows.

Definition 7.4 (Compatible structures) Let $\mathcal{M}_\mathbf{e} = (D_\mathbf{e}, I_\mathbf{e})$ be a $\mathcal{V}_{\mathsf{Exp}}$-structure, and $\mathcal{M} = (D, I)$ a $\mathcal{V}_{\mathsf{Prog}}$-structure. We say that $\mathcal{M}$ *is compatible with* $\mathcal{M}_\mathbf{e}$ when, for every $\tau \in \mathbf{Type}$ and $e^\tau \in \mathbf{Exp}_\tau$, $[\![e^\tau]\!]_\mathcal{M} = [\![e^\tau]\!]_{\mathcal{M}_\mathbf{e}}$ whenever $[\![e^\tau]\!]_{\mathcal{M}_\mathbf{e}} \neq \mathbf{error}$.

The idea is that if some program expression does not evaluate to **error**, then the corresponding assertion or term of the assertion language will evaluate to the same value as the expression. Note that one way to obtain a compatible structure at the logical level is by choosing some designated value of each domain to represent errors, and then *unlifting* the interpretation structure of expressions $\mathcal{M}_\mathbf{e}$, i.e. substituting the designated values for **error** (note that in the presence of a user-provided vocabulary and theory an adequate expansion of this model may be required).

Finally, we introduce a different sort of Hoare triples whose interpretation is modified to include the requirement that execution does not go wrong. We extend the phrase type **Spec** with *safety-sensitive* Hoare triples for partial correctness

$$\mathbf{Spec} \ni S \quad ::= \quad \dots \mid \{\!\!\{\phi\}\!\!\} \, C \, \{\!\!\{\psi\}\!\!\}$$

Informally, $\{\!\!\{\phi\}\!\!\} \, C \, \{\!\!\{\psi\}\!\!\}$ tells us that if ϕ holds in a given state and C is executed in that state, then either the execution of C does not stop, or if it does, the final state is not the error state and ψ will hold in it.

Definition 7.5 (Semantics of safety-sensitive Hoare triples) Let $\mathcal{M}_\mathbf{e} = (D_\mathbf{e}, I_\mathbf{e})$ be a $\mathcal{V}_{\mathsf{Exp}}$-structure and $\mathcal{M} = (D, I)$ a $\mathcal{V}_{\mathsf{Prog}}$-structure such that $\mathcal{M}$ is compatible with $\mathcal{M}_\mathbf{e}$. The semantics of a partial correctness Hoare triple $\{\!\!\{\phi\}\!\!\} \, C \, \{\!\!\{\psi\}\!\!\}$ is given by a function $[\![\{\!\!\{\phi\}\!\!\} \, C \, \{\!\!\{\psi\}\!\!\}]\!]_{\mathcal{M},\mathcal{M}_\mathbf{e}} : \Sigma_D \to \{\mathbf{F}, \mathbf{T}\}$ defined as follows

$$[\![\{\!\!\{\phi\}\!\!\} \, C \, \{\!\!\{\psi\}\!\!\}]\!]_{\mathcal{M},\mathcal{M}_\mathbf{e}}(s) = \mathbf{T} \quad \text{iff} \quad \begin{array}{l} \text{if } [\![\phi]\!]_\mathcal{M}(s) \text{ and } (C, s) \leadsto_{\mathcal{M}_\mathbf{e}} s', \\ \text{then } s' \neq \mathbf{error} \text{ and } [\![\psi]\!]_\mathcal{M}(s') = \mathbf{T} \end{array}$$

We write $\mathcal{M}, \mathcal{M}_\mathbf{e} \models \{\!\!\{\phi\}\!\!\} \, C \, \{\!\!\{\psi\}\!\!\}$ to denote the fact that $[\![\{\!\!\{\phi\}\!\!\} \, C \, \{\!\!\{\psi\}\!\!\}]\!]_{\mathcal{M},\mathcal{M}_\mathbf{e}}(s) = \mathbf{T}$ for all states $s \in \Sigma_D$.

The semantics of total correctness triples can be similarly defined (see Exercise 7.1). In the next section we will design an inference system for this new notion of Hoare triple. We will see that it will be crucial that assertions and program expressions are interpreted in compatible structures.

Before that however we illustrate the semantics introduced in this section for the case of the While^{int} language.

7.1.1 While^{int} *with Errors*

Let us now give an error semantics for While^{int}-expressions. A division by zero expression will be interpreted as **error**, and every expression containing a sub-expression that is interpreted as **error** will also be interpreted as **error**.

Definition 7.6 (Semantics of expressions of While^{int} with error) The functional $[\![\cdot]\!]$ maps every $e \in \mathbf{Exp_{int}}$ to a function $[\![e]\!] : \Sigma \to (\mathbb{Z} \cup \{\mathbf{error}\})$ and every $b \in \mathbf{Exp_{bool}}$ to a function $[\![b]\!] : \Sigma \to (\{\mathbf{F}, \mathbf{T}\} \cup \{\mathbf{error}\})$ as follows

- $[\![e]\!] : \Sigma \to (\mathbb{Z} \cup \{\mathbf{error}\})$ is defined inductively by:

$$[\![n]\!](s) \;=\; n$$

$$[\![x]\!](s) \;=\; s(x)$$

$$[\![-e]\!](s) \;=\; \begin{cases} -[\![e]\!](s) & \text{if } [\![e]\!](s) \neq \mathbf{error} \\ \mathbf{error} & \text{if } [\![e]\!](s) = \mathbf{error} \end{cases}$$

$$[\![e_1 \odot e_2]\!](s) \;=\; \begin{cases} [\![e_1]\!](s) \odot [\![e_2]\!](s) & \text{if } [\![e_1]\!](s) \neq \mathbf{error} \text{ and } [\![e_2]\!](s) \neq \mathbf{error} \\ \mathbf{error} & \text{otherwise} \end{cases}$$
$$\text{for } \odot \in \{+, -, \times\}$$

$$[\![e_1 \mathrel{\mathsf{div}} e_2]\!](s) \;=\; \begin{cases} [\![e_1]\!](s) \div [\![e_2]\!](s) & \text{if } [\![e_1]\!](s) \neq \mathbf{error} \text{ and} \\ & \quad [\![e_2]\!](s) \neq \mathbf{error} \text{ and } [\![e_2]\!](s) \neq 0 \\ \mathbf{error} & \text{otherwise} \end{cases}$$

$$[\![e_1 \mathrel{\mathsf{mod}} e_2]\!](s) \;=\; \begin{cases} [\![e_1]\!](s) \mathrel{\mathsf{mod}} [\![e_2]\!](s) & \text{if } [\![e_1]\!](s) \neq \mathbf{error} \text{ and} \\ & \quad [\![e_2]\!](s) \neq \mathbf{error} \text{ and } [\![e_2]\!](s) \neq 0 \\ \mathbf{error} & \text{otherwise} \end{cases}$$

- $[\![b]\!] : \Sigma \to (\{\mathbf{F}, \mathbf{T}\} \cup \{\mathbf{error}\})$ is defined inductively by:

$$[\![\top]\!](s) \;=\; \mathbf{T}$$

$$[\![\bot]\!](s) \;=\; \mathbf{F}$$

$$[\![\neg b]\!](s) \;=\; \begin{cases} \mathbf{T} & \text{if } [\![b]\!](s) = \mathbf{F} \\ \mathbf{F} & \text{if } [\![b]\!](s) = \mathbf{T} \\ \mathbf{error} & \text{if } [\![b]\!](s) = \mathbf{error} \end{cases}$$

$$[\![b_1 \odot b_2]\!](s) \;=\; \begin{cases} [\![b_1]\!](s) \odot [\![b_2]\!](s) & \text{if } [\![b_1]\!](s) \neq \mathbf{error} \text{ and } [\![b_2]\!](s) \neq \mathbf{error} \\ \mathbf{error} & \text{otherwise} \end{cases}$$
$$\text{for } \odot \in \{=, \neq, <, \leq, >, \geq\}$$

$$[\![b_1 \wedge b_2]\!](s) \;=\; \begin{cases} \mathbf{F} & \text{if } [\![b_1]\!](s) = \mathbf{F} \\ \mathbf{error} & \text{if } [\![b_1]\!](s) = \mathbf{error} \\ [\![b_2]\!](s) & \text{otherwise} \end{cases}$$

$$[\![b_1 \vee b_2]\!](s) \;=\; \begin{cases} \mathbf{T} & \text{if } [\![b_1]\!](s) = \mathbf{T} \\ \mathbf{error} & \text{if } [\![b_1]\!](s) = \mathbf{error} \\ [\![b_2]\!](s) & \text{otherwise} \end{cases}$$

Note that in what concerns terms and assertions nothing changes with respect to what we had in Chap. 5; in particular, division by zero still has some fixed value as result. It is easy to see that the interpretation structure used for $\mathsf{While}^{\mathsf{int}}$ (see Sect. 5.1.2) is compatible with Definition 7.6.

We exemplify the evaluation semantics that results from this interpretation of expressions with errors.

Example 7.7 Consider the evaluation of the $\mathsf{While}^{\mathsf{int}}$-expressions below. Let s be a state such that $s(x) = 10$ and $s(y) = 0$.

$$[\![(x \ \mathsf{div} \ y) > 2]\!](s) = \mathbf{error}, \quad \text{because } [\![y]\!](s) = 0$$

and

$$(\mathbf{if} \ (x \ \mathsf{div} \ y) > 2 \ \mathbf{then} \ C_t \ \mathbf{else} \ C_f, s) \leadsto \mathbf{error}$$

no matter what C_t and C_f are.

7.2 Safety-Sensitive Calculus and VCGen

We now need to adapt the inference system $\mathcal{H}_g$ to cope with this new notion of correctness. Figure 7.2 contains the rules of the inference system $\mathcal{H}_s$ for safety-sensitive Hoare triples. Note that in this system the side conditions of the rules do not depend only on the structure of the command, but also on expressions that may occur in it. In order to be able to infer that a command executes without ever going wrong, we need to have the capacity to describe sufficient conditions guaranteeing that program expressions do not evaluate to **error**. These new side conditions will be called *safety conditions*.

Definition 7.8 Let $\mathsf{safe} : \bigcup_{\tau \in \mathbf{Type}} \mathbf{Exp}_\tau \to \mathbf{Assert}$ be a function that maps every expression into a first-order formula, $\Gamma \subseteq \mathbf{Assert}$, and $\{\!|\phi|\!\} \, C \, \{\!|\psi|\!\} \in \mathbf{Spec}$. We say that $\{\!|\phi|\!\} \, C \, \{\!|\psi|\!\}$ is *derivable in* $\mathcal{H}_s$ *assuming* Γ, denoted $\Gamma \vdash_{\mathcal{H}_s} \{\!|\phi|\!\} \, C \, \{\!|\psi|\!\}$, if there exists a proof of $\{\!|\phi|\!\} \, C \, \{\!|\psi|\!\}$ in $\mathcal{H}_s$ which uses as assumptions (for the side conditions) assertions from Γ.

The idea is that the truth of the assertion $\mathsf{safe}(e^\tau)$ in a given state implies that the evaluation of e^τ in that state will not produce an error—the evaluation is safe. Naturally the soundness of system $\mathcal{H}_s$ depends on this property.

$(skip)$
$$\frac{}{\{\!|\phi|\!\} \,\textbf{skip}\, \{\!|\psi|\!\}} \quad \text{if } \phi \to \psi$$

$(assign)$
$$\frac{}{\{\!|\phi|\!\} \, x := e \, \{\!|\psi|\!\}} \quad \text{if } \phi \to \mathsf{safe}(e) \text{ and } \phi \to \psi[e/x]$$

(seq)
$$\frac{\{\!|\phi|\!\} \, C_1 \, \{\!|\theta|\!\} \qquad \{\!|\theta|\!\} \, C_2 \, \{\!|\psi|\!\}}{\{\!|\phi|\!\} \, C_1 \,;\, C_2 \, \{\!|\psi|\!\}}$$

$(while)$
$$\frac{\{\!|\theta \wedge b|\!\} \, C \, \{\!|\theta|\!\}}{\{\!|\phi|\!\} \,\textbf{while}\, b \,\textbf{do}\, \{\theta\} \, C \, \{\!|\psi|\!\}} \quad \text{if } \phi \to \theta \text{ and } \theta \to \mathsf{safe}(b) \text{ and } \theta \wedge \neg b \to \psi$$

(if)
$$\frac{\{\!|\phi \wedge b|\!\} \, C_t \, \{\!|\psi|\!\} \qquad \{\!|\phi \wedge \neg b|\!\} \, C_f \, \{\!|\psi|\!\}}{\{\!|\phi|\!\} \,\textbf{if}\, b \,\textbf{then}\, C_t \,\textbf{else}\, C_f \, \{\!|\psi|\!\}} \quad \text{if } \phi \to \mathsf{safe}(b)$$

Fig. 7.2 Hoare logic with safety conditions: system $\mathcal{H}_s$

Proposition 7.9 (Soundness of $\mathcal{H}_s$) *Let* $\mathcal{M}_e = (D_e, I_e)$ *be a* V_{Exp}-*structure and* $\mathcal{M} = (D, I)$ *a* V_{Prog}-*structure such that* $\mathcal{M}$ *is compatible with* $\mathcal{M}_e$, *and for every* $s \in \Sigma_D$,

$$[\![e^\tau]\!]_{\mathcal{M}_e}(s) \neq \textbf{\textit{error}} \quad \text{whenever} \quad [\![\mathsf{safe}(e^\tau)]\!]_{\mathcal{M}}(s) = \mathbf{T}$$

Then

$$\text{if } \Gamma \vdash_{\mathcal{H}_s} \{\!|\phi|\!\} \, C \, \{\!|\psi|\!\} \text{ and } \mathcal{M} \models \Gamma, \text{ then } \mathcal{M}, \mathcal{M}_e \models \{\!|\phi|\!\} \, C \, \{\!|\psi|\!\}.$$

Proof By induction on the derivation of $\Gamma \vdash_{\mathcal{H}_s} \{\!|\phi|\!\} \, C \, \{\!|\psi|\!\}$. For the while case we also proceed by induction on the definition of the evaluation relation. □

A safety-sensitive VCGen is given in Fig. 7.3. It is very similar to the original VCGen but includes safety conditions in the set of generated verification conditions. The function that calculates the set of verification conditions is now called VCG^s.

Proposition 7.10 (Adequacy of safety-sensitive VCGen) *Let* $\{\!|\phi|\!\} \, C \, \{\!|\psi|\!\} \in \textbf{Spec}$ *and* $\Gamma \subseteq \textbf{Assert}$. *Then*

$$\Gamma \models \mathsf{VCG}^s(\{\!|\phi|\!\} \, C \, \{\!|\psi|\!\}) \quad \textit{iff} \quad \Gamma \vdash_{\mathcal{H}_s} \{\!|\phi|\!\} \, C \, \{\!|\psi|\!\}$$

Proof ⇒: By induction on the structure of C.
⇐: By induction on the derivation of $\Gamma \vdash_{\mathcal{H}_s} \{\!|\phi|\!\} \, C \, \{\!|\psi|\!\}$. □

$$\mathsf{wp}^{\mathsf{S}}\,(\mathbf{skip}, \psi) \;=\; \psi$$

$$\mathsf{wp}^{\mathsf{S}}\,(x := e, \psi) \;=\; \mathsf{safe}(e) \wedge \psi[e/x]$$

$$\mathsf{wp}^{\mathsf{S}}\,(C_1; C_2, \psi) \;=\; \mathsf{wp}^{\mathsf{S}}\,(C_1, \mathsf{wp}^{\mathsf{S}}\,(C_2, \psi))$$

$$\mathsf{wp}^{\mathsf{S}}\,(\mathbf{if}\ b\ \mathbf{then}\ C_t\ \mathbf{else}\ C_f, \psi) \;=\; \mathsf{safe}(b) \wedge (b \to \mathsf{wp}^{\mathsf{S}}\,(C_t, \psi)) \wedge (\neg b \to \mathsf{wp}^{\mathsf{S}}\,(C_f, \psi))$$

$$\mathsf{wp}^{\mathsf{S}}\,(\mathbf{while}\ b\ \mathbf{do}\ \{\theta\}\, C, \psi) \;=\; \theta$$

$$\mathsf{VC}^{\mathsf{S}}(\mathbf{skip}, \psi) \;=\; \emptyset$$

$$\mathsf{VC}^{\mathsf{S}}(x := e, \psi) \;=\; \emptyset$$

$$\mathsf{VC}^{\mathsf{S}}(C_1; C_2, \psi) \;=\; \mathsf{VC}^{\mathsf{S}}(C_1, \mathsf{wp}^{\mathsf{S}}\,(C_2, \psi)) \cup \mathsf{VC}^{\mathsf{S}}(C_2, \psi)$$

$$\mathsf{VC}^{\mathsf{S}}(\mathbf{if}\ b\ \mathbf{then}\ C_t\ \mathbf{else}\ C_f, \psi) \;=\; \mathsf{VC}^{\mathsf{S}}(C_t, \psi) \cup \mathsf{VC}^{\mathsf{S}}(C_f, \psi)$$

$$\mathsf{VC}^{\mathsf{S}}(\mathbf{while}\ b\ \mathbf{do}\ \{\theta\}\, C, \psi) \;=\; \{\theta \to \mathsf{safe}(b), (\theta \wedge b) \to \mathsf{wp}^{\mathsf{S}}\,(C, \theta), (\theta \wedge \neg b) \to \psi\}$$
$$\cup\, \mathsf{VC}^{\mathsf{S}}(C, \theta)$$

$$\mathsf{VCG}^{\mathsf{S}}(\{\!|\phi|\!\}\, C\, \{\!|\psi|\!\}) \;=\; \{\phi \to \mathsf{wp}^{\mathsf{S}}\,(C, \psi)\} \cup \mathsf{VC}^{\mathsf{S}}(C, \psi)$$

Fig. 7.3 Safety-sensitive VCGen

7.2.1 Safe While^{int} Programs

For the While^{int} language, the function safe can be defined as follows

$$\mathsf{safe} \;:\; (\mathbf{Exp_{int}} \cup \mathbf{Exp_{bool}}) \to \mathbf{Assert}$$
$$\mathsf{safe}(x) \;=\; \top$$
$$\mathsf{safe}(c) \;=\; \top$$
$$\mathsf{safe}(-e) \;=\; \mathsf{safe}(e)$$
$$\mathsf{safe}(e_1 \odot e_2) \;=\; \mathsf{safe}(e_1) \wedge \mathsf{safe}(e_2),$$
$$\text{where } \odot \in \{+, -, \times, =, <, \leq, >, \geq, \neq\}$$
$$\mathsf{safe}(e_1\ \mathsf{div}\ e_2) \;=\; \mathsf{safe}(e_1) \wedge \mathsf{safe}(e_2) \wedge e_2 \neq 0$$
$$\mathsf{safe}(e_1\ \mathsf{mod}\ e_2) \;=\; \mathsf{safe}(e_1) \wedge \mathsf{safe}(e_2) \wedge e_2 \neq 0$$
$$\mathsf{safe}(\neg b) \;=\; \mathsf{safe}(b)$$
$$\mathsf{safe}(b_1 \wedge b_2) \;=\; \mathsf{safe}(b_1) \wedge (b_1 \to \mathsf{safe}(b_2))$$
$$\mathsf{safe}(b_1 \vee b_2) \;=\; \mathsf{safe}(b_1) \wedge (\neg b_1 \to \mathsf{safe}(b_2))$$

It can be easily proved (by induction on the structure of program expressions) that $[\![e^\tau]\!]_{\mathcal{M}_e}(s) \neq \mathbf{error}$ iff $[\![\mathsf{safe}(e^\tau)]\!]_{\mathcal{M}}(s) = \mathbf{T}$. Although in general only an implication is required to guarantee that unsafe programs are rejected, this stronger result implies that no safe program will be rejected.

$$\text{Exp}_{\textbf{array}[N]} \ni a ::= u \mid a[e \rhd e']$$

$$\text{Exp}_{\textbf{int}} \ni e ::= \ldots \mid -1 \mid 0 \mid 1 \mid \ldots \mid x \mid$$
$$-e \mid e_1 + e_2 \mid e_1 - e_2 \mid e_1 \times e_2 \mid e_1 \text{ div } e_2 \mid e_1 \text{ mod } e_2 \mid$$
$$a[e] \mid \text{len}(a)$$

$$\text{Exp}_{\textbf{bool}} \ni b ::= \top \mid \bot \mid \neg b \mid b_1 \wedge b_2 \mid b_1 \vee b_2 \mid e_1 = e_2 \mid e_1 \neq e_2 \mid$$
$$e_1 < e_2 \mid e_1 \leq e_2 \mid e_1 > e_2 \mid e_1 \geq e_2$$

Fig. 7.4 Abstract syntax of program expressions of While$^{\text{array}[N]}$

Note also that in most programming languages evaluation of conjunctive or disjunctive boolean expressions is *short-circuiting*. For instance, in the evaluation of $e_1 \wedge e_2$, if e_1 evaluates to false then e_2 needs not be evaluated at all, since the result will always be false (this is compatible with Definition 5.3). Thus e_2 does not need to be safe. The above definition is consistent with short-circuit evaluation.

Example 7.11 For instance we have

$$\begin{aligned}
\text{safe}((x \text{ div } y) > 2) &= \text{safe}(x) \wedge \text{safe}(y) \wedge y \neq 0 \wedge \text{safe}(2) \\
&= \top \wedge \top \wedge y \neq 0 \wedge \top \\
&\equiv y \neq 0
\end{aligned}$$

$$\begin{aligned}
\text{safe}(7 > x \wedge (x \text{ div } y) > 2) &= \text{safe}(7 > 2) \wedge (7 > x \rightarrow \text{safe}((x \text{ div } y) > 2)) \\
&= \top \wedge \top \wedge (7 > x \rightarrow (\top \wedge \top \wedge y \neq 0 \wedge \top)) \\
&\equiv 7 > x \rightarrow y \neq 0
\end{aligned}$$

Note that we have simplified conjunctions involving $\top$.

7.3 Bounded Arrays: The While$^{\text{array}[N]}$ Language

In Sect. 7.1 we saw how to model execution errors in the language evaluation semantics, and Sect. 7.2 explained how the verification methods of the previous two chapters could be updated to deal with a notion of correctness that implies that programs do not go wrong.

The notion of array introduced in Sect. 5.5 is clearly unrealistic since arrays are virtually infinite, having no lower or upper bound on the set of indexes (even negative indexes are allowed). In this section we modify this aspect of the language and introduce a more realistic notion of bounded arrays. Instead of having a single type **array**, we will have a family of array types $\{\textbf{array}[N]\}_{N \in \mathbb{N}}$. Expressions of type **array**[N] are arrays of length N that admit as valid indexes non-negative integers below N.

The question arises of what to do about operations (lookup and update) involving indexes outside an array's bounds. Since we now have a setup that allows us to adequately address program errors, the natural decision is to treat such operations as runtime errors, and to modify the program logic and VCGen accordingly, to prevent their occurrence. To have access to the length of an array we will have functions $\mathsf{len^{int}} \in \mathcal{F}^{\mathsf{Exp}}$ with $\mathsf{rank}(\mathsf{len^{int}}) = \langle \mathbf{array}[N] \rangle$ for each $N \in \mathbb{N}$. We call this language While^{array[N]} and give in Fig. 7.4 the abstract syntax of its program expressions.

The language semantics is adapted from the While^{array} language with minimal modifications: program states will be the same as in Sect. 5.5, so arrays (of any length) are still modeled by mathematical functions with domain and codomain $\mathbb{Z}$. The presence of the length information allows us to consider only some subset of the domain as valid indexes.

Definition 7.12 (Semantics of expressions of While^{array[N]} with error) The semantics of While^{array[N]} expressions is given by extending the semantics of While^{int} expressions (Definition 7.6) as follows:

- $[\![a]\!] : \Sigma \to ((\mathbb{Z} \to \mathbb{Z}) \cup \{\mathbf{error}\})$ is defined inductively by

$$[\![u]\!](s) = s(u)$$

$$[\![a[e \rhd e']]\!](s) = \begin{cases} [\![a]\!](s)[[\![e]\!](s) \mapsto [\![e']\!](s)] & \text{if } [\![a]\!](s) \neq \mathbf{error} \\ & \text{and } [\![e]\!](s) \neq \mathbf{error} \\ & \text{and } 0 \leq [\![e]\!](s) < [\![\mathsf{len}(a)]\!](s) \\ & \text{and } [\![e']\!](s) \neq \mathbf{error} \\ \mathbf{error} & \text{otherwise} \end{cases}$$

- For integer expressions the definition of $[\![e]\!] : \Sigma \to (\mathbb{Z} \cup \{\mathbf{error}\})$ has the following additional cases:

$$[\![\mathsf{len}(a^{\mathbf{array}[N]})]\!](s) = N$$

$$[\![a[e]]\!](s) = \begin{cases} [\![a]\!](s)([\![e]\!](s)) & \text{if } [\![a]\!](s) \neq \mathbf{error} \text{ and } [\![e]\!](s) \neq \mathbf{error} \\ & \text{and } 0 \leq [\![e]\!](s) < \mathsf{len}(a) \\ \mathbf{error} & \text{otherwise} \end{cases}$$

In the language While^{array[N]} the same syntactic sugar can be used for assignment to array positions as in While^{array}:

$$u[e] := e' \quad \text{is an abbreviation of} \quad u := u[e \rhd e']$$

Example 7.13 Assume u is an array variable with type $\mathbf{array}[50]$. Moreover, suppose s is a state such that $s(x) = 7$ and $s(u) = f$ with $f : \mathbb{Z} \to \mathbb{Z}$. Then we have

$$[\![u[x+8]]\!](s) = f([\![x]\!](s) + [\![8]\!](s)) = f(7+8) = f(15)$$

$$[\![u[10 \times x]]\!](s) = \mathbf{error}, \text{ because } [\![10]\!](s) \times [\![x]\!](s) = 10 \times 7 = 70 \geq \mathsf{len}(u)$$

$$(u[x \times x] := 2, s) \rightsquigarrow s[u \mapsto f[49 \mapsto 2]]$$

$$(u[55] := 2, s) \rightsquigarrow \mathbf{error}$$

Occasionally in examples we will employ the following defined predicates concerning the safety of accesses to an individual array position or a contiguous set of positions.

$$valid\_index(u, i) \overset{\text{def}}{=} 0 \leq i < \text{len}(u) \tag{7.1}$$

$$valid\_range(u, i, j) \overset{\text{def}}{=} 0 \leq i \leq j < \text{len}(u) \vee i > j \tag{7.2}$$

7.3.1 Safe While$^{\text{array[N]}}$ Programs

Recall that safe(e) has the meaning that evaluation of the expression e does not produce an error. For the While$^{\text{array[N]}}$ program expressions the function safe is defined by expanding the definition given in Sect. 7.2.1 with the following additional cases to cope with array expressions:

$$
\begin{aligned}
\text{safe}(u) &= \top \\
\text{safe}(\text{len}(a)) &= \top \\
\text{safe}(a[e]) &= \text{safe}(a) \wedge \text{safe}(e) \wedge 0 \leq e < \text{len}(a) \\
\text{safe}(a[e \rhd e']) &= \text{safe}(a) \wedge \text{safe}(e) \wedge 0 \leq e < \text{len}(a) \wedge \text{safe}(e')
\end{aligned}
$$

Example 7.14 For instance we have

$$
\begin{aligned}
\text{safe}(u[x \text{ div } 2]) &= \text{safe}(u) \wedge \text{safe}(x \text{ div } 2) \wedge 0 \leq x \text{ div } 2 < \text{len}(u) \\
&= \top \wedge \text{safe}(x) \wedge \text{safe}(2) \wedge 2 \neq 0 \wedge 0 \leq (x \text{ div } 2) < \text{len}(u) \\
&= \top \wedge \top \wedge \top \wedge 2 \neq 0 \wedge 0 \leq (x \text{ div } 2) < \text{len}(u) \\
&\equiv 2 \neq 0 \wedge 0 \leq (x \text{ div } 2) < \text{len}(u)
\end{aligned}
$$

$$
\begin{aligned}
\text{safe}(u[3 \rhd 10]) &= \text{safe}(u) \wedge \text{safe}(3) \wedge 0 \leq 3 < \text{len}(u) \wedge \text{safe}(10) \\
&= \top \wedge \top \wedge 0 \leq 3 < \text{len}(u) \wedge \top \\
&\equiv 0 \leq 3 < \text{len}(u)
\end{aligned}
$$

It is easy to see that with this definition of safe, and $\mathcal{M}_e$ given by Definition 7.12, we have in any state s that $[\![a]\!]_{\mathcal{M}_e}(s) \neq \textbf{error}$ iff $[\![\text{safe}(a)]\!]_{\mathcal{M}}(s) = \textbf{T}$, for any $\mathcal{V}_{\text{Prog}}$-structure $\mathcal{M}$ compatible with $\mathcal{M}_e$. Thus the soundness result of $\mathcal{H}_s$ applies.

A rule of system $\mathcal{H}_s$ can be given for array assignment, as a special case of rule (*assign*), by expanding the syntactic sugar:

(*array assign*, $\mathcal{H}_s$)

$$\frac{}{\{\phi\}\, u[e] := e'\, \{\psi\}} \quad \text{if } \phi \rightarrow \text{safe}(u[e \rhd e']) \text{ and } \phi \rightarrow \psi[u[e \rhd e']/u]$$

Clauses of the safety-sensitive VCGen of Fig. 7.3 can also be obtained in the same way:

$$
\begin{aligned}
\text{wp}^s\, (u[e] := e', \psi) &= \text{safe}(u[e \rhd e']) \wedge \psi[u[e \rhd e']/u] \\
\text{VC}^s(u[e] := e', \psi) &= \emptyset
\end{aligned}
$$

$$\mathsf{wp}^\mathsf{S}(C, \theta) \;=\; \mathsf{safe}(u[i] > u[max]) \land$$
$$\big(u[i] > u[max] \to (\mathsf{safe}(i) \land \mathsf{safe}(i+1) \land 1 \le i+1 \le size \land$$
$$0 \le i < i+1 \land \forall a.\, 0 \le a < i+1 \to u[a] \le u[i])\big) \land$$
$$(\neg(u[i] > u[max]) \to (\mathsf{safe}(i+1) \land 1 \le i+1 \le size \land$$
$$0 \le max < i+1 \land \forall a.\, 0 \le a < i+1 \to u[a] \le u[max]))$$

$$\mathsf{wp}^\mathsf{S}(\mathbf{maxarray}, \psi) \;=\; \mathsf{safe}(0) \land \mathsf{safe}(1) \land 1 \le 1 \le size \land 0 \le 0 < 1 \land$$
$$\forall a.\, 0 \le a < 1 \to u[a] \le u[0]$$

$$\mathsf{VC}^\mathsf{S}(\mathbf{maxarray}, \psi) \;=\; \{\, \theta \to \mathsf{safe}(i < size),$$
$$(1 \le i < size \land 0 \le max < i \land \forall a.\, 0 \le a < i \to u[a] \le u[max]) \to \mathsf{wp}^\mathsf{S}(C, \theta),$$
$$(1 \le i = size \land 0 \le max < i \land \forall a.\, 0 \le a < i \to u[a] \le u[max]) \to \psi \,\}$$

$$\mathsf{VCG}^\mathsf{S}(\{\!|\, size \ge 1 \,|\!\} \, \mathbf{maxarray} \, \{\!|\, \psi \,|\!\}) \;=\; \{\, size \ge 1 \to \mathsf{wp}^\mathsf{S}(\mathbf{maxarray}, \psi) \,\} \cup$$
$$\mathsf{VC}^\mathsf{S}(\mathbf{maxarray}, \psi)$$

Fig. 7.5 Generation of verification conditions with safety for **maxarray**

We illustrate the use of the safety-sensitive VCGen through a very simple example, and then apply it the **maxarray** program of Chap. 6. First consider:

$$\mathsf{VCG}^\mathsf{S}(\{\mathsf{len}(u) = 50 \land u[j] > 100\}\, u[i] := 10 \,\{u[j] > 100\})$$
$$= \{\mathsf{len}(u) = 50 \land u[j] > 100 \to \mathsf{safe}(u[i \triangleright 10]) \land (u[j] > 100)[u[i \triangleright 10]/u]\}$$
$$= \{\mathsf{len}(u) = 50 \land u[j] > 100 \to 0 \le i < \mathsf{len}(u) \land (u[i \triangleright 10][j] > 100)\}$$

This verification condition cannot be proved since there is no information about i in the precondition, and moreover it is possible that the values of i and j are equal, thus *read/write axiom 2* of theory $\mathcal{T}_A$ (Sect. 4.7.4) cannot be applied. A precondition like $\mathsf{len}(u) = 50 \land u[j] > 100 \land i \ne j \land i = 70$ would lead to a provable verification condition.

Example 7.15 We consider again the program **maxarray** of Example 6.10. The verification conditions produced by the safety-sensitive VCGen are calculated in Fig. 7.5. Note that most of the safety conditions introduced are trivial. However, the third verification condition contains the following

$$(1 \le i < size \land 0 \le max < i \land \forall a.\, 0 \le a < i \to u[a] \le u[max])$$
$$\to 0 \le i < \mathsf{len}(u) \land 0 \le max < \mathsf{len}(u) \land \cdots$$

which cannot be proved with the available information. We solve this problem by strengthening the precondition with information concerning the range of valid indexes of u as shown in the following modified specification, which states that this program is only guaranteed to work correctly for arrays of length not less than *size*.

$$\{size \ge 1 \land valid\_range(u, 0, size - 1)\}$$
maxarray
$$\{0 \le max < size \land \forall a.\, 0 \le a < size \to u[a] \le u[max]\}$$

The bit of the precondition concerning the range of u remains valid through execution of the program, and is thus a continuous invariant of the loop. We strengthen the loop invariant θ to

$$valid\_range(u, 0, size - 1) \wedge 1 \leq i \leq size \wedge 0 \leq max < i \wedge$$
$$\forall a.\, 0 \leq a < i \rightarrow u[a] \leq u[max]$$

All the safety conditions for the modified annotated program can now be proved. If the program is executed in a context in which u has been allocated with length not smaller than $size$, the precondition $valid\_range(u, 0, size - 1)$ will be trivially satisfied.

7.3.2 An Alternative Formalisation of Bounded Arrays

Example 7.15 illustrated the use of assertions involving the length of arrays in contracts, to allow safety conditions to be discharged. In our treatment of bounded arrays we considered a different type for each array length, and a constant function for each such type that returns the length of arrays of that type—clearly the length of an array did not depend on the state of the program. The existence of this function has similarities with languages like Java, in which programs can freely consult the length of an array. But having the length associated with the array type is less usual.

One alternative would be to have a single **array** type for arrays of any length, and to modify the domain of interpretation to account for the length information. One possibility is

$$[\![a]\!] : \Sigma \rightarrow (((\mathbb{Z} \rightarrow \mathbb{Z}) \times \mathbb{N}) \cup \{\mathbf{error}\})$$

The length of an array is stored in the state explicitly as an integer. The idea is that arrays are still statically allocated and their length remains constant through execution (but note that this is now not forced by the semantics).

In realistic languages array variables are *declared* (globally or locally) before they can be used, which allows to allocate the required amount of memory. From the point of view of this alternative semantics, the declaration of an array variable u would set the value of its length, stored in the state. Suppose we use the syntax (with N a constant)

$$\textbf{array } u[N] \textbf{ in } C$$

to declare a new array variable u of length N to be used in command C. If we treat this declaration as being itself a command, the natural semantics rule for this could be

– If $(C, s[u \mapsto (\overline{0}, N)]) \leadsto_{\mathcal{M}} s'$, then $(\textbf{array } u[N] \textbf{ in } C, s) \leadsto_{\mathcal{M}} s'[u \mapsto s(u)]$,

where $\overline{0}$ denotes the constant function that maps every element of the domain to 0.

For the Hoare calculus we can write a rule like

(*array declaration*)

$$\frac{\{\phi \wedge \mathsf{len}(u) = N\}\, C\, \{\psi\}}{\{\phi\}\, \mathbf{array}\ u[N]\ \mathbf{in}\ C\, \{\psi\}}$$

where u is assumed not to occur free in ϕ or ψ; otherwise it should be substituted by a fresh variable in the premise triple.

7.4 To Learn More

Several works by John Reynolds have inspired this chapter and are very much recommended for further reading. A 1979 research paper [2] set up a basis for reasoning about arrays; the textbook [3] contains many examples involving programs that use arrays, and a more recent book [4] offers a concise discussion of arrays interpreted as total functions whose domains are interval subsets of $\mathbb{Z}$. The latter book also briefly discusses the need for higher-order reasoning, for expressing properties of programs that manipulate arrays. Our treatment of safety in Hoare logic is also inspired by Reynolds' use of enabling predicates [5].

Finally, the survey [1] is an introduction to all the issues related to safety in the context of a modern object-oriented programming language.

7.5 Exercises

7.1. Similarly to what was done in this chapter for partial correctness triples, it makes sense to have a new form of safety-sensitive triples for total correctness. Define the semantics of this form of triple.

7.2. Consider again the program and Hoare triple considered in Examples 6.10 and 7.15. Construct a derivation for this triple in the safety-sensitive system $\mathcal{H}_s$ of Fig. 7.2.

7.3. It may be useful to separate safety verification conditions from other functional VCs. In this case the latter can be generated using the standard VCGen, and the former using the safety-sensitive VCGen, as follows.

$$\mathsf{VCG}^s(\{\phi\}\, C\, \{\psi\}) = \mathsf{VCG}(\{\phi\}\, C\, \{\psi\}) \cup \mathsf{VCG}^s(\{\phi\}\, C\, \{\top\})$$

Prove this result and calculate safety-sensitive VCs for one of the examples in this chapter in this way.

7.4. Certain execution errors are silent in the sense that they do not cause programs to terminate abruptly. They correspond to a behaviour that is wrong but cannot easily be distinguished from a normal behaviour. *Integer overflow* is usually an example of this kind of error: when an arithmetic operation exceeds the storage capacity of the concrete integer type used in the language, a variable may simply take an unexpected value as result. Take for instance 16 bit non-negative integer variables, with values ranging from 0 to 65535. Adding 1 to a variable of this type containing the value 65535 will produce 0 as result.

One way to cope with this is to use the VCGen to statically avoid such errors. To do so, design an error semantics for this language of 16-bit nonnegative integers (such that situations in which the value of an operation exceeds the representation capacity correspond to error states), and define an appropriate safe function for its expressions.

References

1. Hartel, P.H., Moreau, L.: Formalizing the safety of Java, the Java virtual machine, and Java card. ACM Comput. Surv. **33**(4), 517–558 (2001)
2. Reynolds, J.C.: Reasoning about arrays. Commun. ACM **22**(5), 290–299 (1979)
3. Reynolds, J.C.: The Craft of Programming. Prentice-Hall International, Upper Saddle River (1981)
4. Reynolds, J.C.: Theories of Programming Languages. Cambridge University Press, Cambridge (1998)
5. Reynolds, J.C.: Programs that use arrays. Class Notes, Carnegie-Mellon University (February 2004)

Chapter 8
Procedures and Contracts

The presence of sub-routines in the form of procedures or functions is as important as it is challenging from the point of view of verification. The challenges include for instance the treatment of recursive calls (another possible source of non-termination of program execution) and parameters. The presence of functions on the other hand complicates the treatment of expression evaluation, which may now be non-terminating. Also, since expressions may contain multiple function calls, side effects have to be carefully considered.

The reader can find a thorough treatment of recursive procedures in any good textbook on semantics; in this chapter we limit ourselves to a form of program logic for such procedures. We simply claim without proof that it is adequate for reasoning about programs with procedures, and moreover it is adequate for motivating the principles that are used in practice by program verification tools and standard annotation languages, as will be illustrated in Chaps. 9 and 10.

The presence of routines introduces a new level of verification: the material studied in the present chapter covers the *interprocedural* level, but at the *intraprocedural* level we still need to verify the code in the body of procedures; for this we use rules of system $\mathcal{H}_g$. Thus all the inference systems presented in this chapter are meant as extensions of that system. Note that it would also be possible to write them as extensions of the original system $\mathcal{H}$; the main difference is that in system $\mathcal{H}_g$, due to the absence of a consequence rule, the inference rule for procedure invocation needs to consider the issue of adaptation of specifications. We discuss this in a little more detail in the next section. Finally, we note that it would be straightforward to use at the intraprocedural level the safety-sensitive system $\mathcal{H}_s$ instead; in fact some of the examples in this chapter are based on the While$^{\text{array[N]}}$ language.

The chapter starts in Sect. 8.1 with an overview of some of the issues involved in reasoning about procedures. Subsequent sections cover in turn inference rules and verification conditions for programs consisting of mutually recursive parameterless procedures; frame conditions; procedures with parameters; and finally return values and functions.

J.B. Almeida et al., *Rigorous Software Development*,
Undergraduate Topics in Computer Science,
DOI 10.1007/978-0-85729-018-2_8, © Springer-Verlag London Limited 2011

8.1 Procedures and Recursion

In the next section we will formally introduce the setting for the verification of programs based on procedures annotated with contracts, which pretty much corresponds to the basis on which modern verification tools are built. Before that however, we discuss simple recursive (parameterless) procedures and the conceptual modification they require in the presentation of Hoare logic.

Let **proc p** $= C_{\mathbf{p}}$ define the (possibly recursive) procedure **p**; the command $C_{\mathbf{p}}$ is said to be the *body* of **p**. The new command **call p** invokes the procedure, i.e. it transfers execution to the body of **p**. This can be captured by the following natural semantics rule, where **body(p)** denotes the body of procedure **p**.

− If $(\mathbf{body(p)}, s) \leadsto_{\mathcal{M}} s'$, then $(\mathbf{call\,p}, s) \leadsto_{\mathcal{M}} s'$.

How can one reason axiomatically about this new command? If **p** is not recursive, the following rule is sufficient

$$\frac{\{\phi\}\,\mathbf{body(p)}\,\{\psi\}}{\{\phi\}\,\mathbf{call\,p}\,\{\psi\}} \tag{8.1}$$

Example 8.1 The factorial program of Example 5.10 can be readily turned into a procedure as follows.

$$\begin{aligned}
&\mathbf{proc\ fact} = \\
&\quad f := 1;\, i := 1; \\
&\quad \mathbf{while}\ i \le n\ \mathbf{do}\, \{f = fact(i-1) \wedge i \le n+1\}\,\{ \\
&\qquad f := f \times i; \\
&\qquad i := i+1 \\
&\quad \}
\end{aligned}$$

We have the correctness of this procedure for free, since the triple

$$\{n \ge 0 \wedge n = n_0\}\,\mathbf{body(fact)}\,\{f = fact(n) \wedge n = n_0\}$$

was proved correct in Example 6.6, thus the following triple is derived by the above rule:

$$\{n \ge 0 \wedge n = n_0\}\,\mathbf{call\,fact}\,\{f = fact(n) \wedge n = n_0\}$$

This can be seen as a public specification of the **fact** procedure.

Now suppose we want to invoke the procedure in a context characterised by a stronger precondition, say $n = 10$. Of course we could adapt our proof from Chap. 6 to show that **body(fact)** also satisfies this specification, but we would like to be able to reuse the correctness result of that proof without having to reconstruct it, which we can do by simply applying a consequence rule. Following our discussion in Sect. 5.7, in order for this to work properly in the presence of auxiliary variables in the specifications, the version (*conseq-auxvars*) of this rule should be used.

As an example we could derive

$$\frac{\{n \geq 0 \wedge n = n_0\} \, \textbf{call fact} \, \{f = fact(n) \wedge n = n_0\}}{\{n = 10\} \, \textbf{call fact} \, \{f = fact(10)\}} \tag{8.2}$$

with side condition

$$n = 10 \rightarrow \forall n_f, f_f.$$
$$(\forall n_{0f}. \, n \geq 0 \wedge n = n_{0f} \rightarrow f_f = fact(n_f) \wedge n_f = n_{0f}) \rightarrow f_f = fact(10)$$

Rule (8.1) can thus be added to system $\mathcal{H}$ together with (*conseq-auxvars*). An interesting question is how procedure calls could be handled in the context of system $\mathcal{H}_g$—recall that the latter system lacks a consequence rule, thus rule (8.1) would force us to construct a proof of $\{\phi\} \, \textbf{body}(\textbf{p}) \, \{\psi\}$, for every specification (ϕ, ψ) required for some call of $\textbf{p}$. We will address this issue in Sect. 8.1.3 below, by introducing a procedure call rule with an adaptation side condition, that dispenses with the need for a consequence rule.

8.1.1 The ˜ Notation

In practice, modern specification languages avoid the generality allowed by auxiliary variables, and forbid their use in the public specifications of procedures. Instead, a special notation is introduced that allows postconditions to refer to the initial value of program variables.

In this book we will use the special variable $x\tilde{}$ to denote the value of x in the pre-state. Using this notation, the specification of the factorial procedure can be written instead as

$$\{n \geq 0\} \, \textbf{call fact} \, \{f = fact(n) \wedge n = n\tilde{}\} \tag{8.3}$$

A suitable version of the consequence rule to deal with such specifications is the following

(*conseq˜*)

$$\frac{\{\phi\} \, C \, \{\psi\}}{\{\phi'\} \, C \, \{\psi'\}} \quad \text{if} \quad \phi' \rightarrow \forall \overline{x_f}. \, (\phi \rightarrow \lfloor \psi[\overline{x_f}/\overline{x}] \rfloor) \rightarrow \psi'[\overline{x_f}/\overline{x}]$$

where $\overline{x}$ are the program variables of C, $\overline{x_f}$ are fresh variables, and $\lfloor \psi[\overline{x_f}/\overline{x}] \rfloor$ denotes the result of substituting in $\psi[\overline{x_f}/\overline{x}]$ every variable $x\tilde{}$ occurring in it by the corresponding x.

The goal triple of (8.2) would now be derived as follows.

$$\frac{\{n \geq 0\} \, \textbf{call fact} \, \{f = fact(n) \wedge n = n\tilde{}\}}{\{n = 10\} \, \textbf{call fact} \, \{f = fact(10)\}}$$

with side condition

$$n = 10 \rightarrow \forall n_f, f_f. \, (n \geq 0 \rightarrow f_f = fact(n_f) \wedge n_f = n) \rightarrow f_f = fact(10)$$

It remains to discuss how the Hoare triple (8.3) could be derived, since it does not follow from rule (8.1). We need a new version of the latter rule to cope with the $\tilde{}$ notation:

$$\frac{\{\phi \wedge x_1 = x_1\tilde{} \wedge \cdots \wedge x_n = x_n\tilde{}\} \, \textbf{body(p)} \, \{\psi\}}{\{\phi\} \, \textbf{call p} \, \{\psi\}} \tag{8.4}$$

where $x_1, \ldots, x_n$ are the program variables of **body(p)**. For Example 8.1 this would yield (note that the validity of the premise has been established before)

$$\frac{\{n \geq 0 \wedge n = n\tilde{}\} \, \textbf{body(fact)} \, \{f = fact(n) \wedge n = n\tilde{}\}}{\{n \geq 0\} \, \textbf{call fact} \, \{f = fact(n) \wedge n = n\tilde{}\}}$$

Example 8.2 Suppose the following has been derived for some procedure **p**:

$$\{x > 0 \wedge y > 0 \wedge x = x\tilde{} \wedge y = y\tilde{} \wedge z = z\tilde{}\} \, \textbf{body(p)} \, \{z = x + y \wedge x = x\tilde{}\}$$

and thus by rule (8.4) we have the following specification for **p**

$$\{x > 0 \wedge y > 0\} \, \textbf{call p} \, \{z = x + y \wedge x = x\tilde{}\}$$

We are granted that if the precondition is met, the final value of variable z will be the sum of (the final values of) x and y, and moreover the value of x is unchanged with respect to its initial value. Now suppose we want to derive $\{\phi\} \, \textbf{call p} \, \{\psi\}$ where

$$\begin{aligned} \phi \quad &\text{is} \quad x > 0 \wedge y > x \wedge x = x\tilde{} + 100 \\ \psi \quad &\text{is} \quad z = x + y \wedge x = x\tilde{} + 100 \end{aligned}$$

Observe that $x\tilde{}$ in ϕ and ψ refers to the value of x in the initial state of execution of the caller procedure, whereas $x\tilde{}$ in the postcondition of **p** refers to the initial value of x from the point of view of the callee **p**.

The rule (*conseq*$\tilde{}$) can be applied with side condition

$$x > 0 \wedge y > x \wedge x = x\tilde{} + 100 \rightarrow \forall x_f, y_f, z_f.$$
$$(x > 0 \wedge y > 0 \rightarrow z_f = x_f + y_f \wedge x_f = x) \rightarrow z_f = x_f + y_f \wedge x_f = x\tilde{} + 100$$

which holds. Note that without the $x = x\tilde{}$ part of the specification of **p**, it would not have been possible to establish the second part of ψ, which is a consequence of the fact that the procedure does not modify the value of x.

8.1.2 Recursive Procedures

In the presence of recursion, **body(p)** may itself contain one or more **call p** commands, in which case we may be led to the construction of infinite derivations if we use rule (8.1). For this reason, we need the following rule proposed by Hoare, that uses a notion familiar from propositional logic: reasoning under assumptions.

$$[\{\phi\} \, \textbf{call p} \, \{\psi\}]$$
$$\vdots$$
$$\{\phi\} \, \textbf{body}(\textbf{p}) \, \{\psi\}$$
$$\rule{4cm}{0.4pt}$$
$$\{\phi\} \, \textbf{call p} \, \{\psi\}$$

(8.5)

The meaning of this rule is that if, assuming $\{\phi\} \, \textbf{call p} \, \{\psi\}$ can be derived, a derivation for $\{\phi\} \, \textbf{body}(\textbf{p}) \, \{\psi\}$ can be constructed, then $\{\phi\} \, \textbf{call p} \, \{\psi\}$ can be derived without assumptions (that is why the assumption is canceled, which is denoted by putting it inside square brackets). This may seem a little confusing at first, since (unlike, say, the introduction rule for implication of propositional logic) the assumption that is canceled coincides with the conclusion of the rule, but this is precisely the recursive aspect at play—an axiomatic counterpart of fixpoint induction.

Finally, note that the presence of recursive procedures creates a new source of non-termination in addition to loops. To handle total correctness, one would have to introduce some measure such that procedure invocations that do not lead to a strict decrease of this measure are forbidden. In the rest of the chapter we will consider partial correctness only.

Example 8.3 Consider the following recursive version of factorial.

$$\textbf{proc factr} =$$
$$\textbf{if } n = 0 \textbf{ then } f := 1$$
$$\textbf{else } \{$$
$$\quad n := n - 1 \, ;$$
$$\quad \textbf{call factr} \, ;$$
$$\quad n := n + 1 \, ;$$
$$\quad f := n \times f$$
$$\}$$

This is a classic example that has been used in many research papers to illustrate the adaptation problem in the presence of auxiliary variables, and several solutions ranging from adding more structural rules to specialized so-called *rules of adaptation*. But in the presence of rule (*conseq-aux-vars*) the triple $\{n \geq 0 \wedge n = n_0\} \, \textbf{call factr} \, \{f = fact(n) \wedge n = n_0\}$ can be proved correct simply using rule (8.5), see Exercise 8.3.

8.1.3 Procedure Calls in System $\mathcal{H}_g$

In this book we have used a goal-directed inference system, without consequence rule, as a basis for the generation of verification conditions. In this chapter we have been looking at how procedure calls can be dealt with by introducing rules that work with system $\mathcal{H}$. We now consider how this would be done in system $\mathcal{H}_g$, which raises the problem of adaptation of specifications, in the absence of a consequence rule.

We saw in Chap. 6 that in system $\mathcal{H}_g$ side conditions were distributed by several rules; the way to deal with adaptation in this system is to modify the procedure call rule to include an adaptation condition, as follows

$$\frac{\{\phi\}\,\mathbf{body}(\mathbf{p})\,\{\psi\}}{\{\phi'\}\,\mathbf{call}\,\mathbf{p}\,\{\psi'\}} \quad \text{if } \phi' \to \forall \overline{x_f}.\ (\forall \overline{y_f}.\ \phi[\overline{y_f}/\overline{y}] \to \psi[\overline{y_f}/\overline{y}, \overline{x_f}/\overline{x}]) \to \psi'[\overline{x_f}/\overline{x}]$$

$$(8.6)$$

$$\text{where} \quad \begin{aligned} &\overline{y} \text{ are the auxiliary variables of } \{\phi\}\,\mathbf{body}(\mathbf{p})\,\{\psi\}\\ &\overline{x} \text{ are the program variables of } \mathbf{body}(\mathbf{p})\\ &\overline{x_f} \text{ and } \overline{y_f} \text{ are fresh variables} \end{aligned}$$

The idea of this rule is that the body of $\mathbf{p}$ is proved correct with respect to (ϕ, ψ) once and for all, thus this specification should be as strong as possible, and the above rule can be used to adapt the procedure to weaker specifications.

Note that extended in this way, system $\mathcal{H}_g$ maintains its characteristic absence of ambiguity, and adaptation conditions are applied only when strictly necessary, i.e. in procedure calls. This stands in contrast with the use of rule (*conseq-aux-vars*) that always introduces adaptation conditions.

In the next section we will abandon the use of auxiliary variables and of a consequence rule, and move to a setting in which programs consist of sets of mutually recursive procedures annotated with *contracts*. A contract is a public specification with respect to which a procedure is proved correct once and for all. Auxiliary variables will be forbidden in contracts, and replaced by the use of the ˜ notation to refer to the value of a variable in the initial state. Rule (8.6) will be modified accordingly. A mutual recursion rule will also be introduced, as a generalisation of rule (8.5).

8.2 Contracts and Mutual Recursion

To this point there was no distinction in our simple language between programs and commands. We will now formally add procedures to the language, and see programs as sets of procedures. In our definition procedures will have no parameters; as such, a program is a collection of procedures that communicate through a set of global variables. Naturally, the procedures defined in a program are allowed to freely invoke each other; they are said to be *mutually recursive*. Note that this view also offers an analogue of classes in object-oriented programming languages: classes consist of sets of methods that share a pool of instance variables (but of course, other aspects of object-oriented languages, such as inheritance, are completely left out of this analogy).

Let **PN** denote the set of procedure names and let $\mathbf{p}, \mathbf{q}, \ldots$ range over **PN**. The programming language syntax is extended as follows, where **Proc** and **Prog** are two new phrase types, for *procedure definitions* and *programs* respectively, and **Pspec** is a phrase type for program (rather than command) correctness formulas.

$$\mathbf{Comm} \ni C \quad ::= \quad \ldots \mid \mathbf{call\,p}$$

$$\mathbf{Proc} \ni \Phi \quad ::= \quad \mathbf{pre}\,\phi\,\mathbf{post}\,\psi\,\mathbf{proc\,p} = C$$

$$\mathbf{Prog} \ni \Pi \quad ::= \quad \Phi \mid \Pi\,\Phi$$

$$\mathbf{Pspec} \ni S_p \quad ::= \quad \{\Pi\}$$

For each variable $x \in \mathbf{Var}$, we now require a special variable $x^\sim$ (that will only be allowed to occur in contract postconditions). We let $\mathbf{Var}^\sim = \{x^\sim \mid x \in \mathbf{Var}\}$ and denote by $\mathbf{Var}(\phi)$ the subset of **Var** consisting of all variables occurring free in ϕ, and by $\mathbf{Var}^\sim(\phi)$ the subset of **Var** consisting of all variables x such that $x^\sim$ occurs free in ϕ. For convenience, we will occasionally consider that these operators return a *sequence* rather than a set of variables.

We will denote by $\lfloor \theta \rfloor$ the formula that results from substituting in θ every $x^\sim$ variable occurring in it by the corresponding x i.e., $\lfloor \theta \rfloor = \theta[x_1/x_1^\sim, \ldots, x_n/x_n^\sim]$ with $\mathbf{Var}^\sim(\theta) = \{x_1, \ldots, x_n\}$. The effect of this operator can be seen as "dropping the $\sim$".

Each procedure definition associates with a procedure name (i) a precondition; (ii) a postcondition; and (iii) a procedure body (a command). A program consists of a sequence of such definitions.

We will let $\mathbf{PN}(\Pi)$ denote the set of names of the procedures defined in the program Π. The following notation will be used for representing the different parts of the procedures of a program. For $\mathbf{p} \in \mathbf{PN}(\Pi)$, $\mathbf{pre}_\Pi(\mathbf{p})$, $\mathbf{post}_\Pi(\mathbf{p})$, and $\mathbf{body}_\Pi(\mathbf{p})$ denote, respectively, the precondition, postcondition, and body command of the procedure named $\mathbf{p}$ in Π (the program name will be dropped when it is clear from context). Thus given in program Π the procedure definition

$$\mathbf{pre}\,\phi\,\mathbf{post}\,\psi\,\mathbf{proc\,p} = C$$

we will have

$$\mathbf{pre}_\Pi(\mathbf{p}) = \phi$$
$$\mathbf{post}_\Pi(\mathbf{p}) = \psi$$
$$\mathbf{body}_\Pi(\mathbf{p}) = C$$

A program is said to be *well-defined* if each name is unique (i.e. the same name is not used for more than one procedure), the program is closed with respect to procedure invocation (i.e. if $\mathbf{p} \in \mathbf{PN}(\Pi)$ and $\mathbf{q}$ is invoked inside $\mathbf{p}$ then $\mathbf{q} \in \mathbf{PN}(\Pi)$), and moreover for every $\mathbf{p} \in \mathbf{PN}(\Pi)$, $\mathbf{pre}(\mathbf{p})$ and $\mathbf{post}(\mathbf{p})$ contain no occurrences of auxiliary variables (all variables are program variables or logical variables used with quantifiers), and additionally **pre(p)** contains no occurrences of $\tilde{\ }$, i.e. $\mathbf{Var}^\sim(\mathbf{pre(p)}) = \emptyset$.

From the point of view of operational semantics, in addition to the rule for the **call** command given in Sect. 8.1, a special procedure should be designated as an entry point into the program. This is the procedure that will first be executed when the program begins to run (the role of the `main` function in C for instance). This issue is not however relevant from the axiomatic point of view.

A *contract triple* is a Hoare triple of the form $\{\mathbf{pre(p)}\}\,\mathbf{call\,p}\,\{\mathbf{post(p)}\}$ for some procedure **p**. A program Π is said to be *correct*, written as the formula $\{\Pi\}$, if all its procedures are correct with respect to their corresponding specifications.

Definition 8.4 Let $\mathcal{M} = (D, I)$ be a $\mathcal{V}_{\mathsf{Prog}}$-structure. The semantics of program correctness formula $\{\Pi\}$ is given by a function $[\![\{\Pi\}]\!]_{\mathcal{M}} : \Sigma_D \to \{\mathbf{F}, \mathbf{T}\}$ defined as follows

$$[\![\{\Pi\}]\!]_{\mathcal{M}}(s) = \mathbf{T} \quad \text{iff} \quad \text{for all } \mathbf{p} \in \mathbf{PN}(\Pi),$$
$$[\![\{\mathbf{pre}_\Pi(\mathbf{p}) \wedge x_1 = x_1^\sim \wedge \cdots \wedge x_k = x_k^\sim\}$$
$$\mathbf{call\,p}\,\{\mathbf{post}_\Pi(\mathbf{p})\}]\!]_{\mathcal{M}} = \mathbf{T}$$

8.2.1 Programming with Contracts

Specifications associated to procedures (or methods in the object-oriented context) are usually called *contracts*. A whole methodology for software development prescribes writing contracts for procedures, that can be dynamically checked in runtime. The idea is that each procedure **p** establishes a contract with the other procedures in the program, in the sense that if **p** is called in a state in which the precondition **pre(p)** is valid, then if **p** terminates, the postcondition **post(p)** will be valid when control is passed back to the caller.

In this kind of development based on contracts, a special compiler introduces extra code in the procedures that will check whether **pre(p)** is true when **p** is called and **post(p)** is true when its execution finishes. This is of course very useful in the development stage, for debugging and testing, and what is more it helps programmers abandon the defensive programming habit of including code that checks that certain (often implicit) preconditions are valid when a procedure is called. At a mature stage, the runtime checks can simply be turned off.

There are two important reasons for adopting contracts in the context of static program verification, as we do in the present chapter. The first is that the contract of a procedure **p** is a specification that can be proved once and then used as an interface for reasoning externally about invocations of **p**. Thus it should describe the behaviour of **p** as precisely as possible, in order to be adapted to specific requirements for each invocation, in the spirit of the discussion of Sect. 5.7.

A second reason is that the use of contracts has become very popular in recent years; standard interface specification languages have been put forward and are used in practice (some programming languages even include contracts natively, and not through a complementary annotation language). Contracts are used by a wide range

(*mutual recursion—parameterless*)

$$[\{\Pi\}] \qquad\qquad\qquad\qquad [\{\Pi\}]$$
$$\vdots \qquad\qquad\qquad\qquad\qquad \vdots$$

$$\frac{\{\widetilde{\mathbf{pre}_\Pi(\mathbf{p}_1)}\}\,\mathbf{body}_\Pi(\mathbf{p}_1)\,\{\mathbf{post}_\Pi(\mathbf{p}_1)\} \quad\cdots\quad \{\widetilde{\mathbf{pre}_\Pi(\mathbf{p}_n)}\}\,\mathbf{body}_\Pi(\mathbf{p}_n)\,\{\mathbf{post}_\Pi(\mathbf{p}_n)\}}{\{\Pi\}}$$

where

$$\mathbf{PN}(\Pi) \;=\; \{\mathbf{p}_1,\ldots,\mathbf{p}_n\}$$

$$\widetilde{\mathbf{pre}_\Pi(\mathbf{p}_i)} \;=\; \mathbf{pre}_\Pi(\mathbf{p}_i) \wedge x_1 = \tilde{x_1} \wedge \cdots \wedge x_k = \tilde{x_k}$$
$$\text{with } \mathbf{Var}\tilde{\ }(\mathbf{post}_\Pi(\mathbf{p}_i)) = \{x_1,\ldots,x_k\}$$

(*procedure call—parameterless*)

$$\frac{\{\Pi\}}{\{\phi\}\,\mathbf{call}\,\mathbf{p}\,\{\psi\}} \quad \text{if } \phi \rightarrow \forall \overline{x_f}.\,(\mathbf{pre}_\Pi(\mathbf{p}) \rightarrow \lfloor \mathbf{post}_\Pi(\mathbf{p})[\overline{x_f}/\overline{x}]\rfloor) \rightarrow \psi[\overline{x_f}/\overline{x}]$$

where

$$p \;\in\; \mathbf{PN}(\Pi)$$

$$\overline{x} \;=\; \mathbf{Var}(\mathbf{post}_\Pi(\mathbf{p})) \cup \mathbf{Var}(\psi)$$

$$\overline{x_f} \quad \text{are fresh variables}$$

Fig. 8.1 Inference rules for programs with mutually recursive parameterless procedures

of tools, for tasks ranging from runtime checking to test case generation. Thus it is a matter of opportunity for static verification tools to adopt when possible the same principles and specification languages already used in other software engineering tasks.

8.2.2 Inference System for Parameterless Procedures

Figure 8.1 contains two inference rules that together with system $\mathcal{H}_g$ allow us to reason about programs with mutually recursive procedures. The first rule is a generalisation of rule (8.5) and establishes the simultaneous correction of a set of procedures. Note that it is not possible to derive procedure correctness with an empty set of assertions (i.e. without assuming the correctness of other procedures) individually for each procedure. But if the correctness of the body of each procedure in the set can be derived assuming only the correctness of the procedures in that same set, then all the procedures (and therefore the program) are indeed correct.

The second rule concerns the procedure call command, and has as premise the correctness of the program. It looks up the procedure's contract, and the side condition establishes the adaptation between the actual required specification and the contract. No consequence rule is required in this system.

We remark that the rules are compatible with modular reasoning and the principles underlying contract-based development—each procedure is individually proved correct with respect to its specification, assuming every procedure invoked by it is correct with respect to its public contract; if this succeeds for every procedure then the program is correct.

8.2.3 Verification Conditions for Parameterless Procedures

It should be straightforward to understand how the VCGen of Fig. 6.5 (or the safety-sensitive version of Fig. 7.3) can be extended to cope with procedures.

First we need clauses for the **call** command. Its weakest precondition is extracted directly from the corresponding rule of Fig. 8.1, and there are no additional verification conditions:

$$
\begin{aligned}
\text{wp} \, (\textbf{call p}, \psi) &= \forall \overline{x_f}. \, (\textbf{pre(p)} \to \lfloor \textbf{post(p)}[\overline{x_f}/\overline{x}] \rfloor) \to \psi[\overline{x_f}/\overline{x}] \\
\text{VC}(\textbf{call p}, \psi) &= \emptyset
\end{aligned}
$$

Similarly to what has been done for Hoare triples, we must now consider what the set of verification conditions for a program correctness formula $\{\Pi\}$ should be. This is simply defined as the union of all the sets of verification conditions required for its constituent procedures:

$$
\text{VCG}(\{\Pi\}) = \bigcup_{p \in \text{PN}(\Pi)} \text{VCG}(\{\widetilde{\textbf{pre}_\Pi(\textbf{p})}\} \, \textbf{body}_\Pi(\textbf{p}) \, \{\textbf{post}_\Pi(\textbf{p})\})
$$

where we have as before the set of verification conditions of a triple given by

$$
\text{VCG}(\{\phi\} \, C \, \{\psi\}) = \{\phi \to \text{wp}\,(C, \psi)\} \cup \text{VC}(C, \psi)
$$

Example 8.5 Let Π be the following program consisting of procedures p_1 and p_2:

$$
\begin{aligned}
&\textbf{pre} \, x > 0 \wedge y > 0 \\
&\textbf{post} \, x = x^{\sim} \wedge y = 2 \times y^{\sim} \wedge z = x + y \wedge z > 2 \\
&\textbf{proc} \, p_1 = \\
&\quad y := 2 \times y; \\
&\quad z := x + y
\end{aligned}
$$

$\mathrm{VCG}(\{\widetilde{\mathbf{pre}(p_1)}\}\,\mathbf{body}(p_1)\,\{\mathbf{post}(p_1)\})$
$\begin{aligned}
= \quad & \mathrm{VCG}(\{x > 0 \wedge y > 0 \wedge x = x^\sim \wedge y = y^\sim \wedge z = z^\sim\} \\
& \quad y := 2 \times y;\; z := x + y \\
& \quad \{x = x^\sim \wedge y = 2 \times y^\sim \wedge z = x + y \wedge z > 2\}) \\
= \quad & \{x > 0 \wedge y > 0 \wedge x = x^\sim \wedge y = y^\sim \wedge z = z^\sim \rightarrow \\
& \quad x = x^\sim \wedge 2 \times y = 2 \times y^\sim \wedge x + 2 \times y = x + 2 \times y \wedge x + 2 \times y > 2\}
\end{aligned}$

$\mathrm{VCG}(\{\widetilde{\mathbf{pre}(p_2)}\}\,\mathbf{body}(p_2)\,\{\mathbf{post}(p_2)\})$
$\begin{aligned}
= \quad & \mathrm{VCG}(\{x > 0 \wedge x = x^\sim \wedge y = y^\sim \wedge z = z^\sim\} \\
& \quad y := x + 100;\; \mathbf{call}\,p_1 \\
& \quad \{z = 3 \times x^\sim + 200\}) \\
= \quad & \{x > 0 \wedge x = x^\sim \wedge y = y^\sim \wedge z = z^\sim \rightarrow \\
& \quad \mathrm{wp}\,(\mathbf{call}\,p_1, z = 3 \times x^\sim + 200)[x + 100/y]\}
\end{aligned}$

where
$\mathrm{wp}\,(\mathbf{call}\,p_1, z = 3 \times x^\sim + 200)$
$\begin{aligned}
= \quad & \forall x_f, y_f, z_f.\,(x > 0 \wedge y > 0 \rightarrow \\
& \quad \lfloor x_f = x^\sim \wedge y_f = 2 \times y^\sim \wedge z_f = x_f + y_f \wedge z_f > 2\rfloor) \rightarrow z_f = 3 \times x^\sim + 200
\end{aligned}$

thus
$\mathrm{VCG}(\{\widetilde{\mathbf{pre}(p_2)}\}\,\mathbf{body}(p_2)\,\{\mathbf{post}(p_2)\})$
$\begin{aligned}
= \quad & \{x > 0 \wedge x = x^\sim \wedge y = y^\sim \wedge z = z^\sim \rightarrow \forall x_f, y_f, z_f.\,(x > 0 \wedge x + 100 > 0 \rightarrow \\
& \quad x_f = x \wedge y_f = 2 \times (x + 100) \wedge z_f = x_f + y_f \wedge z_f > 2) \rightarrow z_f = 3 \times x^\sim + 200\}
\end{aligned}$

Fig. 8.2 Calculation of verification conditions for a program with two procedures

$$\begin{aligned}
&\mathbf{pre}\,x > 0 \\
&\mathbf{post}\,z = 3 \times x^\sim + 200 \\
&\mathbf{proc}\,p_2 = \\
&\quad y := x + 100; \\
&\quad \mathbf{call}\,p_1
\end{aligned}$$

The verification conditions for this program are calculated in Fig. 8.2; both VCs hold, thus the program is correct.

We end the section with an example that illustrates the generation of verification conditions for a procedure in which the $\sim$ notation is used with an array variable (note that for this example we consider that the rules of Fig. 8.1 and the resulting VCGen clauses are used with the safety-sensitive calculus and VCGen of Chap. 7).

Example 8.6 Let us consider again the program **factab** of Example 6.11. We will now write it as a procedure, and additionally modify it in order to cope with safety. In particular, the precondition will be strengthened to state that the relevant range of both the input and the output arrays is safe. Furthermore the external loop's invariant must also be strengthened with these safety conditions (the internal loop does not access any array positions).

pre $size \geq 0 \wedge$
   $valid\_range(in, 0, size - 1) \wedge$
   $valid\_range(out, 0, size - 1) \wedge$
   $\forall a.\ 0 \leq a < size \rightarrow in[a] \geq 0$

post $\forall a.\ 0 \leq a < size \rightarrow out[a] = fact(in\tilde{}[a]) \wedge size = size\tilde{}$

proc factab $=$
  $k := 0;$
  while $k < size$ **do** $\{\theta_2^0 \wedge 0 \leq k \leq size \wedge$
                            $\forall a.\ 0 \leq a < k \rightarrow out[a] = fact(in[a])\}$
  $\{$
    $f := 1;\ i := 1;\ n := in[k];$
    while $i \leq n$ **do** $\{\theta_1^0 \wedge 1 \leq i \leq n + 1 \wedge f = fact(i - 1)\}\{$
      $f := f \times i;$
      $i := i + 1$
    $\}$
    $out[k] := f;$
    $k := k + 1$
  $\}$

where
θ_2^0 is $size \geq 0 \wedge valid\_range(in, 0, size - 1) \wedge valid\_range(out, 0, size - 1) \wedge$
   $\forall a.\ 0 \leq a < size \rightarrow in[a] \geq 0 \wedge in = in\tilde{} \wedge size = size\tilde{}$
θ_1^0 is $\theta_2^0 \wedge n = in[k] \wedge 0 \leq k < size \wedge \forall a.\ 0 \leq a < k \rightarrow out[a] = fact(in[a])$

Fig. 8.3 **factab** as a procedure

The new definition is shown in Fig. 8.3. To check the procedure, one is led to calculate the verification conditions for the triple

   $\{size \geq 0 \wedge valid\_range(in, 0, size - 1) \wedge valid\_range(out, 0, size - 1) \wedge$
   $\forall a.\ 0 \leq a < size \rightarrow in[a] \geq 0 \wedge$
   $in = in\tilde{} \wedge size = size\tilde{}\}$
   body(factab)
   $\{\forall a.\ 0 \leq a < size \rightarrow out[a] = fact(in\tilde{}[a]) \wedge size = size\tilde{}\}$

which we leave to the reader to complete. The reader is also invited to write a procedure that invokes **factab** and calculate the verification conditions for the program consisting of both procedures.

8.3 Frame Conditions

The reader will have noticed that after the execution of a procedure call command **call p**, nothing can be assumed about the value of any variable, unless the variable is mentioned in the postcondition **post(p)**. This may be very annoying, since the program may deal with a large set of variables, of which **p** assigns to only a small subset, and it is not possible to infer that the values of the unassigned variables is preserved. This is a consequence of the design of the procedure call rule,

$$\frac{\{\Pi\}}{\{\phi\}\,\textbf{call p}\,\{\psi\}} \quad \text{if } \phi \to \forall \overline{x_f}.\,(\textbf{pre(p)} \to \lfloor \textbf{post(p)}[\overline{x_f}/\overline{x}]\rfloor) \to \psi[\overline{x_f}/\overline{x}]$$

where $\overline{x} = \textbf{Var}(\textbf{post(p)}) \cup \textbf{Var}(\psi)$. The problem lies in fact in the definition of $\overline{x}$, which correctly prescribes that any variable occurring in the current postcondition ψ must be considered in the post-state (i.e. it is substituted by a fresh variable), since it may have been assigned by **p**. Thus any relation between the values of the variable in the pre-state and in the post-state has to be explicitly stated in the contract's postcondition **post(p)**.

If it was known exactly what variables **p** modifies, or at least some safe approximation of this set, then the definition could be modified to

$$\overline{x} = \textbf{frame(p)}$$

where **frame(p)** denotes the set of variables *possibly assigned* by **p** (i.e. variables not listed are guaranteed to have their values preserved by the procedure). Thus a variable x occurring in ψ but not assigned by **p** will in the side condition of the procedure call rule be considered in the pre-state, as desirable—even if it occurs in **post(p)**. The effect will be the same as if **post(p)** contained the condition $x = x\tilde{\ }$.

Consider the following procedure similar to that of Example 8.2:

$$\textbf{pre } x > 0 \wedge y > 0$$
$$\textbf{post } z = x + y$$
$$\textbf{frame } z$$
$$\textbf{proc p} = \cdots$$

Instead of explicitly stating in the postcondition that the value of x is preserved by executions of **p**, the contract of **p** states that the procedure only assigns to z. If **p** is called in a state in which the following precondition ϕ is true:

$$x > 0 \wedge y > x \wedge x = x\tilde{\ } + 100$$

and we wish to prove that the following postcondition ψ will hold after the execution of **call p**,

$$z = x + y \wedge x = x\tilde{\ } + 100$$

the side condition of the procedure call rule, modified as explained above, would now be

$$x > 0 \wedge y > x \wedge x = x\tilde{\ } + 100 \to \forall z_f.\,(x > 0 \wedge y > 0 \to z_f = x + y) \to$$
$$z_f = x + y \wedge x = x\tilde{\ } + 100$$

It remains to discuss how frame conditions are themselves checked. Let **frame(p)** denote the set of assigned variables declared in the contract. Consider the function $assigned : \textbf{Comm} \rightarrow \mathcal{P}(\textbf{Var})$ defined inductively as follows:

$$assigned(\textbf{skip}) = \emptyset$$

$$assigned(x := e) = \{x\}$$

$$assigned(u[e] := e') = \{u\}$$

$$assigned(C_1; C_2) = assigned(C_1) \cup assigned(C_2)$$

$$assigned(\textbf{if } b \textbf{ then } C_t \textbf{ else } C_f) = assigned(C_t) \cup assigned(C_f)$$

$$assigned(\textbf{while } b \textbf{ do } \{\theta\} C) = assigned(C)$$

$$assigned(\textbf{call p}) = \textbf{frame(p)}$$

Note that this is a just an approximation; in particular it would be much more useful to be able to list sets of array positions modified by a procedure, rather than just array variables.

The declared frame condition of each procedure **p** should include the set $assigned(\textbf{p})$, so $\textbf{VCG}(\{\Pi\})$ should include for each $\textbf{p} \in \textbf{PN}(\Pi)$ the condition $assigned(\textbf{p}) \subseteq \textbf{frame(p)}$ (encoded as a first order formula). We omit the details here.

Lists of assigned variables explicitly included in contracts are usually called *frame conditions* and are contemplated in standard annotation languages for realistic languages, like JML (for Java), ACSL (for C, see Chap. 9), and SPARK. Both JML and ACSL include optional frame conditions in contracts along the lines discussed above (and the absence of such a condition is interpreted as the routine possibly modifying the entire global state). In SPARK frame conditions (in the form of *global* definitions) are part of the dataflow annotations that must be given by the programmer for every routine. Every read and assigned variable has to be identified, so in fact writing precise frame conditions in SPARK is compulsory.

8.4 Procedures with Parameters

In this section we consider procedures with parameters. This raises some standard issues (related in particular to scoping rules and evaluation strategies) that have relevant consequences in the discussion of axiomatic reasoning involving such procedures. We will not give the operational semantics of this extended language, or prove soundness results for the inference and VCGen rules presented. Our goal in this section is to explain the main issues involved, which will in turn facilitate the use of standard specification languages as exemplified later in the book.

The first step is to extend the programming language syntax by allowing procedures to take a list of formal arguments. In the following a ranges over **Var** and ε denotes the empty sequence.

$$\textbf{Arglist} \quad \ni \quad \lambda \quad ::= \quad a, \lambda \mid \textbf{var}\, a, \lambda \mid \varepsilon$$

$$\textbf{Proc} \quad \ni \quad \Phi \quad ::= \quad \textbf{pre}\, \phi \; \textbf{post}\, \psi \; \textbf{proc}\, \textbf{p}(\lambda) = C$$

$$\textbf{Comm} \quad \ni \quad C \quad ::= \quad \ldots \mid \textbf{call}\, \textbf{p}(\overline{e})$$

where the length of $\overline{e}$ coincides with the arity of **p**. For a procedure to be well-defined the corresponding list of formal parameters is required to contain no repeated variables. Following the Pascal syntax, parameters passed by reference are identified by the keyword **var** in the procedure definition. This parameter mode corresponds to effectively passing to the procedure a reference to a variable of the caller procedure; local assignments to this parameter are reflected in the value of the caller's variable.

For $\textbf{p} \in \textbf{PN}(\Pi)$, we let $\textbf{param}_\Pi(\textbf{p})$ denote the list of formal parameters of **p** passed by value and $\textbf{varparam}_\Pi(\textbf{p})$ denote the list of formal parameters passed by reference. Usually we drop the program name when it is clear from the context. For instance given the definition

$$\textbf{pre}\, \theta$$
$$\textbf{post}\, \rho$$
$$\textbf{proc}\, \textbf{p}(x, \textbf{var}\, y, z, \textbf{var}\, w) = C$$

one has

$$\textbf{param}(\textbf{p}) \;=\; x, z$$
$$\textbf{varparam}(\textbf{p}) \;=\; y, w$$

With the introduction of parameters, program variables become clearly divided in two separate classes: in addition to global variables there are now parameter variables, which have local scope. Any program variable occurring in the body of a parameter and not in its parameter list is a global variable.

The procedure's contract ($\textbf{pre}(\textbf{p})$ and $\textbf{post}(\textbf{p})$) may contain occurrences of the formal parameter variables x, y, z, and w, which allows us to impose conditions on the procedure's input as part of the precondition, and on the other hand to specify the output of the procedure in relation to the parameters' values. In addition to local parameters, a procedure's contract may of course still contain occurrences of global variables.

Parameters can be freely renamed, i.e. they can be substituted by *fresh* variables in the procedure's code *and contract*. By fresh we mean that they do not occur free in either the procedure's body, precondition, or postcondition. For instance assuming x', y', z', and w' do not occur free in C, θ, or ρ, then the definition of **p** above is equivalent to the following

$$\textbf{pre } \theta[x'/x, y'/y, z'/z, w'/w]$$
$$\textbf{post } \rho[x'/x, y'/y, z'/z, w'/w]$$
$$\textbf{proc p}(x', \textbf{var } y', z', \textbf{var } w') = C[x'/x, y'/y, z'/z, w'/w]$$

If the same variable x is employed as a global variable and as a parameter, the effect of this is that the global variable is not visible inside the procedure. Moreover, *it is not visible in the procedure's contract*: all occurrences of x in the annotated precondition and postcondition are interpreted as referring to the local parameter. On the other hand a procedure's parameters are never visible inside other procedures, and they may not occur in another procedure's contracts. *Static scoping* is assumed, so when a procedure is called the values of the caller's local parameters do not affect the callee.

Let us start by considering the procedure call rule of Fig. 8.1, and what needs to be changed in the presence of local parameters, say to derive the following

$$\frac{\{\Pi\}}{\{\phi\} \textbf{ call p}(e_1, e_2, e_3, e_4) \, \{\psi\}}$$

As in the parameterless case, the side condition of this rule will be a first-order formula involving ϕ, $\textbf{pre(p)}$, $\textbf{post(p)}$, and ψ. The first key point to understand is that occurrences of the formal parameter variables x, y, z, w in $\textbf{pre(p)}$ and $\textbf{post(p)}$ have to be substituted by the actual parameters e_1, e_2, e_3, e_4. Note that if the formal parameters occur also in ϕ or in ψ they must be interpreted as distinct variables that have nothing to do with $\textbf{p}$'s parameters and will of course not be substituted.

Now we face a difficulty. The variables occurring in the actual arguments e_1 to e_4, and also in the assertions ϕ and ψ, may be either global, or *local parameters of the caller procedure*—and note that there is no way to know to which category a given variable occurrence belongs. Thus even though $\textbf{pre(p)}$ and $\textbf{post(p)}$ do not contain occurrences of the caller's parameter variables, for instance the assertion $\textbf{pre(p)}[e_1/x, e_2/y, e_3/z, e_4/w]$ may well contain such occurrences. What is dangerous here is that if the caller uses some parameter name, say x, that also occurs as a global variable in $\textbf{p}$'s contract, then in the side condition we risk having occurrences of x (those in e_1 to e_4, ϕ and ψ) that refer to the caller's parameter, while others (those in $\textbf{pre(p)}$ and $\textbf{post(p)}$) refer to the global variable. The latter occurrences are captured by the binding of parameters of the caller procedure.

This confusion arises from the use of locally-scoped variables (of which, with the exception of Exercise 5.10, parameters are the first instance in this book), and cannot be detected at the level of the procedure call. We will eliminate the problem by resorting to the following

Variable convention: The variables that are used globally in a program Π are not used as parameters in any procedure $\textbf{p} \in \textbf{PN}(\Pi)$.

This convention will prevent the problem just described from occurring in the derivations constructed as premises of the mutual recursion rule: since a variable cannot be used simultaneously as local and global, there is no way that a global variable can be captured by an application of the procedure call rule. Programs that do not meet this requirement can of course have their procedures's parameters conveniently renamed, as outlined above. The variable convention implies that the set of program variables is partitioned in two subsets: those used as global variables and those used as parameters. We assume that $\mathbf{Var}(\phi)$ and $\mathbf{Var}^\frown(\phi)$ return sets of global variables.

We will now look in more detail at what the procedure call and mutual recursion rules should look like, first for parameters passed by value and then for parameters passed by reference.

8.4.1 Parameters Passed by Value

For the sake of simplicity we first consider a procedure **p** with a single formal parameter a passed by value, and a version of the procedure call rule that does not perform adaptation between the specification required by the invocation and the contract:

$$\frac{\{\Pi\}}{\{\phi\}\,\mathbf{call}\,\mathbf{p}(e)\,\{\psi\}} \quad \text{if } \phi \to \mathbf{pre(p)}[e/a] \text{ and } \mathbf{post(p)}[e/a] \to \psi \qquad (8.7)$$

As stated above, occurrences of parameter variables in the procedure's contract hide homonymous variables if these exist. If the caller procedure also has a parameter a, occurrences of a in ϕ and $\mathbf{pre(p)}$ refer to the caller's parameter and to the callee's parameter respectively (the variable convention implies that no global variable of the same name may exist). The substitution of the actual parameter e for a ensures that all occurrences of a in $\phi \to \mathbf{pre(p)}[e/a]$ actually refer to the caller's variable, which may also of course occur in e. Naturally, the same is true of ψ and $\mathbf{post(p)}$.

It is a matter of language design whether parameters passed by value can be assigned to inside the procedure's body, and used as local auxiliary variables (this is not a big issue since the same effect could be obtained by using locally declared variables, following Exercise 5.10). If parameters cannot be assigned, they act as local constants and are also said to be passed by *constant value*; this is the case of *in* parameters in the Ada programming language. If a is a constant value parameter then it has the same value in the initial and final states of execution of the procedure.

If parameters passed by value can be assigned, then how should occurrences of a in $\mathbf{post(p)}$ be interpreted? Since the internal final value of the parameter variable is totally irrelevant to the caller, we will take such occurrences as referring to the initial value of a. This is in accordance with the interpretation prescribed by standard specification languages like JML.

This discussion affects the form of the mutual recursion rule. With parameters passed by constant value the rule of Fig. 8.1 could be used with no modifications. We will consider instead that parameter variables can be assigned, which requires the rule to be modified. The branch concerning our procedure **p** will be as shown below.

$$[\{\Pi\}]$$
$$\vdots$$

$$\dots \quad \{\mathbf{pre(p)} \wedge a = a\tilde{\ }\}\,\mathbf{body(p)}\,\{\mathbf{post(p)}[a\tilde{\ }/a]\} \quad \dots \qquad (8.8)$$
_____
$$\{\Pi\}$$

Thus a is simply treated in **post(p)** in the same way as $a\tilde{\ }$. Note that using $a\tilde{\ }$ in **post(p)** becomes unnecessary since a can be used to the same effect.

Example 8.7 Following Example 8.1, let us now write factorial as a one-parameter procedure. This is straightforward: variable n is a natural candidate to being promoted to a parameter. f and i are still global variables.

> **pre** $n \geq 0$
> **post** $f = fact(n)$
> **fact**$(n) =$
> $f := 1; i := 1;$
>   **while** $i \leq n$ **do** $\{f = fact(i-1) \wedge i \leq n+1\}\{$
> $f := f \times i;$
> $i := i+1$
> $\}$

Note that there is no need to use $n\tilde{\ }$ in the postcondition. The following triple can be derived by applying rule (8.7). x can be either a global variable or a parameter of the caller procedure.

$$\{x \geq -10\}\,\mathbf{call\,fact}(x+20)\,\{f = fact(x+20)\}$$

with the following valid side conditions:

1. $x \geq -10 \to n \geq 0\,[x+20/n]$
2. $f = fact(n)\,[x+20/n] \to f = fact(x+20)$

Let us turn to the general case: procedures may now have more than one parameter, all passed by valued, that can be freely assigned to in the procedure's code, and the annotated postcondition **post(p)** may contain occurrences of $\tilde{\ }$ to refer to the values of arbitrary variables in the pre-state. The appropriate rules are shown in Fig. 8.4. The mutual recursion rule is just a generalisation of (8.8) that copes with the use of $\tilde{\ }$ both for global variables and to treat occurrences of parameters in the postconditions.

In the procedure call rule, the actual parameters $\bar{e}$ are substituted for the formal parameters $\bar{a}$ in the contract's conditions. There may be occurrences in $\bar{e}$ of global variables that also occur in **post(p)**. Occurrences that come from $\bar{e}$ are not substituted by fresh variables, because they should be interpreted in the pre-state, contrary

(mutual recursion—pbv)

$$[\{\Pi\}] \qquad\qquad\qquad [\{\Pi\}]$$
$$\vdots \qquad\qquad\qquad\qquad \vdots$$

$$\frac{\{\widetilde{\mathbf{pre}_\Pi(\mathbf{p}_1)}\}\,\mathbf{body}_\Pi(\mathbf{p}_1)\,\{\widetilde{\mathbf{post}_\Pi(\mathbf{p}_1)}\} \quad \cdots \quad \{\widetilde{\mathbf{pre}_\Pi(\mathbf{p}_n)}\}\,\mathbf{body}_\Pi(\mathbf{p}_n)\,\{\widetilde{\mathbf{post}_\Pi(\mathbf{p}_n)}\}}{\{\Pi\}}$$

where

$$\mathbf{PN}(\Pi) \;=\; \{\mathbf{p}_1,\ldots,\mathbf{p}_n\}$$

$$\widetilde{\mathbf{pre}_\Pi(\mathbf{p}_i)} \;=\; \mathbf{pre}_\Pi(\mathbf{p}_i)\wedge x_1 = x_1{}^{\sim}\wedge\cdots\wedge x_k = x_k{}^{\sim}$$
$$\text{with } \mathbf{Var}\tilde{}(\mathbf{post}_\Pi(\mathbf{p}_i))\cup\mathbf{param}_\Pi(\mathbf{p}_i) = \{x_1,\ldots,x_k\}$$

$$\widetilde{\mathbf{post}_\Pi(\mathbf{p}_i)} \;=\; \mathbf{post}_\Pi(\mathbf{p}_i)[a_1{}^{\sim}/a_1\ldots a_m{}^{\sim}/a_m] \quad \text{with } \mathbf{param}_\Pi(\mathbf{p}_i)=\{a_1,\ldots,a_m\}$$

(procedure call—pbv)

$$\frac{\{\Pi\}}{\{\phi\}\,\mathbf{call}\,\mathbf{p}(\overline{e})\,\{\psi\}} \quad \text{if } \phi\to\forall\overline{x_f}.\,(\mathbf{pre}_\Pi(\mathbf{p})[\overline{e}/\overline{a}]\to\lfloor\mathbf{post}_\Pi(\mathbf{p})[\overline{e}/\overline{a},\overline{x_f}/\overline{x}]\rfloor)\to\psi[\overline{x_f}/\overline{x}]$$

where

$$p \;\in\; \mathbf{PN}(\Pi)$$

$$\overline{a} \;=\; \mathbf{param}_\Pi(\mathbf{p})$$

$$\overline{x} \;=\; \mathbf{Var}(\mathbf{post}_\Pi(\mathbf{p}))\cup\mathbf{Var}(\psi)$$

$$\overline{x_f} \quad \text{are fresh variables}$$

Fig. 8.4 Inference rules for programs with mutually recursive procedures with parameters passed by value

to occurrences that were originally in **post(p)**. This explains the use of simultaneous substitution of $\overline{a}$ and $\overline{x}$. Note that $\overline{x}$ does not contain any parameter variables—in particular it does not contain parameters of the caller procedure. Thus parameters are not substituted by fresh variables, which is coherent with the fact that **p** cannot access parameter variables of the caller procedure, whose values are thus preserved by the call to **p**.

Example 8.8 Consider a variant of the procedure **p** of Example 8.2 with the following definition, where a is a parameter passed by value.

$$\textbf{pre } a > 0 \wedge y > 0$$
$$\textbf{post } z = a + y$$
$$\textbf{proc } \mathbf{p}(a) = \cdots$$

If the precondition is met, the final value of variable z will be the sum of the initial value of the parameter a and the final value of y. Now consider a call to this procedure in a state in which the following precondition ϕ is true:

$$x > 0 \wedge y > x$$

and we wish to prove that the following postcondition ψ will hold after the execution of $\textbf{call p}(2 \times x + 1)$:

$$z = 2 \times x + 1 + y$$

Note that, in ϕ and ψ, y is necessarily a global variable, since it is a global variable in the definition of $\mathbf{p}$. Let us assume that x is also a global variable. Assuming the program is correct, the side condition of the procedure call rule will be

$$x > 0 \wedge y > x \rightarrow \forall x_f, y_f, z_f.$$
$$(2 \times x + 1 > 0 \wedge y > 0 \rightarrow z_f = 2 \times x + 1 + y_f) \rightarrow z_f = 2 \times x_f + 1 + y_f$$

This is clearly not valid: the contract provides a value for z_f (z in the final state) in terms of the value of x at call time, but the desired postcondition expresses z_f in terms of its value after execution of $\textbf{call p}(2 \times x + 1)$, and there is no way to infer that the value of x has been preserved by the call. If on the other hand x is a local parameter of the caller procedure, the side condition will be instead

$$x > 0 \wedge y > x \rightarrow \forall y_f, z_f.$$
$$(2 \times x + 1 > 0 \wedge y > 0 \rightarrow z_f = 2 \times x + 1 + y_f) \rightarrow z_f = 2 \times x + 1 + y_f$$

This is valid, in accordance with the fact that $\mathbf{p}$ cannot modify the value of a parameter of the caller procedure.

If x is global, we can of course include in the contract of $\mathbf{p}$ explicit information regarding the preservation of its value:

$$\textbf{pre } a > 0 \wedge y > 0$$
$$\textbf{post } z = a + y \wedge x = x\tilde{\ }$$
$$\textbf{proc } \mathbf{p}(a) = \cdots$$

the side condition becomes

$$x > 0 \wedge y > x \rightarrow \forall x_f, y_f, z_f.$$
$$(2 \times x + 1 > 0 \wedge y > 0 \rightarrow$$
$$z_f = 2 \times x + 1 + y_f \wedge x_f = x) \rightarrow z_f = 2 \times x_f + 1 + y_f$$

Finally, consider the situation in which a is both a parameter of $\mathbf{p}$ and of the caller procedure, and it may thus occur in ϕ and/or ψ, for instance:

$$\text{let } \phi \text{ be } x > 0 \wedge y > x \wedge a = a\tilde{\ }$$
$$\text{and } \psi \text{ be } z = 2 \times x + 1 + y \wedge a = a\tilde{\ }$$

The side condition becomes simply

$$x > 0 \wedge y > x \wedge a = a^\sim \rightarrow \forall x_f, y_f, z_f.$$
$$(2 \times x + 1 > 0 \wedge y > 0 \rightarrow z_f = 2 \times x + 1 + y_f \wedge x_f = x) \rightarrow$$
$$z_f = 2 \times x_f + 1 + y_f \wedge a = a^\sim$$

which is coherent with the fact that **p** cannot modify the value of the parameter a of the caller.

8.4.2 Parameters Passed by Reference

The difference with respect to parameters passed by value is that assignments to a parameter passed by reference will affect the final value of the actual parameter (necessarily a variable) when the procedure's execution terminates. To illustrate this, let the procedure **p** be defined as

$$\begin{aligned} &\textbf{pre } \phi \\ &\textbf{post } \psi \\ &\textbf{proc p}(\textbf{var } a) = C \end{aligned}$$

Then after execution of the command **call p**(y), the value of y will be whatever value was last assigned to the parameter a inside C. Note that, as was the case when a was passed by value, y can be either a global variable or a local parameter of the caller procedure.

Parameter-passing by reference can be implemented by passing the memory address of (i.e. a reference to) the actual parameter variable to the callee, so that all assignments to the formal parameter variable will immediately be reflected in the value of the actual variable used in the call. It can also be implemented *by copy*: the value of the actual parameter is copied to the formal parameter when the procedure is called, and it is copied back in the reverse direction upon exit (this mode is also called *copy-restore* or *value-return*, and is used for instance in the Ada programming language, with *in/out parameters*).

In any case it is clear that such a parameter a *can* be assigned inside the procedure's body, and occurrences of a in **post(p)** must now be interpreted as referring to the value of the parameter variable in the final state of execution of the procedure. Similarly to global variables, it should be possible to refer in the procedure's postcondition to the initial value of a, using the notation $a^\sim$. As in the parameterless case, the corresponding premise in the mutual recursion rule uses a strengthened precondition to set the initial value of $a^\sim$:

$$[\{\Pi\}]$$
$$\vdots$$
$$\cdots \quad \{\textbf{pre(p)} \wedge a = a^\sim\} \, \textbf{body(p)} \, \{\textbf{post(p)}\} \quad \cdots \qquad (8.9)$$
$$\overline{}$$
$$\{\Pi\}$$

With respect to the procedure call rule, it suffices to understand that the call **call p**(y) should be equivalent to a call to a parameterless procedure that works directly on y instead of a. Consider the virtual procedure

$$\mathbf{pre}\ \phi\,[y/a]$$
$$\mathbf{post}\ \psi\,[y/a, y\tilde{\ }/a\tilde{\ }]$$
$$\mathbf{proc}\ \mathbf{p}' = C\,[y/a]$$

Then the inference rule for the command **call p**(y) should be the same as for **call p′**, and can thus be synthesized from the rule of Fig. 8.1:

$$\frac{\{\Pi\}}{\{\phi\}\,\mathbf{call}\,\mathbf{p}(y)\,\{\psi\}}\quad \text{if}\quad \begin{array}{l}\phi \to \forall y_f.\\ (\mathbf{pre}(\mathbf{p})[y/a] \to \mathbf{post}(\mathbf{p})[y_f/a, y/a\tilde{\ }]) \to \psi[y_f/y]\end{array}$$

$$(8.10)$$

Note that an important difference with respect to parameters passed by value is that the call to procedure **p** *can* now modify the value of parameter variables of the caller, if they are passed by reference to **p**. Thus in the above rule it is indifferent whether y is a global variable or a formal parameter of the caller procedure; it must in any case be substituted by a fresh variable since its value may have been modified by the call to **p**.

Figure 8.5 contains mutual recursion and procedure call rules for the general case of procedures having parameters both passed by value and passed by reference. We remark that the substitution of fresh variables for variables whose value may have been modified by **p** takes place after the substitution of actual parameters passed by reference for $\overline{a_r}$, and simultaneously with the substitution of parameters passed by value for $\overline{a_v}$.

Note also that occurrences of a given fresh variable x_f in the side condition of the procedure call rule may have different origins: x may occur directly in **post**(**p**) and also in $\overline{e_r}$ (possibly even more than once). All these occurrences will be substituted by the same x_f. This, in fact, is when aliasing happens—the final assertion contains occurrences of x corresponding to two or more different ways of referring to the same global variable. The aliasing phenomenon is so important that it deserves to be discussed in a separate section, but first let us illustrate the use of the inference rule for procedure calls in the absence of aliasing.

Example 8.9 To illustrate the procedure call rule with parameters passed by reference we continue developing variants of Example 8.8. The procedure now returns the calculated value in a parameter passed by reference:

$$\mathbf{pre}\ a > 0 \wedge y > 0$$
$$\mathbf{post}\ z = a + y \wedge x = x\tilde{\ }$$
$$\mathbf{proc}\ \mathbf{p}(a, \mathbf{var}\,z) = \cdots$$

where x and y are (necessarily) global variables. Let us consider the invocation **call p**($2 \times x + 1, w$), with the precondition $x > 0 \wedge y > x$ and postcondition $w =$

(*mutual recursion—pbvr*)

$$[\{\Pi\}] \qquad\qquad\qquad [\{\Pi\}]$$
$$\vdots \qquad\qquad\qquad\qquad \vdots$$

$$\{\widetilde{\mathbf{pre}_\Pi}(\mathbf{p}_1)\}\,\mathbf{body}_\Pi(\mathbf{p}_1)\,\{\widetilde{\mathbf{post}_\Pi}(\mathbf{p}_1)\} \quad\cdots\quad \{\widetilde{\mathbf{pre}_\Pi}(\mathbf{p}_n)\}\,\mathbf{body}_\Pi(\mathbf{p}_n)\,\{\widetilde{\mathbf{post}_\Pi}(\mathbf{p}_n)\}$$

$$\overline{\rule{10cm}{0.4pt}}$$

$$\{\Pi\}$$

where

$$\mathbf{PN}(\Pi) \;=\; \{\mathbf{p}_1,\ldots,\mathbf{p}_n\}$$

$$\widetilde{\mathbf{pre}_\Pi}(\mathbf{p}_i) \;=\; \mathbf{pre}_\Pi(\mathbf{p}_i) \wedge x_1 = x_1^{\tilde{}} \wedge \cdots \wedge x_k = x_k^{\tilde{}} \text{ with}$$
$$\mathbf{Var}^{\tilde{}}(\mathbf{post}_\Pi(\mathbf{p}_i)) \cup \mathbf{param}_\Pi(\mathbf{p}_i) \cup \mathbf{varparam}_\Pi(\mathbf{p}_i) = \{x_1,\ldots,x_k\}$$

$$\widetilde{\mathbf{post}_\Pi}(\mathbf{p}_i) \;=\; \mathbf{post}_\Pi(\mathbf{p}_i)[a_1^{\tilde{}}/a_1 \ldots a_m^{\tilde{}}/a_m] \quad \text{with } \mathbf{param}_\Pi(\mathbf{p}_i) = \{a_1,\ldots,a_m\}$$

(*procedure call—pbvr*)

$$\frac{\{\Pi\}}{\{\phi\}\,\mathbf{call}\,\mathbf{p}(\overline{e})\,\{\psi\}} \quad \text{if} \quad \begin{array}{l} \phi \to \forall \overline{x_f}.\,(\mathbf{pre}_\Pi(\mathbf{p})[\overline{e}/\overline{a}] \to \\[4pt] \lfloor \mathbf{post}_\Pi(\mathbf{p})[\overline{e_r}/\overline{a_r},\overline{e_r^{\tilde{}}}/\overline{a_r^{\tilde{}}}][\overline{e_v}/\overline{a_v},\overline{x_f}/\overline{x}]\rfloor) \to \psi[\overline{x_f}/\overline{x}] \end{array}$$

where

$$\mathbf{p} \;\in\; \mathbf{PN}(\Pi)$$

$\overline{e_v},\overline{e_r}$ are the subsequences of $\overline{e}$ corresponding to parameters passed by value and by reference respectively

$$\overline{a_v} \;=\; \mathbf{param}_\Pi(\mathbf{p})$$

$$\overline{a_r} \;=\; \mathbf{varparam}_\Pi(\mathbf{p})$$

$$\overline{x} \;=\; \mathbf{Var}(\mathbf{post}_\Pi(\mathbf{p})) \cup \mathbf{Var}(\psi) \cup \overline{e_r}$$

$\overline{x_f}$ are fresh variables

Fig. 8.5 Inference rules for programs with mutually recursive procedures with parameters, general case

$2 \times x + 1 + y$. Assuming that the program is correct, the side condition of the procedure call rule would be

$$x > 0 \wedge y > x \to \forall x_f, y_f, w_f.\,(2 \times x + 1 > 0 \wedge y > 0 \to$$
$$w_f = 2 \times x + 1 + y_f \wedge x_f = x) \to w_f = 2 \times x_f + 1 + y_f$$

Note that w can be either a formal parameter of the caller procedure or a global variable. The side condition is the same in both cases.

Next we consider that the initial value of z is used instead of y:

$$\textbf{pre } a > 0 \wedge z > 0$$
$$\textbf{post } z = a + \tilde{z} \wedge x = \tilde{x}$$
$$\textbf{proc p}(a, \textbf{var } z) = \cdots$$

Suppose the invocation is still **call p**$(2 \times x + 1, w)$, with the precondition $x > 0 \wedge w > x \wedge w = 100$ and postcondition $w = 2 \times x + 1 + 100$. This would result in the side condition

$$x > 0 \wedge w > x \wedge w = 100 \rightarrow \forall x_f, w_f. (2 \times x + 1 > 0 \wedge w > 0 \rightarrow$$
$$w_f = 2 \times x + 1 + w \wedge x_f = x) \rightarrow w_f = 2 \times x_f + 1 + 100$$

If z is also a parameter of the caller procedure it may be passed to substitute the formal parameter z (this may even be a recursive call of **p** from its own body). This is illustrated by the command **call p**$(2 \times x + 1, z)$, with precondition $x > 0 \wedge z > x \wedge z = 100$ and postcondition $z = 2 \times x + 101$. The side condition will be instead

$$x > 0 \wedge z > x \wedge z = 100 \rightarrow \forall x_f, z_f. (2 \times x + 1 > 0 \wedge z > 0 \rightarrow$$
$$z_f = 2 \times x + 1 + z \wedge x_f = x) \rightarrow z_f = 2 \times x_f + 101$$

8.4.3 Aliasing

Let us now explain the devastating consequences that the aliasing phenomenon can have on the kind of reasoning that we have been promoting. Consider a procedure

$$\textbf{pre } \top$$
$$\textbf{post } a = 10 \wedge b = 20$$
$$\textbf{proc p}(\textbf{var } a, \textbf{var } b) =$$
$$a := 10;$$
$$b := 20$$

The reader will agree that the body of the procedure conforms to the contract:

$$\{a = \tilde{a} \wedge b = \tilde{b}\} \, \textbf{body}(\textbf{p}) \, \{a = 10 \wedge b = 20\}$$

However, consider now a call of the form **call p**(x, x) where x is either a global variable or local to the caller procedure. The procedure invocation rule of Sect. 8.4.2 allows us to derive the following, which clearly does not hold:

$$\{\top\} \, \textbf{call p}(x, x) \, \{x = 10 \wedge x = 20\}$$

The problem here has to do with the fact that the same global variable is being locally referred through two different parameters a and b. But aliasing may occur even if the procedure has a single parameter; consider the following definition, where y is global.

$$\textbf{pre}\ \top$$
$$\textbf{post}\ c = 10 \wedge y = 20$$
$$\textbf{proc}\ \textbf{q}(\textbf{var}\ c) =$$
$$\quad c := 10;$$
$$\quad y := 20$$

The validity of the following triple is straightforward to establish

$$\{c = \tilde{c} \wedge y = \tilde{y}\}\ \textbf{body}(\textbf{q})\ \{c = 10 \wedge y = 20\}$$

Consider now the invocation $\textbf{call}\ \textbf{q}(y)$; the variable y will be referred both directly, and indirectly through the parameter c. Again the procedure invocation rule of the previous section would derive a triple that does not hold:

$$\{\top\}\ \textbf{call}\ \textbf{q}(y)\ \{y = 10 \wedge y = 20\}$$

So clearly the procedure call rule given in Sect. 8.4.2 is not sound. How could this problem be fixed in our inference system? Roughly, we could consider additional side conditions in the rule,

— to prevent the same actual parameter variable (passed by reference) to occur more than once in an invocation; and
— to prevent global variables that are assigned inside a procedure to be used as actual parameters (passed by reference) in invocations of that procedure. One way to do this would be to include in the procedure's contract a list of the assigned variables (see Sect. 8.3).

It is worth considering how two practical verification platforms handle the problem. In SPARK/Ada, a language used mostly in the development of safety-critical software, programs in which aliasing may occur are nor considered valid programs, because this is seen as a bad practice, not admissible in the safety-critical context. In SPARK all procedures must contain *dataflow* annotations that list the variables written by the procedure, and the SPARK Examiner (a tool that validates SPARK programs through syntactic and flow analysis) can use this information to reject programs with aliasing. In particular both examples given above would be rejected. The SPARK VCGen, also implemented as part of the SPARK Examiner, does not need to consider the possibility of aliasing, since it only takes as input valid SPARK programs. Thus a procedure call rule like the one of Sect. 8.4.2 can be applied with no restrictions.

The C programming language on the other hand does of course allow aliasing, but consider how it is created. Parameter-passing in C is always by value; call-by-reference is implemented through the use of pointer variables (which are themselves passed by value). The first example given above would look like

```
p (int *ap, int *bp) {
    *ap = 10;
    *bp = 20;
}
```

Note that the parameters are now memory addresses (pointer variables), whose contents `*ap` / `*bp` correspond to the values of the variables being passed by reference. So the parameters have access to both the values of the variables and to their addresses `ap` / `bp`. Aliasing can be treated as part of a more general issue: *separation of memory regions*. The assertion language may for instance contain a predicate to assert that two memory locations belong to disjoint memory regions. In our example an assertion could then be added to the precondition of `p`, stating that `ap` and `bp` point to separate locations, which would prevent the possibility of aliasing taking place. For instance the call `p(&x, &x)` would not satisfy the precondition. The Frama-C / Jessie VCGen (see Chap. 10) can be used in a mode that assumes that all data structures accessed through pointers are stored in separate memory regions.

8.5 Return Values and Pure Functions

The name "function" when applied to a form of routine/sub-program is often used with very different meanings. In two extremes we find the use of the word in the C programming language, in which all routines are called functions, and in SPARK/Ada, where the name is used for a *pure* routine that does not have any side effects. Assignment to global variables is forbidden, and only parameters passed by (constant) value are allowed, thus the global state is never modified. In fact in SPARK function execution depends only on the values of the input parameters and the global state—the notion is as close as possible to mathematical functions. It is the latter notion that will be considered in this section.

Let us start by seeing what can be simplified in the procedure call and mutual recursion rules for such pure routines. In the procedure call rule the absence of side effects and ˜ notation mean that there is no need to substitute fresh variables to account for the new values of variables in the post-state, so the rule will resemble the following, where $\bar{a} = \mathbf{param}(\mathbf{p})$.

$$\frac{\{\Pi\}}{\{\phi\}\,\mathbf{call}\,\mathbf{p}(\bar{e})\,\{\psi\}} \quad \text{if } \phi \to (\mathbf{pre}(\mathbf{p})[\bar{e}/\bar{a}] \to \mathbf{post}(\mathbf{p})[\bar{e}/\bar{a}]) \to \psi$$

Note the side condition of the procedure call rule, instead of the stronger

$$\phi \to \mathbf{pre}(\mathbf{p})[\bar{e}/\bar{a}] \quad \text{and} \quad \mathbf{post}(\mathbf{p})[\bar{e}/\bar{a}] \to \psi$$

The fact that the procedure is pure does not mean that it would not be useful to relate the values of variables before and after the call; it means instead that these values are the same. The latter side condition does not take this into account, but the former does, since it allows for information to be transported from ϕ to ψ.

As it is, this procedure call rule is however completely useless: a pure procedure with all parameters passed by constant value cannot communicate with the caller procedure. In fact, all uses of the term "function" have in common the fact that functions have a *return value* that is passed back to the caller. Let us then extend the programming language syntax by allowing function definitions and a command for designating return values as follows

$$\textbf{Proc} \quad \ni \quad \Phi \quad ::= \quad \dots \mid \textbf{pre}\,\phi\,\textbf{post}\,\psi\,\textbf{fun}\,\textbf{f}(\overline{x}) = C$$

$$\textbf{Comm} \quad \ni \quad C \quad ::= \quad \dots \mid \textbf{return}\,e \mid x := \textbf{fcall}\,\textbf{f}(\overline{e})$$

where   the length of $\overline{e}$ coincides with the arity of $\textbf{f}$. We let $\textbf{FN}(\Pi)$ denote the set of names of the functions defined in the program Π. Return values are designated using a special **return** command, as in the C programming language.

So for instance the following defines a function with precondition ϕ, postcondition ψ, parameter list $\overline{a}$, and body C.

$$\textbf{pre}\,\phi$$
$$\textbf{post}\,\psi$$
$$\textbf{fun}\,\textbf{f}(\overline{a}) = C$$

The **return** command will in fact be treated as syntactic sugar as follows:

$$\textbf{return}\,e \quad \text{is an abbreviation of} \quad result := e$$

We use a special (reserved) program variable *result* to deal with return values. Occurrences of the variable in the contract are limited to postconditions, to make possible specifying how the return value of a function is related to the values of its parameters. ˜ notation does not apply to this variable. With the exception of its use in postconditions, the return value variable *result* is only accessible to the programmer through the syntactic sugar.

Note that while in C the **return** statement has an *abrupt exit* semantics, allowing it to return control from the callee function to the caller at any point in the function code, we exclude that possibility here (see Exercise 8.7). If the return value is assigned more than once in the function, it is the last such value that is returned to the caller.

Function calls on the other hand can only be made with an assignment command, whose effect is to call the function and then copy the return value to the assigned variable. Again this is done through syntactic sugar:

$$y := \textbf{fcall}\,\textbf{f}(\overline{e}) \quad \text{is an abbreviation of} \quad \textbf{call}\,f(\overline{e})\,;\; y := result$$

(*mutual recursion*)

for each $\mathbf{p} \in \mathbf{PN}(\Pi)$ and $\mathbf{f} \in \mathbf{FN}(\Pi)$

$$[\{\{\Pi\}\}]$$
$$\vdots$$
$$\{\widetilde{\mathbf{pre}_\Pi(\mathbf{p})}\}\,\mathbf{body}_\Pi(\mathbf{p})\,\{\widetilde{\mathbf{post}_\Pi(\mathbf{p})}\}$$

$$[\{\{\Pi\}\}]$$
$$\vdots$$
$$\{\mathbf{pre}_\Pi(\mathbf{f})\}\,\mathbf{body}_\Pi(\mathbf{f})\,\{\mathbf{post}_\Pi(\mathbf{f})\}$$

$$\overline{}$$
$$\{\Pi\}$$

(*function call*)

$$\frac{\{\Pi\}}{\{\phi\}\,y := \mathbf{fcall}\,\mathbf{f}(\overline{e})\,\{\psi\}} \quad \text{if } \phi \to (\mathbf{pre}_\Pi(\mathbf{f})[\overline{e}/\overline{a}] \to \mathbf{post}_\Pi(\mathbf{f})[\overline{e}/\overline{a}]) \to \psi[result/y]$$

where

$$\mathbf{f} \in \mathbf{FN}(\Pi)$$

$$\overline{a} = \mathbf{param}_\Pi(\mathbf{f})$$

(*return*)

$$\frac{}{\{\phi\}\,\mathbf{return}\,e\,\{\psi\}} \quad \text{if } \phi \to \psi[e/result]$$

Fig. 8.6 Inference rules for programs containing functions and procedures

Since the program may consist of a mix of pure functions and impure procedures, some premises of the mutual recursion rule correspond to procedures, and some to functions. The latter are much simplified, since all parameters are passed by constant value, and no uses of the ~ notation are allowed. The full set of rules is shown in Fig. 8.6.

The VCGen definition can be extended by simply giving rules for **return** and function call as follows (where $\overline{a}$ is the list of formal parameters of $\mathbf{f}$, i.e. $\overline{a} = \mathbf{param}(\mathbf{f})$)

$$\begin{aligned}
\mathsf{wp}\,(y := \mathbf{fcall}\,\mathbf{f}(\overline{e}),\,\psi) &= (\mathbf{pre}(\mathbf{f})[\overline{e}/\overline{a}] \to \mathbf{post}(\mathbf{f})[\overline{e}/\overline{a}]) \to \psi[result/y] \\
\mathsf{wp}\,(\mathbf{return}\,e,\,\psi) &= \psi[e/result]
\end{aligned}$$

$$\begin{aligned}
\mathsf{VC}(y := \mathbf{fcall}\,\mathbf{f}(\overline{e}),\,\psi) &= \emptyset \\
\mathsf{VC}(\mathbf{return}\,e,\,\psi) &= \emptyset
\end{aligned}$$

And the set of verification conditions generated for a program Π will have to contain, for every $\mathbf{f} \in \mathbf{FN}(\Pi)$, the following verification conditions concerning the correctness of $\mathbf{f}$ with respect to its contract:

$$\mathsf{VCG}(\{\mathbf{pre}_\Pi(\mathbf{f})\}\, \mathbf{body}_\Pi(\mathbf{f})\, \{\mathbf{post}_\Pi(\mathbf{f})\})$$

Local Variables in Functions Since functions are not allowed to modify the global state, it may be necessary to employ variables that are local to their bodies. In the examples in this section we consider a form of local variables that are particularly easy to handle: they are declared and initialised at the top of the function's body, and we simply require that the same variable is not declared locally more than once in a given function, and that local variables do not coincide with variables that occur in the routine's contract (either local parameters or global variables). With these restrictions, we have that

$$\mathsf{wp}\,(\mathbf{local}\ x := e\ \mathbf{in}\ C, \psi) \quad = \quad \mathsf{wp}\,(x := e\,;\, C, \psi)$$

i.e. the VCGen can simply treat their initialisations as ordinary assignments, as if they were global. If they do occur in the contract, those occurrences surely refer to global variables, and the local variables can be renamed before verification conditions are generated. Naturally, this discussion also applies in general to variables declared locally inside procedures.

Example 8.10 Factorial gives us a good opportunity to illustrate the verification of a recursive function, which in itself also illustrates reasoning with function calls. Let us write the obvious recursive definition of factorial as a function, and prove it (partially) correct with respect to the factorial axioms.

> **pre** $n \geq 0$
> **post** $result = fact(n)$
> **fun factrec**$(n) =$
>   **local** $f := 1$ **in**
>   **if** $n = 0$ **then skip**
>   **else** $\{\ f := \mathbf{fcall\ factrec}(n - 1)\,;$
> $f := n \times f$
> $\}\,;$
>   **return** f

Let C_{else} be $f := \mathbf{fcall\ factrec}(n - 1)\,;\ f := n \times f$. The verification conditions are calculated as shown in Fig. 8.7. Both resulting VCs can be discharged using the factorial axioms.

Example 8.11 Of course we could have written instead an iterative version of the factorial function:

$\text{wp} (\textbf{if } n = 0 \textbf{ then skip else } C_{else}\,; \textbf{ return } f, result = fact(n))$

$=\quad \text{wp} (\textbf{if } n = 0 \textbf{ then skip else } C_{else}, f = fact(n))$

$=\quad (n = 0 \to \text{wp} (\textbf{skip}, f = fact(n))) \land (\neg(n = 0) \to \text{wp} (C_{else}, f = fact(n)))$

$=\quad (n = 0 \to f = fact(n)) \land$
$\qquad (\neg(n = 0) \to \text{wp} (f := \textbf{fcall factrec}(n - 1), n \times f = fact(n)))$

$=\quad (n = 0 \to f = fact(n)) \land$
$\qquad (\neg(n = 0) \to (\textbf{pre(factrec)}[n - 1/n] \to \textbf{post(factrec)}[n - 1/n])$
$\qquad\qquad\quad \to (n \times f = fact(n))[result/f])$

$=\quad (n = 0 \to f = fact(n)) \land$
$\qquad (\neg(n = 0) \to (n - 1 \geq 0 \to result = fact(n - 1))$
$\qquad\qquad\quad \to (n \times result = fact(n)))$

$\text{wp} (\textbf{body(factrec)}, result = fact(n))$

$=\quad \text{wp} (\textbf{local } f := 1 \textbf{ in if } n = 0 \textbf{ then skip else } C_{else}\,; \textbf{ return } f,$
$\qquad\qquad result = fact(n))$

$=\quad \text{wp} (f := 1, (n = 0 \to f = fact(n)) \land (\neg(n = 0) \to$
$\qquad\qquad (n - 1 \geq 0 \to result = fact(n - 1)) \to (n \times result = fact(n))))$

$=\quad (n = 0 \to 1 = fact(n)) \land (\neg(n = 0) \to$
$\qquad\qquad (n - 1 \geq 0 \to result = fact(n - 1)) \to (n \times result = fact(n)))$

$\text{VCG}(\{\textbf{pre(factrec)}\}\,\textbf{body(factrec)}\,\{\textbf{post(factrec)}\})$

$=\quad \{ n \geq 0 \to \text{wp} (\textbf{body(factrec)}, result = fact(n)) \} \cup$
$\qquad \text{VC}(\textbf{body(factrec)}, result = fact(n))$

$=\quad \{ n \geq 0 \to (n = 0 \to 1 = fact(n)) \land$
$\qquad\qquad (\neg(n = 0) \to (n - 1 \geq 0 \to result = fact(n - 1)) \to$
$\qquad\qquad\qquad (n \times result = fact(n))) \}$

Fig. 8.7 Calculating the verification conditions for the recursive definition of factorial

$\textbf{pre } n \geq 0$

$\textbf{post } result = fact(n)$

$\textbf{fun factf}(n) =$
$\quad \textbf{local } f := 1 \textbf{ in}$
$\quad \textbf{local } i := 1 \textbf{ in}$
$\quad \textbf{while } i \leq n \textbf{ do} \{ f = fact(i - 1) \land i \leq n + 1 \} \{$
$\qquad f := f \times i\,;$
$\qquad i := i + 1$
$\quad \}\,;$
$\quad \textbf{return } f$

The function will introduce the following verification conditions:

$$\text{VCG}(\{n \geq 0\}\,\textbf{body(factf)}\,\{result = fact(n)\})$$

$\textbf{pre } size \geq 0 \wedge$
$\quad valid\_range(in, 0, size - 1) \wedge$
$\quad valid\_range(out, 0, size - 1) \wedge$
$\quad \forall a.\ 0 \leq a < size \rightarrow in[a] \geq 0$

$\textbf{post } \forall a.\ 0 \leq a < size \rightarrow out[a] = fact(in\tilde{\ }[a]) \wedge size = size\tilde{\ }$

$\textbf{proc factab} =$
$\quad k := 0;$
$\quad \textbf{while } k < size \textbf{ do } \{\theta_2^0 \wedge 0 \leq k \leq size \wedge$
$\qquad\qquad\qquad\qquad \forall a.\ 0 \leq a < k \rightarrow out[a] = fact(in[a])\}$
$\quad \{$
$\qquad out[k] := \textbf{fcall factf}(in[k]);$
$\qquad k := k + 1$
$\quad \}$

Fig. 8.8 Procedure **factab** with an auxiliary factorial function

Now notice that **body(factf)** is **body(fact)** ; **return** f, where **body(fact)** is the code of Examples 5.10 and 6.9, and we have

$$\textsf{wp}\,(\textbf{body}(\textbf{factf}), result = fact(n)) = \textsf{wp}\,(\textbf{body}(\textbf{fact}), f = fact(n)), \quad \text{and}$$

$$\textsf{VC}(\textbf{body}(\textbf{factf}), result = fact(n)) = \textsf{VC}(\textbf{body}(\textbf{fact}), f = fact(n))$$

The correctness of the function is then immediate to establish.

Example 8.12 Our final example in this chapter is a program with bounded arrays, consisting of the factorial function **factf** together with a version of the **factab** procedure that calls that function, see Fig. 8.8. The reader will have no difficulty in generating the set of verification conditions for **factab**. Naturally, only the contract of **factf** is relevant for this purpose, not its implementation.

Admittedly, our function call construction is quite limited since it only allows for function calls to occur directly as the value assigned to a variable. Note however that the use of functions in arbitrary expressions could be dealt with in our framework as syntactic expansions. To exemplify this consider some command C containing the expression $e = 5 \times \textbf{fcall f}(e_1) + 10 \times \textbf{fcall g}(e_2)$. This can be transformed into

$$\textbf{local } y_1 := \textbf{fcall f}(e_1) \textbf{ in}$$
$$\textbf{local } y_2 := \textbf{fcall g}(e_2) \textbf{ in}$$
$$C[5 \times y_1 + 10 \times y_2/e]$$

Since we are dealing with functions without side effects, the order in which the function calls are evaluated is irrelevant: neither of them affects the global state. In the presence of side effects the evaluation order would be relevant for the above expansion.

Note however that even with pure functions the program obtained is not necessarily operationally equivalent to the initial. As an example, consider the boolean expression $(x > 10) \vee (\textbf{fcall } \textbf{f}(x) < 100)$. Short-circuit evaluation implies that the second operand of the disjunction operator will only be evaluated if the first operand does not evaluate to true. As such, **fcall** $\textbf{f}(x)$ will be executed or not depending on the value of x. The transformation is however sound from the axiomatic point of view.

8.6 To Learn More

A discussion of the semantics of procedure calls, including locally-defined procedures and the differences between static and dynamic scoping, can be found in [7].

Hoare discussed the axiomatic treatment of recursive procedures and introduced the rule of recursive invocation in [2]. Our treatment of adaptation, essential in the study of procedure calls, is inspired by Kleymann's work [4]. Subsequent works that consider systems of mutually recursive procedures include [8, 9], and [1]. Authors who have in the past made advances in this topic and proposed different notions of adaptation rules include Apt, Olderog, America and de Boer, Cartwright and Oppen, and Morris. See any of the previous papers for full references. Homeier and Martin [3] have extended their previous work on the mechanical verification of a VCGen, to cope with mutually recursive procedures.

The idea of programming with contracts was introduced in the design of the Eiffel programming language [6] by Meyer. The article [5] is a short introduction to the topic. Interface specification languages have played an important role in reasoning about procedures annotated with contracts. See references at the end of Chap. 9.

8.7 Exercises

8.1. Write a total correctness specification of the factorial procedure of Example 8.1, and derive the triple

$$[n = 10] \, \textbf{call fact} \, [f = fact(10)]$$

using the rule ($conseq\text{-}auxvars\text{-}total$) of Sect. 5.7.

8.2. Write a version of the ($conseq^{-}$) rule suitable for deriving total correctness triples. You should seek inspiration in the $conseq\text{-}auxvars\text{-}total$ rule.

8.3. Prove the correctness of the recursive factorial procedure of Example 8.3 using rules (8.5) and ($conseq\text{-}aux\text{-}vars$).

8.4. Consider again Example 8.5, in which we calculated the verification conditions for a program Π with two parameterless procedures. Now obtain a derivation of $\{\Pi\}$ in the inference system of Fig. 8.1.

8.5. Section 8.4 discussed procedures with parameters and proposed inference rules to reason about their correctness. Give appropriate VCGen rules to match the inference systems defined.

8.6. Our examples in Sect. 8.4 considered only parameters of type **int**. Rewrite the **factab** procedure of Fig. 8.8 to receive both the input and output arrays as parameters. Discuss the use of the inference rules given, for parameters of array type, and calculate the verification conditions for this modified program.

8.7. Discuss, from the point of view of writing a VCGen algorithm, the difficulties that are raised by having an abruptly terminating **return** command that may occur at any point (and possibly more than once) in a function definition.

8.8. Safety conditions have been left out when discussing procedure and function calls. Discuss what would have to be modified in the different sets of inference rules studied in this chapter, to cope with safety.

References

1. Back, R.-J., Preoteasa, V.: Reasoning about recursive procedures with parameters. In: MERLIN '03: Proceedings of the 2003 ACM SIGPLAN Workshop on Mechanized Reasoning about Languages with Variable Binding, pp. 1–7. ACM, New York (2003)
2. Hoare, C.A.R.: Procedures and parameters: An axiomatic approach. In: Proceedings of Symposium on Semantics of Algorithmic Languages. Lecture Notes in Mathematics, vol. 188. Springer, Berlin (1971)
3. Homeier, P.V., Martin, D.F.: Secure mechanical verification of mutually recursive procedures. Inf. Comput. **187**(1), 1–19 (2003)
4. Kleymann, T.: Hoare logic and auxiliary variables. Form. Asp. Comput. **11**(5), 541–566 (1999)
5. Meyer, B.: Applying "design by contract". Computer **25**(10), 40–51 (1992)
6. Meyer, B.: Eiffel: The Language. Prentice Hall, Hemel Hempstead (1992)
7. Nielson, H.R., Nielson, F.: Semantics with Applications: An Appetizer. Undergraduate Topics in Computer Science. Springer, Berlin (2007)
8. Nipkow, T.: Hoare logics for recursive procedures and unbounded nondeterminism. In: Bradfield, J.C. (ed.) CSL. Lecture Notes in Computer Science, vol. 2471, pp. 103–119. Springer, Berlin (2002)
9. von Oheimb, D.: Hoare logic for mutual recursion and local variables. In: Pandu Rangan, C., Raman, V., Ramanujam, R. (eds.) FSTTCS. Lecture Notes in Computer Science, vol. 1738, pp. 168–180. Springer, Berlin (1999)

Chapter 9
Specifying C Programs

In this chapter we shift the focus of our study from the programming language framework that we have been considering and extending since Chap. 5, to programs written in a realistic programming language. In particular, we introduce the *ANSI/ISO C Specification Language* (ACSL), which is an annotation language for C programs.

Our choice of programming language leads us to make the following very important remark: the distance between the languages that we have been using and a language like C is huge, and the verification of many aspects of the language is still a matter of active research, with different tools proposing different approaches. This is due in particular to the presence of an aspect that we have not treated in Chaps. 5 to 8: *heap memory allocation* and *indirect referencing* through pointer variables, which in turn allow for the manipulation of linked data-structures. Although they are often avoided in safety-critical software, these are of course very important features of programming languages in general. From the point of view of deductive verification, these aspects are addressed by constructing a *memory model* that allows for reasoning about data structures allocated in a program's heap. Although the discussion of memory models is out of the scope of this book, and we do not illustrate the verification of programs that employ linked data structures, the model assumed by ACSL is implicitly used in this chapter in the treatment of arrays.

In Sect. 8.2 we have introduced contracts and the principles of design by contract. ACSL adheres to these principles: each C function in a program is annotated with an ACSL specification—the function's contract. Verification of a program consisting of a number of mutually-recursive functions is completely modular: each function is verified against its own contract assuming that all other functions are correct; the program is correct if all functions are correct.

ACSL also has scope for many other aspects of behavioural specification, some of which have been covered in previous chapters. For instance, functions may also carry frame conditions (as discussed in Sect. 8.3), and loops can be annotated with loop invariants and variants. Predicates and logic functions can be either defined or specified by a set of axioms, following the discussion of Chap. 5. Other aspects have not been covered before; for instance ACSL offers a *state label* mechanism. This is a generalisation of the˜notation that allows assertions to refer to the value of

J.B. Almeida et al., *Rigorous Software Development*, 229
Undergraduate Topics in Computer Science,
DOI 10.1007/978-0-85729-018-2_9, © Springer-Verlag London Limited 2011

```
int size, u[], max;

/*@ requires size >= 1 && \valid_range(u,0,size-1);
  @ ensures  0 <= max < size &&
  @          (\forall int a; 0 <= a < size ==> u[a] <= u[max]);
  @*/
void maxarray() {
  int i = 1;
  max = 0;

  /*@ loop invariant
    @    1 <= i <= size && 0 <= max < i &&
    @    (\forall int a; 0 <= a < i ==> u[a] <= u[max]);
    @ loop variant
    @    size-i;
    @*/
  while (i < size) {
    if (u[i] > u[max]) max = i;
    i = i+1;
  }
}
```

Fig. 9.1 maxarray in C with ACSL annotations

a program expression in any state, identified through C program labels (the pre-state is just a particular case).

ACSL annotations are included in the program files as special comments identified by the character @. This is a standard approach in specification languages for source code: the annotations are ignored by the compiler but not by the verification tools.

9.1 An Introduction to ACSL

In what follows we consider a few examples of annotated code, based on programs already studied in Chaps. 5 to 8. They are here rewritten in the C programming language and annotated using the ACSL specification language.

9.1.1 Array-Based Programs

As a first example of a C function annotated with an ACSL specification, consider again the **maxarray** code of Examples 6.10 and 7.15. We write this as a C function in Fig. 9.1. Both the input (size, u) and output (max) variables are global. Note that in ACSL contracts can only be associated to functions, not arbitrary blocks of code.

The comment lines that precede the function contain the function's precondition (identified by the keyword `requires`) and postcondition (identified by the keyword `ensures`). The comment lines that precede the while loop contain an invariant and a variant. All annotations result from a straightforward translation of Example 6.10 into ACSL syntax, and strengthening the precondition with a safety annotation. Apart from a few keywords, learning the ACSL syntax should be straightforward for C programmers, since the specification language builds on the syntax of the programming language expressions.

C arrays are of course quite different from the arrays introduced in Sect. 5.5. They can be allocated either statically or dynamically, accessed through pointer arithmetics, and passed (by reference) as argument to functions. Annotating programs that manipulate arrays is however not so different from what we had in our simple language; in particular the index and range validity predicates used for safety annotations have the same intuitive meaning (even though in ACSL their meaning is based on the underlying memory model).

9.1.2 Using Axiomatics

Consider now the **factab** program of Examples 6.11 and 8.6. The safety-aware version of Fig. 8.3 is adapted to C and ACSL in Fig. 9.2. The annotated code corresponds almost exactly to that in Fig. 8.3. There are two notable differences:

1. In Chap. 7 procedure parameters could only be of type **int**, so arrays had to be kept global. In Fig. 9.2 the input and output arrays are passed (by reference, naturally) as parameters to the `factab` function. Note that this does not imply any modification in the safety assertions that must be included as preconditions.
2. The `\valid_range` safety preconditions, which remain true through execution of the program, are continuous invariants that had to be included as part of the loop invariants in our initial version. In ACSL however, continuous invariants are implicit and can be omitted in the loop invariants. Thus the θ_1^0 and θ_2^0 continuous invariants of Example 6.11 can be omitted. This substantially simplifies the annotated invariants.

This example also illustrates the use of a predicate and a function at the logical level. In ACSL logical declarations are included in *axiomatics*. In this example the `factorial` axiomatic contains the two following declarations.

```
predicate isfact(integer n, integer r);
logic integer fact (integer n);
```

The first declares a new predicate that takes two integer arguments, and the second a function that takes an integer argument. The axiomatic also contains the axioms for factorial, following the discussion of page 136. Note the use of the logic type `integer` instead of the C type `int` in the axiomatic; this makes the axiomatic more abstract, which means it can be used with any concrete integer type employed

```
/*@ axiomatic factorial {
  @
  @ predicate isfact(integer n, integer r);
  @ axiom isfact0:
  @   isfact(0,1);
  @ axiom isfactn:
  @   \forall integer n, integer f;
  @   n>0 ==> isfact (n-1,f) ==> isfact(n,f*n);
  @
  @ logic integer fact (integer n);
  @ axiom fact1:
  @   \forall integer n; isfact (n,fact(n));
  @ axiom fact2:
  @   \forall integer n, integer f; isfact (n,f) ==>
  @   f == fact(n);}
  @*/


/*@ requires
  @   size>=0 &&
  @   \valid_range(inp,0,size-1) &&
  @   \valid_range(outp,0,size-1) &&
  @   \forall int a; 0 <= a < size ==> inp[a] >= 0;
  @ assigns outp[0..size-1];
  @ ensures
  @   \forall int a ; 0 <= a < size ==>
  @   outp[a] == fact (inp[a]);*/
void factab (int inp[], int outp[], int size)
{
  int k = 0 ;

  /*@ loop invariant 0 <= k <= size &&
    @   \forall int a; 0 <= a < k ==>
    @   outp[a] == fact (inp[a]);
    @ loop variant size-k;
    @*/
  while (k < size) {
    int f = 1, i = 1, n = inp[k] ;

    /*@ loop invariant 1 <= i <= n+1 && f == fact(i-1);
      @ loop variant n+1-i;
      @*/
    while (i <= n) {
      f *= i;
      i++;
    }
    outp[k++] = f ;
  }
}
```

Fig. 9.2 factab in C with ACSL annotations

in the program. In the example the logical `fact` function occurs in the postcondition and loop invariants, where it is passed arguments of `int` C type.

Finally, the example also illustrates the use of a *frame condition* (see Sect. 8.3). The clause

```
assigns outp[0..size-1];
```

in the function's contract lists all the memory locations assigned by the function. These may in general be parts of the global state or else of arguments passed to the function. The `factab` function only assigns to positions of the parameter array `outp`, and the clause identifies the range of indexes assigned, from 0 to `size-1`.

The presence of this `assigns` clause will cause specific verification conditions to be generated, to ensure that the function does not assign to any memory locations not listed in the clause. This implies reasoning about the memory locations written by loops, which of course requires the use of loop invariants. ACSL provides a `loop assigns` annotation to this effect; for instance one could add the following to the external loop in `factab`:

```
loop assigns outp[0..size-1];
```

which should be read as follows: if at the beginning of a given iteration being executed, no locations other than `outp[0..size-1]` have been written, then after execution of the iteration no locations other than `outp[0..size-1]` will have been written. `loop assigns` clauses are often omitted, in which case they can be assumed by the VCGen to be equal to the frame condition of the routine.

A frame condition is an important part of a routine's contract when reasoning about calls to that routine, since it immediately implies the preservation of the values contained in all locations not mentioned.

9.1.3 Function Calls

We give in Fig. 9.3 a version of `factab` that uses an auxiliary function `factf` for calculating factorial. This is a C / ACSL adaptation of the program shown previously in Fig. 8.8.

This example illustrates the idea of programming with contracts: to emphasize the modularity allowed by contracts, we do not give a C implementation for the function `factf` (the user is asked to propose one in Exercise 9.2). Instead, a prototype is given for that function, together with a contract. The idea is that we do not care about the way in which factorial is calculated, we simply rely on the function `factf` to do so. It should be possible to verify the function `factab` independently of any implementation of `factf`. Note also the use of `\result` in a postcondition to refer to the return value of a function, and the `assigns \empty` clause stating that the function does not write to any memory locations visible externally, i.e. it is free of side effects. Such functions are usually called *pure*.

A second important remark concerns the way in which the safety of array accesses is handled in this example. This version of the `factab` function works on

```
/*@ requires n >= 0;
  @ assigns \empty;
  @ ensures \result == fact(n);
  @*/
int factf (int n);



#define LENGTH 1000
int inp[LENGTH], outp[LENGTH];

/*@ requires
  @   0 <= size <= LENGTH &&
  @   \forall int a; 0 <= a < size ==> inp[a] >= 0;
  @ assigns outp[0..size-1];
  @ ensures
  @   \forall int a; 0 <= a < size ==>
  @     outp[a] == fact (inp[a]);
  @*/
void factab (int size)
{
  int k = 0 ;

  /*@ loop invariant 0 <= k <= size &&
    @   \forall int a; 0 <= a < k ==>
    @     outp[a] == fact (inp[a]);
    @ loop variant size-k;
    @*/
  while (k < size) {
    outp[k] = factf(inp[k]) ;
    k++;
  }
}
```

Fig. 9.3 factab in C using a fact function (axiomatisation omitted)

global input and output arrays, which are defined in the same file. The frame condition is the same as before, but it now applies to a global array instead of a parameter.

Thus, since the length of the arrays is known, it is not necessary to include safety \valid_range assertions in preconditions. Instead, it suffices to ensure (and this *must* be included as a precondition) that the array size considered by the procedure is not greater than the allocated length of the arrays. The safety verification conditions generated will be satisfied by the context information extracted from the global array definitions.

9.1.4 State Labels and Behaviours

Figure 9.4 contains the C code for a function that implements the insertion sort algorithm. Naturally, the contract of this function can be used for any sorting algorithm,

```
/*@ predicate Sorted{L}(int t[], integer i, integer j) =
  @    \forall int k; i <= k < j ==>
  @      \at(t[k],L) <= \at(t[k+1],L);
  @*/

/*@ requires N >= 1  && \valid_range(A,1,N);
  @ assigns A[1..N];
  @ ensures Sorted{Here}(A,1,N);
  @*/
void insertion_sort(int A[], int N) {
  int i, j, key;

  /*@ loop invariant
    @    2 <= j <= N+1 &&
    @    Sorted{Here} (A,1,j-1);
    @ loop variant (N-j);
    @*/
  for (j=2; j<=N; j++) {
    key = A[j];
    i = j-1;
    /*@ loop invariant
      @  ...
      @ loop variant
      @  ...
      @*/
    while (i>0 && A[i]>key) {
      A[i+1] = A[i];
      i--;
    }
    A[i+1] = key;
  }
}
```

Fig. 9.4 Insertion sort in C

since it only expresses *what* the algorithm does; the loop invariants on the other hand describe *how* this algorithm achieves what it does, and are specific to each particular sorting algorithm.

When execution terminates, the array is sorted between indexes 1 and N. To state this, the postcondition makes use of a defined predicate Sorted, whose definition is also included in the figure. The predicate takes as arguments an array and two indexes that define the range in which the array is said to be sorted. Additionally, the predicate also takes a state label L, with the meaning that the array is sorted in a certain program state. Note the use of the operator \at to get the value of an expression in a given program state, corresponding in this case to the parameter label L. The predicate simply states that (in the state corresponding to L) the contents of adjacent positions of the array are in the right order.

The use of state labels makes it possible to refer, even within the same assertion, to the value of an expression in different states. Two special labels are particularly useful: Here denotes the current state to which the assertion refers; Old denotes the pre-state of the function. In this example only the Here label is employed: the predicate Sorted is used with this label in the postcondition, where Here refers to the final state of the program, as well as in the invariant of the for loop, where it refers to an arbitrary state in which an iteration of the loop is starting.

Insertion sort is an incremental sorting algorithm: each iteration of the external loop adds one element of the array to the initial segment which is already sorted. The invariant of the for loop simply expresses this fact. The invariant of the inner while loop is left for the reader to work out (see Exercise 9.3).

The reader may have noticed that our specification of insertion sort is incomplete. A sorting algorithm is not simply an algorithm that produces sorted sequences: it must do so while preserving the elements contained in the initial sequence received as input. More precisely, the *multiset* of elements in the array is preserved, i.e. not only the elements are the same but the number of occurrences of each element is preserved. Another way of stating this is that there exists a bijection on the set of indices that establishes a permutation between the two arrays. This is an essential property to prove for sorting algorithms.

It is not easy to formalise this property. As a first attempt, consider the following first-order formula

$$\forall k.\ p \leq k \leq r \rightarrow (\exists l.\ p \leq l \leq r \rightarrow A[k] = B[l] \wedge A[l] = B[k])$$

This indeed implies that B is a permutation of A between indexes p and r. It could be written in ACSL as the following post-condition, in which the \at operator is used, this time with the Old label, which allows us to refer to the value contained in a given array position in the function's pre-state (i.e. when the function was invoked).

```
@   (\forall int k; p <= k <= r =>
@      (\exists int l; p <= l <= r =>
@        A[k] == \at(A[l],Old) && A[l] == \at(A[k],Old)))
```

In fact the use of the Old label in postconditions is so common that there is a short-cut notation in ACSL via the \old operator (which resembles the ~ notation used in Chap. 8). The above could be written instead as

```
@   (\forall int k; p <= k <= r =>
@      (\exists int l; p <= l <= r =>
@        A[k] == \old(A[l]) && A[l] == \old(A[k])))
```

The problem is that this property is *stronger* than what is desired: it only covers the cases in which B is directly obtained from A by swapping pairs of elements (i.e. each element may only be swapped once—in mathematical terms the bijection between A and B is said to be an involution). It is indeed easy to see that a sequence of swaps produces an array that is no longer related to the original in this way, but is still a permutation of the initial array:

$$1\,\underline{2}\,3\,4 \longrightarrow 2\,1\,\underline{3}\,\underline{4} \longrightarrow 2\,1\,\underline{4}\,3 \longrightarrow 2\,4\,1\,3$$

```
@ predicate Swap{L1,L2}(int a[], integer i, integer j) =
@    \at(a[i],L1) == \at(a[j],L2) &&
@    \at(a[j],L1) == \at(a[i],L2) &&
@    \forall integer k; k!=i && k!=j ==>
@       \at(a[k],L1) == \at(a[k],L2);
@
@ axiomatic permutation {
@
@ predicate Permuta{L1,L2}(int a[], integer l, integer h);
@
@ axiom Permuta_refl{L}:
@    \forall int a[], integer l, h; Permuta{L,L}(a, l, h) ;
@ axiom Permuta_sym{L1,L2}:
@    \forall int a[], integer l, h;
@      Permuta{L1,L2}(a, l, h) ==> Permuta{L2,L1}(a, l, h) ;
@ axiom Permuta_trans{L1,L2,L3}:
@    \forall int a[], integer l, h;
@      Permuta{L1,L2}(a, l, h) && Permuta{L2,L3}(a, l, h) ==>
@      Permuta{L1,L3}(a, l, h) ;
@ axiom Permuta_swap{L1,L2}:
@    \forall int a[], integer l, h, i, j;
@      l <= i <= h && l <= j <= h && Swap{L1,L2}(a, i, j) ==>
@      Permuta{L1,L2}(a, l, h) ;
@ }
```

Fig. 9.5 The permutation axiomatic

Now consider the following alternative formalisation, which states that every element in each array must occur in the other array.

$$\forall k.\ p \le k \le r \to (\exists l.\ p \le l \le r \to A[k] = B[l])$$
$$\wedge$$
$$\forall k.\ p \le k \le r \to (\exists l.\ p \le l \le r \to B[k] = A[l])$$

Admittedly, this property is *too weak* since it does not take into account the number of occurrences i.e. it accounts for the preservation of the set of elements, not the multiset. For instance the arrays 1 2 2 and 2 1 1 satisfy the condition. This would be adequate if we were dealing with sequences containing no repeated elements, but not in the general case.

One possibility to treat permutations is to see them as sequences of pairwise swaps. Figure 9.5 contains an axiomatic that captures this idea. First note that the defined predicate Swap has as parameters an array, two indexes, and two state labels. Swap{L1,L2}(a,i,j) has the meaning that the contents of array a in states L1 and L2 are the same, with the exception of indexes i and j, which are swapped.

Predicate Permuta is not defined: a set of axioms is given instead. The assertion Permuta{L1,L2}(a, l, h) has the meaning that the contents of array a in the range from index l to index h in state L2 is a permutation of its contents in that

```
/*@ requires 0 <= p <= r  && \valid_range(A,p,r);
  @ assigns A[p..r];
  @ behavior sorted:
  @    ensures
  @       Sorted{Here}(A,p,r);
  @ behavior permutation:
  @    ensures
  @       Permuta{Old,Here}(A,p,r);
  @*/
void sort(int A[], int p, int r) ;
```

Fig. 9.6 Complete Specification of a sorting algorithm in ACSL

range in state L1. Note that we could give the predicate a different signature, taking two arrays as parameters, but in practice, in the specification of an in-place sorting algorithm we want to express a relation between the contents of the same array in two different states.

Axiom Permuta_swap states that a pairwise swap (involving two indexes within the appropriate range) is a permutation; in fact it is the most elementary permutation. Note that the state labels handled by an axiom must be made explicit.

The remaining axioms correspond to reflexivity (a sequence is a permutation of itself, or in the "in place" setting, an array in a given state contains a permutation of itself in the same state), symmetry (if A is a permutation of B then B is a permutation of A), and transitivity (if A is a permutation of B and B is a permutation of C then A is a permutation of C) of the permutation relation.

Armed with the permutation axiomatic we may now write a complete specification of a sorting algorithm, see Fig. 9.6.

Contracts can be structured in a number of behaviours, to allow the verification process to be broken in smaller parts. Moreover the generated verification conditions can be labeled with the behavior they relate to, which facilitates tracing them back. The specification of Fig. 9.6 makes use of two behaviours, sorted and permutation, each with its own postcondition, in addition to a common precondition.

9.2 To Learn More

The ACSL reference manual [1] is available online.[1] A tutorial [2] is also available. The ACSL specification language is linked to the Frama-C software analysis tool; the VCGen functionality of this tool will be the topic of the next chapter.

ACSL is closely related to the *Java Modeling Language* (JML), and in fact these are very similar specification languages as far as the imperative subset is concerned. JML covers object-oriented specification aspects, and moreover it can be

---

[1]From http://frama-c.com/.

used equally for deductive verification and for dynamic (runtime) verification. The paper [4] is a tutorial on JML. One important aspect of ACSL that we have not covered (which is also present in JML) is *ghost code* and *ghost variables*. These are essentially data and code that exist only at the specification level, and are not part of the program itself. See the previous references for details.

Finally, [3] offers a general discussion of interface specification languages.

9.3 Exercises

9.1 Consider again Exercise 5.1. Write ACSL specifications for functions corresponding to (a), (b), and (c) in that exercise.

9.2 Give an implementation of the `factf` function that satisfies the contract of Fig. 9.3.

9.3 Consider again the insertion sort algorithm of Fig. 9.4. The role of the inner `while` loop is to ensure the preservation of the invariant of the external `for` loop from one iteration to the next, i.e. to extend the sorted prefix of the array by one position. How is this achieved? Write an invariant for the `while` loop that expresses how it works. Write also a variant for the same loop.

9.4 Take your favourite sorting algorithm and adapt it to meet the specification of Fig. 9.6. Think of adequate invariants and variants for every loop.

9.5 Consider a defined type `Stack` to model stacks of integers in C. Write prototypes and contracts for the usual stack functions, including *init*, *push*, *pop*, *top*, *isEmpty*. The contracts should work for any concrete definition of `Stack`.

9.6 Consider now the concrete definition `typedef int Stack [SIZE];` give ACSL-annotated implementations for the functions in the previous exercise.

References

1. Baudin, P., Cuoq, P., Filliâtre, J.-C., Marché, C., Monate, B., Moy, Y., Prevosto, V.: ACSL: ANSI/ISO C Specification Language. CEA LIST and INRIA (2009–2010)
2. Burghardt, J., Gerlach, J., Hartig, K., Soto, J., Weber, C.: ACSL By Example. DEVICE-SOFT project publication. Fraunhofer FIRST Institute (January 2010)
3. Hatcliff, J., Leavens, G.T., Leino, K.R.M., Müller, P., Parkinson, P.: Behavioral interface specification languages. Technical Report CS-TR-09-01, School of EECS, University of Central Florida (2009)
4. Leavens, G.T.: Tutorial on JML, the Java modeling language. In: Stirewalt, R.E.K., Egyed, A., Fischer, B. (eds.) Proceedings of ASE'07, p. 573. ACM, New York (2007)

Chapter 10
Verifying C Programs

The previous chapter contains many ACSL examples; in this chapter we will study a verification tool for ACSL-annotated C programs. In fact the tool used here, Frama-C, is closely linked to the development of ACSL. Both are being developed jointly by CEA-LIST (Software Reliability Laboratory) and INRIA-Saclay (ProVal team), and supported by a number of collaborative projects. More information can be obtained from the homepage of Frama-C.[1] We are convinced that ACSL will in the near future be adopted by other verification tools for C.

The first thing that must be said about Frama-C is that it is much more than a verification tool: it is a general, plug-in-based program analysis tool, designed to be used in practice in industrial projects. The architecture of the tool allows each plug-in to use services provided by other plug-ins. In addition to the verification capabilities described in this chapter, Frama-C provides, among others, plug-ins for value analysis, impact analysis, scope and data-flow browsing, variable occurrence browsing, semantic constant folding, slicing, and spare code removal.

The program verification functionality of Frama-C used in the present chapter is provided by the *Jessie* plug-in. Jessie in turn relies on the use of a VCGen tool called *Why*. The use of this tool has two major advantages:

1. It is *multi-prover*. This means that it can export verification conditions as proof obligations for many different proof tools. These include for instance the Simplify, Alt-Ergo, Z3 automatic provers, and other SMT solvers, and proof assistants like Coq, Isabelle, PVS and HOL. The benefits of this possibility should be clear by now: one can take profit of the best features of each automatic prover and leave harder conditions to be interactively proved later.
2. It provides a graphical front-end that allows to monitor in real-time which obligations have been discharged by which provers.

Instead of applying Jessie to the annotated examples of the previous chapter, our approach in the present chapter will be to start from an algorithm for which only an informal specification is given. We will annotate the C code of this function as we

---

[1] http://frama-c.com.

J.B. Almeida et al., *Rigorous Software Development*,
Undergraduate Topics in Computer Science,
DOI 10.1007/978-0-85729-018-2_10, © Springer-Verlag London Limited 2011

Fig. 10.1 partition in C

```
int partition (int A[], int p, int r)
{
    int x = A[r];
    int tmp, j, i = p-1;

    for(j=p; j<r; j++)
      if (A[j] <= x) {
         i++;
         tmp = A[i];
         A[i] = A[j];
         A[j] = tmp;
      }
    tmp = A[i+1];
    A[i+1] = A[r];
    A[r] = tmp;
    return i+1;
}
```

go along, starting with the minimal annotations required for verification of safety (Sect. 10.1), followed later by other functional properties (Sects. 10.2 and 10.3).

This process corresponds to one typical use of a program verification tool: instead of being used on code that was developed based on a formal specification, verification engineers often have to work with code for which only informal specifications exist, and whose formal contracts must be retroactively written. When the verification fails, it is not clear that the program is wrong: it may well be the case that it is the formal specification that does not yet capture exactly what the code does.

10.1 Safety Verification

We will start by considering the use of Frama-C for verifying the safe execution of programs. We will illustrate this with a function containing no contracts or invariants, and adding the minimal annotations required for safety verification.

Consider the partition function in Fig. 10.1. It implements an algorithm that separates the elements contained in positions p to r of array A into those that are smaller than or equal to a pivot element, and those that are greater than the pivot. We take the pivot to be the element in the last position, x = A[r]. Later we will write this as an ACSL specification and prove that the algorithm indeed partitions the array in this way.

For now we concentrate on something more basic: that the partition function is free of problems that may cause the program to behave abnormally. We start by running the Frama-C VCGen on this function as follows:

```
> frama-c -jessie -jessie-atp simplify partition.c
```

The -jessie switch instructs Frama-C to run the Jessie plugin to generate verification conditions; the -jessie-atp switch indicates our choice of automatic theorem prover (we have picked the Simplify prover) that should be used to discharge them. The output of this will be something like

```
total   : 20
valid   : 10 ( 50%)
invalid :  0 (  0%)
unknown : 10 ( 50%)
timeout :  0 (  0%)
failure :  0 (  0%)
```

Note the 5 possible outcomes for each VC: the prover may be able to establish that it is *valid* or *invalid*, but it may also fail to do so (*unknown*) or stop because of a *timeout*. Finally, *failure* indicates a problem in the use of the proof tool.

Omitting the `-jessie-atp` switch will cause Frama-C to invoke the multi-prover graphic interface Gwhy, which has the advantage of allowing the user to interactively invoke any proof tool available locally, individually for each verification condition.

The outcome may seem surprising at first: if the function is not annotated with a contract, why were 20 VCs generated, and how should we interpret the fact that half of these could not be discharged? The VCGen is simply behaving in the same way as the safety-sensitive VCGens of Chap. 7: even though the precondition and postcondition are omitted (which corresponds to being interpreted as true), verification conditions for safe execution were generated.

Gwhy shows a label for each verification condition that identifies the type of condition. In this example 10 VCs are labeled *check arithmetic overflow* and the remaining 10 are labeled *pointer dereferencing*.

10.1.1 Arithmetic Overflow Safety

A source of execution errors that we have not explored in the previous chapters of this book is *integer overflow*. For both signed and unsigned integer variables, there is a minimum and maximum value that can be stored in it (regardless of whether it is of type short or long integer). Arithmetic operations may result in values that do not fit within the range that can be stored, which is known as an overflow situation. This does not cause the program to stop with an exception, but leads instead to unexpected results. Most programmers do not worry about this in everyday programming, but in certain contexts, notably when developing safety-critical applications, such matters must be taken into account.

Verification conditions can be easily visualized in Gwhy: clicking a particular VC causes the relevant line in the code to be highlighted in a window, and the VC itself can be read in another window. Consider the `i` = `p-1` initialisation. It originates two verification conditions, corresponding to the lower and upper bound on the numbers that can be stored in variable `i`. These VCs are displayed as follows

```
1. -2147483648 <= integer_of_int32(p) - 1
2. integer_of_int32(p) - 1 <= 2147483647
```

The verification conditions are quite intuitive to understand. The numbers refer to the range of a signed long integer (32 bits). The variable p is cast from an implementation 32 bit integer type to an abstract (logical) integer type by the `integer_of_int32` operator. It is quite natural that VC 1 cannot be discharged, whereas VC 2 can. If p fits within that range, then p-1 may exceed the lower bound but not the upper. In what follows the program will be annotated in a way that will naturally allow all the overflow-related VCs to be discharged.

This discussion assumes the so-called *bounded integer model*, used by default, which forbids overflow. If overflow is not a concern in a given verification context, the *exact integer model* can be used instead. This can be done by inserting a *pragma* in the program file like this:

```
# pragma JessieIntegerModel(exact)
```

This model assumes unbounded integers; overflow checking is thus disabled. Alternative values for `JessieIntegerModel` include `strict` and `modulo`, which both model machine arithmetics. The former, used by default, forbids overflow and generates VCs to ensure that; in the latter overflow is allowed, with the same behaviour as in program execution.

10.1.2 Safety of Array Access

Arrays in C are passed by reference. The function `partition` of Fig. 10.1 accesses the array for both reading and writing, and this generates verification conditions to ensure that these accesses are within the array's allocated range. For instance consider the initialization of x with the value of A[r]. This array access results in the following two VCs.

3. `offset_min(int_P_A_1_alloc_table, A) <= integer_of_int32(r)`
4. `integer_of_int32(r) <= offset_max(int_P_A_1_alloc_table,A)`

Naturally, none of these can be proved, since there is no information in the context concerning the array's range. `offset_min(int_P_A_1_alloc_table, A)` and `offset_max(int_P_A_1_alloc_table, A)` represent the initial and final valid indexes in the allocated range of A. The first parameter corresponds to an allocation table.

This can be remedied in part by adding a precondition annotation to the program stating that the range of the array between indexes p and r can be safely accessed, and also the facts that p and r are not negative, and p is not greater than r. We add the following at the beginning of the program:

```
/*@ requires 0 <= p <= r && \valid_range(A,p,r)
    @*/
```

Note the use of && for conjunction; the assertion `0 <= p <= r` is equivalent to `0 <= p && p <= r`. Adding the precondition will allow the VCs 3 and 4 shown above to be discharged automatically, but there are still 7 conditions left unproved.

Incidentally note that concerning overflow, VC 1, that had been left unproved, has now also been proved since p is known to be non-negative, thus p-1 is guaranteed to fit within the representable bounds of an integer variable.

10.1.3 Adding Loop Invariants

The remaining VCs are all generated from instructions occurring inside the loop or after the loop. Consider for instance the array access contained in the boolean condition A[j] <= x inside the loop. This results in the VC

```
offset_min(int_P_A_1_alloc_table, A) <= integer_of_int32(j0)
```

Overflow VCs concerning instructions inside or after the loop are also still left unproved, for instance the i++ instruction generates

```
integer_of_int32(i0) + 1 <= 2147483647
```

What is missing now is a loop invariant. Note that Frama-C propagates continuous invariants in a transparent way for the user, so these do not have to be explicitly included in the loop invariants. This is the case for instance of the assertion \valid_range(A,p,r), which remains valid throughout the program and is available in the context when reasoning about instructions inside the loop.

The reason why the VCs shown above cannot be proved is that there is no information in the context concerning the values of i and j. A first attempt would be to add the following annotation immediately before the loop.

```
@ invariant p <= j <= r && p-1 <= i <= r
```

The first thing to notice is that the introduction of a loop invariant causes verification conditions to be displayed by Gwhy grouped in two categories, or behaviours: those concerning safety are as before grouped under the *safety* behaviour, but a new set of VCs is shown, grouped under *default*. This default behaviour includes VCs that are not automatically created in the absence of annotations—they are generated from user-provided assertions annotated in the code. In the present example there are now 24 safety VCs and 12 default VCs, labeled either "initialization of loop invariant" or "preservation of loop invariant".

Some of these VCs have still not been discarded automatically. In fact, even though the invariant we added is valid, it is not sufficiently strong to allow all safety VCs to be discarded. Moreover the preservation of the i < j part of the invariant (a *default* VC) could not be proved. Notice however that although the relation between the values of i and j is not immediate from the program, a tighter upper bound exists for i: the invariant can be strengthened to i <= r.

Figure 10.2 shows the partition function with the annotations required for safety verification. All verification conditions are now discharged automatically (including those concerning overflow).

```
/*@ requires 0 <= p <= r && \valid_range(A,p,r)
  @*/
int partition (int A[], int p, int r)
{
  int x = A[r];
  int tmp, j, i = p-1;

  /*@ invariant p <= j <= r && p-1 <= i < j
    @ variant (r-j)
    @*/
  for(j=p; j<r; j++)
    if (A[j] <= x) {
       i++;
       tmp = A[i];
       A[i] = A[j];
       A[j] = tmp;
    }
  tmp = A[i+1];
  A[i+1] = A[r];
  A[r] = tmp;
  return i+1;
}
```

Fig. 10.2 partition in C, annotated for safety verification

10.1.4 Termination Checking and Loop Variants

Note that the obvious loop variant has also been added in Fig. 10.2; this has resulted in the generation of 4 VCs labeled "variant decrease" under the safety behaviour, all proved automatically. Consequently this is a terminating function. Frama-C has three different policies for termination checking; the user can specify that one of these should be used by inserting a pragma in the program file like this:

```
# pragma JessieTerminationPolicy(user)
```

If the user policy is used, Frama-C treats variants in the way suggested in Sect. 5.6: termination is proved independently for each loop for which a variant is given, so it is possible to prove the termination of only a few loops in the program; if all loops are adequately annotated with variants then the verification will, if successful, ensure total correctness.

The two other possible values for this pragma are never and always. The former means that a partial correctness setting should be assumed (no verification conditions for termination are generated); the latter specifies a total correctness setting, which means that every loop must be annotated with an appropriate variant. If this is not the case, verification conditions will be generated using a constant variant, and the verification will fail.

```
/*@ requires \valid_index(t,i) && \valid_index(t,j);
  @*/
void swap(int t[], int i, int j) {
  int tmp = t[i];
  t[i] = t[j];
  t[j] = tmp;
}

/*@ requires 0 <= p <= r && \valid_range(A,p,r);
  @*/
int partition (int A[], int p, int r)
{
  int x = A[r];
  int tmp, j, i = p-1;

  /*@ loop invariant p <= j <= r && p-1 <= i < j;
    @ loop variant (r-j);
    @*/
  for (j=p; j<r; j++)
    if (A[j] <= x) {
      i++;
      swap(A,i,j);
    }
  swap(A,i+1,r);
  return i+1;
}
```

Fig. 10.3 `partition` using a `swap` function, annotated for safety verification

10.1.5 Safety of Function Calls

Consider now that the "swap the contents of two array positions" bit of partition is outsourced to a separate function, invoked twice by partition. The new code is given in Fig. 10.3. Since swap accesses the array, it needs to be annotated with a precondition concerning the safety of those accesses. The function only reads and writes the contents of two indexes, thus two \valid_index assertions are sufficient—the range of the array may be unknown, as long as i and j are valid positions.

The generated VCs are grouped in three behaviours: *partition safety*, *swap safety*, and *partition default*. No VCs are generated for *swap default*. The behaviour partition safety now includes a set of VCs labeled *"precondition for user call"*. These are created because of the two calls to swap inside partition, and correspond to the satisfaction of the latter's precondition. It should be clear after Chap. 8 that the verification of each function is independent of the other; for instance

- If swap accesses other positions than i or j and this is not reflected in the precondition, its verification will fail, but the VCs generated from partition will still be proved, since all function calls in the latter function conform to the contract of swap.

– If the call inside the loop is modified to, say, swap(A,i,j+100), of course the safety of swap can still be proved; it is one of the VCs in the partition safety behaviour that will fail to be proved.

10.2 Functional Correctness: Array Partitioning

It should be clear that we have in this chapter been doing something quite different from what we had done in Chap. 9: we took a piece of code that was not specified formally, and we started by adding all the annotations required for its safety verification. This is a typical activity for engineers working with legacy code for which only an informal specification exists.

Let us now turn to the specification of the functional behaviour of the partition algorithm. Recall the informal specification of the function's behaviour given at the beginning of Sect. 10.1:

> It implements an algorithm that separates the elements contained in positions p to r of array A into those that are smaller than or equal to a pivot element, and those that are greater than the pivot. We take the pivot to be the element in the last position, x = A[r].

How can this be described formally? One has to look at the code in order to understand precisely what the algorithm does. In particular, note that at the end of execution of the loop, the array will consist of a sequence (ending in index i) of elements that are not greater than the pivot x, followed by a sequence of elements that are greater than x, and finally the pivot will be unchanged in the position with the last index (r) in the range considered. But note that an additional swap operation takes place after the loop, that will bring the pivot to the position with index i+1, which means that it will now be placed between the two sequences. Moreover, the return value is precisely i+1, i.e. the final index of the pivot. This can be referred to using \result in the postcondition, which can now be written as follows.

```
@ ensures
@    p <= \result <= r &&
@    (\forall int l; p <= l < \result => A[l] <= A[\result]) &&
@    (\forall int l; \result < l <= r => A[l] >  A[\result]) &&
@    A[\result] == \old(A[r])
@
```

Note the use of the \old operator in the last formula. Without this last postcondition, the specification would state that some pivot is used to calculate the partition, and its current index is returned by the function, but it would not actually specify that the element initially contained in the last position of the array was used as pivot.

Of course the same has to be done for the swap function, since its current specification merely consists of a safety precondition. partition relies on swap, and contract-based verification implies that, from the point of view of the verification of partition, all that matters is the latter function's contract, not its code. The relevant postcondition is

```
@ ensures t[i] == \old(t[j]) && t[j] == \old(t[i]) &&
@    \forall integer k; k != i && k != j ==>
      t[k] == \old(t[k]);
```

Note again the fundamental role played by the \old operator, and also the fact that it is not sufficient to mention what changes in the array from the pre-state to the post-state; it is also necessary to state that the contents of the remaining array positions are unchanged. In fact, this last bit could alternatively be achieved through a frame condition, by including an assigns clause in the function's contract as follows.

```
@ requires \valid_index(t,i) &&  \valid_index(t,j);
@ assigns t[i], t[j];
@ ensures t[i] == \old(t[j]) && t[j] == \old(t[i]);
```

This clause lists the set of array positions that may be assigned by the function. Following the principles of contracts, in the presence of an assigns clause for swap Frama-C will

1. generate an additional verification condition in the default behaviour for swap corresponding to the truth of the frame condition;
2. generate hypotheses that can be used to prove the verification conditions generated for partition.

We remark that the frame condition is *stronger* than, and not equivalent to, the postcondition

```
\forall integer k; k != i && k != j ==> t[k] == \old(t[k])
```

The former implies that the function does not write any other positions of the array passed as argument, *or anything else* in the global state. We choose to keep also the latter postcondition in the specification, since it is more explicit and may facilitate automatic proofs. The resulting contract thus contains some redundancy.

Running Frama-C on the code annotated with the postconditions discussed above will generate many additional verification conditions. In particular, a default behaviour is created for swap, consisting of 3 VCs labeled "postcondition". All 3 are automatically discharged. The default behaviour of partition will also contain new VCs labeled "postcondition", and some of these cannot be discharged. One could think that this is because something still needs to be added to the precondition of partition, but in fact that is not the case: the precondition required for safety verification is all that is needed (which means that the function should work for any array, provided it is safely allocated). What is missing is that we still need to strengthen the loop invariant, which for now contains no information that can be used to prove the postcondition.

At any point of execution the initial segment of the array ending in index $j-1$ will be organised as two adjacent sequences containing elements respectively smaller than and strictly greater than x, with i the last index in the first sequence. Writing this in ACSL yields the annotated code shown in Fig. 10.4. Frama-C produces 56 verification conditions from it, all of which are automatically discharged.

```
/*@ requires \valid_index(t,i) &&  \valid_index(t,j);
  @ assigns t[i], t[j];
  @ ensures t[i] == \old(t[j]) && t[j] == \old(t[i]) &&
  @    \forall integer k; k != i && k != j ==>
  @      t[k] == \old(t[k]);
  @*/
void swap(int t[],int i,int j) {
  int tmp = t[i];
  t[i] = t[j];
  t[j] = tmp;
}


/*@ requires 0 <= p <= r && \valid_range(A,p,r);
  @ assigns A[p..r];
  @ ensures
  @   p <= \result <= r &&
  @   (\forall integer l; p <= l < \result ==>
  @     A[l] <= A[\result]) &&
  @   (\forall integer l; \result < l <= r ==>
  @     A[l] >  A[\result]) &&
  @   A[\result] == \old(A[r]) ;
  @*/
int partition (int A[], int p, int r)
{
  int x = A[r];
  int tmp, j, i = p-1;

  /*@ loop invariant
    @   p <= j <= r && p-1 <= i < j &&
    @   (\forall integer k; (p <= k <= i) ==> A[k] <= x) &&
    @   (\forall integer k; (i <  k <  j) ==> A[k] >  x) &&
    @   A[r] == x;
    @ loop variant (r-j);
    @*/
  for (j=p; j<r; j++)
    if (A[j] <= x) {
      i++;
      swap(A,i,j);
    }
  swap(A,i+1,r);
  return i+1;
}
```

Fig. 10.4 `partition` annotated for safety and partial functional specification

10.3 Functional Correctness: Multiset Preservation

Even though the partition function does not sort arrays, the discussion of the example of Sect. 9.1.4 also applies in the present context: the functional property described and verified so far does not correspond to a complete specification of what the function does. It is important to also state that is preserves the multiset of elements initially contained in the array, or in other words that the final contents of the array parameter is a permutation of its initial contents. We simply follow the example of Sect. 9.1.4 and specify two different behaviours for the function: one for partitioning and another for the permutation aspect.

Figure 10.5 contains the complete specification of partition. Note that in addition to using the predicate `Permuta` for the permutation behaviour, we have also used the defined predicate `Swap` (see Fig. 9.5) in the postcondition of function `swap`. The example also illustrates the use of loop invariants associated to particular behaviours. Such invariants will be ignored if the corresponding behaviour is not being considered for verification.

There are now 6 different behaviours:

1. Default behaviour for `partition`
2. Partition behaviour for `partition`
3. Permutation behaviour for `partition`
4. Safety behaviour for `partition`
5. Default behaviour for `swap`
6. Safety behaviour for `swap`

These originate a total of 60 VCs (with the exact integers model used). In our installation, Simplify reports only two unknown conditions (in the `permutation` behaviour), with the remaining all valid, while the Ergo prover times out after 60 seconds for 4 conditions, all related to `assigns` clauses. Since the unproved conditions do not coincide for both tools, the program is successfully checked. This confirms the interest of using a multi-prover VCGen.

10.4 A Word of Caution

We finish the section with a word of caution: we would not like to induce the reader into thinking that automatic proof always works. In fact, in this particular example the complete proof was automated because it was possible to write the specification in a way that facilitated the proof process. In particular, note that in abstract terms what partition does is a sequence of swap operations, and the permutation axiomatic is written in accordance with this. But for other algorithms the correspondence may not be so direct. Consider again the insertion sort algorithm of Fig. 9.4. Clearly it is correct with respect to the permutation behaviour, however, since the algorithm is not implemented by directly swapping elements, it is hopeless trying to automatically establish the permutation aspect.

Let us take the current example to illustrate another feature of ACSL: the language allows for the user to write lemmas written in a similar way to axioms, as the following example shows.

```
/*@ requires \valid_index(t,i) && \valid_index(t,j);
  @ assigns t[i],t[j];
  @ ensures Swap{Old,Here}(t,i,j);
  @*/
void swap(int t[],int i,int j) {
  int tmp = t[i];
  t[i] = t[j];
  t[j] = tmp;
}


/*@ requires 0 <= p <= r && \valid_range(A,p,r);
  @ assigns A[p..r];
  @ behavior partition:
  @   ensures
  @     p <= \result <= r &&
  @     (\forall int l; p <= l < \result ==>
  @       A[l] <= A[\result]) &&
  @     (\forall int l; \result < l <= r ==> A[l] > A[\result]) &&
  @     A[\result] == \old(A[r]) ;
  @ behavior permutation:
  @   ensures
  @     Permuta{Old,Here}(A,p,r) &&
  @     \forall integer a; (a<p || a>r) ==>
  @       A[a] == \at(A[a],Old);
  @*/
int partition (int A[], int p, int r)
{
  int x = A[r];
  int tmp, j, i = p-1;

  /*@ loop invariant
    @   p <= j <= r && p-1 <= i < j;
    @ for partition:
    @   loop invariant
    @     (\forall int k; (p <= k <= i) ==> A[k] <= x) &&
    @     (\forall int k; (i <  k <  j) ==> A[k] >  x) &&
    @     A[r] == x;
    @ for permutation:
    @   loop invariant
    @     Permuta{Pre,Here}(A,p,r) &&
    @     \forall integer a; (a<p || a>=j) ==>
    @       A[a] == \at(A[a],Pre);
    @ loop variant (r-j);
    @*/
  for (j=p; j<r; j++)
    if (A[j] <= x) {
      i++;
      swap(A,i,j);
    }
  swap(A,i+1,r);
  return i+1;
}
```

Fig. 10.5 `partition`: complete specification (predicate definitions and axiomatics omitted)

```
@ lemma Permut_swap_sequence{L1,L2,L3}:
@   \forall int a[], integer l, h, i, j;
@   Permuta{L1,L2}(a, l, h) ==>
@   (l <= i <= h && l <= j <= h && Swap{L2,L3}(a, i, j)) ==>
@   Permuta{L1,L3}(a, l, h) ;
```

Like axioms, lemmas will be used by the provers in their attempts to discharge the verification conditions; but unlike axioms, lemmas need to be proved. This means that Frama-C will generate a proof obligation for each lemma (grouped under the label "user goals").

What is interesting about the above lemma is that in its presence Simplify will conclude the proof in a completely automatic way, with all verification conditions discharged (but not the lemma). It is a fairly easy lemma to prove interactively using axioms `Permuta_swap` and `Permuta_trans`. But it makes all the difference to Simplify.

This points us to a general strategy for discharging verification conditions: sometimes writing a few (possibly difficult) lemmas may have a dramatic effect, causing most verification conditions to be automatically discharged, and leaving only the lemmas to be proved. The difficult results, encapsulated in the lemmas, can then be proved interactively if required, or simply certified by a human as trusted—this may sound really bad after we have invested so much effort into accomplishing rigorous verification methods, but when software certification, rather than complete verification, is being carried out, it may be an option. The SPARK toolset for instance, which is widely used to develop safety critical applications, explicitly contemplates "user-approved" proof obligations.

10.5 Pointer Variables and Parameters Passed by Reference

The following function is a translation to C of an example from Sect. 8.4.3.

```
/*@ requires \valid(a) && \valid(b);
  @ ensures *a == 10 && *b == 20;
  @*/
void p(int *a, int *b)
{
   *a = 10;
   *b = 20;
}
```

The treatment of pointer variables was of course implicit in our use of arrays in the previous sections; here they are used explicitly to allow for the use of parameters (which are not of array type) passed by reference. Note the use of the \valid predicate in the precondition to signal the fact that pointers a and b can be safely dereferenced.

Following our discussion of aliasing in Sect. 8.4.3, the reader should be surprised to observe that the function's verification can be successfully performed. The reason for this is that Jessie is considering a "separated regions" assumption, under which the parameters a and b are assumed to point to non-overlapping memory regions. In this case, they cannot point to the same integer, thus the verification is successful.

This assumption can be turned off by inserting the following pragma:

```
# pragma SeparationPolicy (none)
```

In this case the postcondition $*a == 10$ can no longer be proved. In general, parameters can point to the root of some dynamically allocated structures like trees, and one has to consider carefully whether it is reasonable to rely on the separated regions assumption, which can greatly simplify things. The possibility of overlapping heap structures and aliasing contribute to making reasoning about C programs very challenging, thus the motto "precise static analyzers despite the pitfalls of C", used by the Frama-C developing team.

10.6 To Learn More

The examples in this chapter were checked with the Beryllium version of Frama-C. The Frama-C user manual is available from http://frama-c.com/. A tutorial on the Jessie plugin [6] is also available.

Jessie relies on the Why tool. Why is a generic, multi-prover VCGen for a simple ML-like language (also called Why). It is generic in the sense that it is really not meant to be simply a VCGen for the Why language; the idea is that one can obtain a VCGen for an arbitrary language by encoding this target language into Why. The paper [5] describes the tool as well as two VCGens obtained by using it: Caduceus for C and Krakatoa for the Java programming languages. [4] contains a detailed case study of the verification of a C program using Caduceus.

Why's input language is a very basic while language tailored for program verification. In spite of its simplicity, its associated logic (annotation) language is sufficiently powerful to allow for the definition of functions and predicates and the introduction of axioms concerning these predicates. Thus it is possible to model abstract data types. The lecture notes [3] explain in detail how Why can be used.

Boogie [2] is another generic tool that can be used for the same purpose; it has been used in particular as the basis for the Spec# [1] system for verification of C# programs. Other experimental tools for the verification of C code include Havoc[2] and VCC.[3] The former is particularly suited for the verification of programs that use linked heap-allocated data structures; the latter is suited for the verification of concurrent programs. Both rely on Boogie.

---

[2]http://research.microsoft.com/en-us/projects/havoc/.

[3]http://research.microsoft.com/en-us/projects/vcc/.

10.7 Exercises

10.1. Chapter 9 contains many examples of programs with ACSL annotations. You are now invited to verify any of them using Frama-C. Try to make small modifications to the specification and/or the code to increase your understanding of both the specification language, the tool, and the proof obligations generated.

10.2. Write C functions matching the specifications of Exercise 9.1, and use Frama-C to check their correctness.

10.3. Consider the following function annotated with a postcondition.

```
#define N 10

/*@ ensures *x == \old( *x );
  @*/
void ex_one (int *x, int *y)
{
  int i = 0;
  while (i < N) {
    *y = i;
    i++;
  }
}
```

Use Frama-C / Jessie to verify its total correctness. Add any preconditions and additional annotations that may be required.

10.4. Verify the program of the previous exercise but now include a pragma to instruct Jessie that it should not assume separated regions. Carefully observe the new proof obligations that are created and add annotations that allow them to be discharged.

10.5. Now strengthen the postcondition in the previous exercise as follows.

```
@ ensures *x == \old( *x ) && *y == N-1;
```

What other annotations are required to allow this postcondition to be proved?

References

1. Barnett, M., Leino, K.R.M., Schulte, W.: The Spec# programming system: An overview. In: CASSIS: Construction and Analysis of Safe, Secure, and Interoperable Smart Devices, vol. 3362, pp. 49–69. Springer, Berlin (2004)
2. Barnett, M., Chang, B.-Y.E., DeLine, R., Jacobs, B., Leino, K.R.M.: Boogie: A modular reusable verifier for object-oriented programs. In: de Boer, F.S., Bonsangue, M.M., Graf, S., de Roever, W.P. (eds.) FMCO. Lecture Notes in Computer Science, vol. 4111, pp. 364–387. Springer, Berlin (2005)
3. Filliâtre, J.-C.: Program verification using coq—introduction to the why tool. Lecture Notes TYPES Summer School (2005)

4. Filliâtre, J.-C.: Queens on a chessboard: An exercise in program verification (2007). http://why.lri.fr/queens/
5. Filliâtre, J.-C., Marché, C.: The Why/Krakatoa/Caduceus platform for deductive program verification. In: Damm, W., Hermanns, H. (eds.) CAV. Lecture Notes in Computer Science, vol. 4590, pp. 173–177. Springer, Berlin (2007)
6. Moy, Y., Marché, C., Jessie Plugin Tutorial. LRI (February 2010). Beryllium Version

Index

J.B. Almeida et al., *Rigorous Software Development*,
Undergraduate Topics in Computer Science,
DOI 10.1007/978-0-85729-018-2, © Springer-Verlag London Limited 2011

6084